Multifunctional Coordination Materials for Green Energy Technologies

As an emerging material platform, multifunctional coordination materials offer many advantages such as remarkable porosity, structural flexibility, crystallinity, and modifiable functionalities that render them highly suited to generate and store green energy. This book covers the design and fabrication approaches of multifunctional coordination materials for green energy-related technologies, including batteries, supercapacitors, solar cells, and nanogenerators.

- Discusses fundamentals of multifunctional coordination materials.
- Explains vital synthesis and design techniques as well as theoretical modeling.
- Offers a comprehensive overview of preparation, structural and morphological properties, and applications in a wide variety of energy production, energy storage, and energy device technologies.
- Assesses environmental impacts, recycling, challenges, and future perspectives.

Multifunctional Coordination Materials for Green Energy Technologies is an ideal reference for advanced students and researchers working in materials engineering, including new catalyst development, battery design, and related areas.

Emerging Materials and Technologies

Series Editor
Boris I. Kharissov

The *Emerging Materials and Technologies* series is devoted to highlighting publications centered on emerging advanced materials and novel technologies. Attention is paid to those newly discovered or applied materials with the potential to solve pressing societal problems and improve quality of life, corresponding to environmental protection, medicine, communications, energy, transportation, advanced manufacturing, and related areas.

The series takes into account that, under present strong demands for energy, material, and cost savings, as well as heavy contamination problems and worldwide pandemic conditions, the area of emerging materials and related scalable technologies is a highly interdisciplinary field, with the need for researchers, professionals, and academics across the spectrum of engineering and technological disciplines. The main objective of this book series is to attract more attention to these materials and technologies and invite conversation among the international R&D community.

Computational Studies
From Molecules to Materials
Edited by Ambrish Kumar Srivastava

2D Semiconductors for Environmental Remediation
Edited by Honey John, Nisha T Padmanabhan, Sona Stanly and Jith C Janardhanan

Materials from Natural Sources
Structure, Properties, and Applications
Edited by Ramesh Gardas, Neha Patni, and Amita Chaudhary

Dielectric Materials for Capacitive Energy Storage
Edited By Haibo Zhang and Hua Tan

Multifunctional Coordination Materials for Green Energy Technologies
Edited by Ghulam Yasin, Anuj Kumar, Sajjad Ali, Tuan Anh Nguyen and Saira Ajmal

Advancements in Nanomaterials for Energy Conversion and Storage
Edited by Piyush Kumar Sonkar and Vellaichamy Ganesan

For more information about this series, please visit: www.routledge.com/Emerging-Materials-and-Technologies/book-series/CRCEMT

Multifunctional Coordination Materials for Green Energy Technologies

Edited by
Ghulam Yasin, Anuj Kumar, Sajjad Ali, Tuan Anh Nguyen and Saira Ajmal

CRC Press is an imprint of the
Taylor & Francis Group, an informa business

Designed cover image: Shutterstock

First edition published 2025
by CRC Press
2385 NW Executive Center Drive, Suite 320, Boca Raton FL 33431

and by CRC Press
4 Park Square, Milton Park, Abingdon, Oxon, OX14 4RN

CRC Press is an imprint of Taylor & Francis Group, LLC

ISBN: 978-1-032-38555-6 (hbk)
ISBN: 978-1-032-38617-1 (pbk)
ISBN: 978-1-003-34588-6 (ebk)

DOI: 10.1201/9781003345886

Typeset in Times LT Std
by Apex CoVantage, LLC

Contents

PART II Multifunctional Coordination Materials for Energy Storage Technologies

Preface

In recent decades, the urgency to mitigate carbon emissions and transition towards clean and sustainable energy sources has catalyzed a monumental shift in our approach to materials science today. Owing to the pressing global challenges like climate change, dwindling fossil fuel reserves, and environmental degradation, the imperative for sustainable, eco-friendly, and inexpensive energy solutions has never been more urgent. The journey towards sustainable energy is multifaceted, necessitating interdisciplinary bonding and innovation across related scientific domains. As we embark on this exploration, it is decisive to realize the importance of the transition towards green and sustainable energy technologies. It not only promises to mitigate the hostile effects of climate change and reduce our dependence on limited resources, but it also enables economic opportunities, job creation, driving innovation, and sustainable growth. Moreover, building green energy technologies fosters energy security, resilience, and equitable access to clean energy resources, thereby advancing global development goals and social welfare. Nevertheless, the journey towards a clean and sustainable energy future is not without its challenges. Among them, technical hurdles, scalability concerns, and economic constraints pose formidable barriers to the widespread adoption of green energy systems. Addressing these challenges demands concerted efforts from researchers, industry, policymakers, and society at large.

The transition towards a greener, more sustainable future demands novel scalable approaches and transformative technologies. By harnessing the potential of a new family of multifunctional coordination materials and fostering collaboration across disciplines, we can overcome these challenges and accelerate the transition towards a greener, more sustainable energy paradigm. Multifunctional coordination materials, with their unique chemical structures and versatile properties, exhibit remarkable characteristics tailored for a diverse array of clean energy technologies. These materials, often based on metal-organic frameworks (MOFs), covalent organic frameworks (COFs), and other coordination porous polymers and related structures, possess a unique combination of porosity, structural versatility, and tunable chemical functionality, making them ideally suited for uses ranging from energy conversion and storage to various catalysis and sensing devices. Furthermore, from photovoltaics and thermoelectric systems to fuel cells and energy storage technologies, these materials present solutions that transcend conventional boundaries, paving the way, in particular, for enhanced efficiency, durability, and environmental friendliness in energy conversion and storage devices, ascertaining the promising materials that hold the potential to revolutionize clean and sustainable energy technologies.

This book, *Multifunctional Coordination Materials for Green Energy Technologies* delves into the intricate family of these materials, exploring their significance, functionalities, and uses in the realm of sustainable energy. This also begins with a basic introduction and overview of the fundamental principles underlying coordination chemistry and the potential design strategies utilized in the construction of multifunctional coordination materials. Through a comprehensive investigation of their characteristics, preparation approaches, characterization, and performance in various energy devices, this book aims to provide a comprehensive understanding of the pivotal role that multifunctional coordination materials play in shaping the landscape of clean energy technologies. Drawing on the expertise of leading researchers from around the globe, each chapter delves into a specific aspect of this rapidly growing field, providing both a broad overview of contemporary state-of-the-art information and in-depth insights into cutting-edge research.

Throughout these chapters, we highlight not only the remarkable achievements of the past but also the challenges and opportunities that lie ahead. From the synthesis of innovative coordination nanoarchitectures to the engineering of materials with tailored properties, there is still much to be explored and revealed in this dynamic avenue. Moreover, as we strive towards a more sustainable future, it is crucial to consider not only the practical viability of new materials and technologies but also their scalability, economic viability, and environmental impact.

It is our hope that this book will serve as a valuable reference for researchers, students, professionals, and engineers seeking to comprehend the contribution of multifunctional coordination materials in shaping the future of green and sustainable energy technologies. The readers who are fresh in this field will be motivated to learn fundamental illustrations that may well deliver a vivid devotion earlier to stern reading. Meanwhile, the updated citations in all chapters ought to facilitate readers to rapidly probe the exciting field with evidence on the hottest advances. So, this book, *Multifunctional Coordination Materials for Green Energy Technologies,* is an indispensable latest reference on coordination nanomaterials for energy conversion and storage systems to teachers, researchers, scientists, engineers, and students in the domain of materials science, nanotechnology, chemical science, and electrochemistry. Academic experts can utilize this book to immediately swot the contemporary advances to broaden their understanding of the new family of coordination nanomaterials with emerging innovative technologies for green energy conversion and storage systems.

Finally, we want to express our earnest and sincere gratitude to Allison Shatkin and their coworkers at Taylor & Francis Group for their tolerant and substantial support and cooperation throughout the accomplishment of this book. We would also like to acknowledge our all collaborators, authors, and associates who contributed to this book. Last, but not least, we are thankful to our families for their incessant love, tolerance, and generous support.

Dr. Ghulam Yasin
School of Environment and Civil Engineering
Dongguan University of Technology
Guangdong, China

About the Editors

Dr. Ghulam Yasin is a researcher in the School of Environment and Civil Engineering at Dongguan University of Technology, Guangdong, China. His expertise covers the design and development of hybrid devices and technologies of carbon nanostructures and advanced nanomaterials for real-world impact in energy-related and other functional applications.

Dr. Anuj Kumar is Assistant Professor at GLA University, Mathura, India. For outstanding contribution in his research field, he has been awarded "Best Young Scientist Award 2021" from the Tamil Nadu Association of Intellectuals and Faculty (TAIF) and GRBS Educational Charitable Trust, India and "Young Researcher Award 2020" by Central Education Growth and Research (CEGR), India.

Dr. Sajjad Ali received his PhD in Materials Science and Engineering from the University of Chinese Academy of Science, China. He is Researcher (Assistant Professor) at Energy, Water and Environmental Lab, Prince Sultan University, Saudi Arabia.

Dr. Tuan Anh Nguyen completed his PhD in chemistry from Paris Diderot University, France in 2003. In 2012, he was appointed as Head of the Microanalysis Department at the Institute for Tropical Technology, Vietnam Academy of Science and Technology. He is the Editor-in-Chief, Kenkyu Journal of Nanotechnology and Nanoscience. Currently, he is the senior principal research scientist at the Vietnam Academy of Science and Technology.

Dr. Saira Ajmal received her PhD from Fudan University, China in environmental science and engineering with a specialization in environmental chemistry. Dr. Saira is Researcher (Academic) at the School of Mechanical Engineering, Dongguan University of Technology, Dongguan, Guangdong, China. Her research focuses on the design and development of advanced materials for green energy technologies.

Contributors

Khalid Aziz
Institute of Chemical Sciences
Bahauddin Zakariya University
Multan, Pakistan

Muhammad Arif
Institute of Chemical and Environmental Engineering
Khwaja Fareed University of Engineering and Information Technology
Rahim Yar Khan, Punjab, Pakistan
and
Centre for Thermal and Renewable Energy Research
Khwaja Fareed University of Engineering and Information Technology
Rahim Yar Khan, Punjab, Pakistan

Mahnoor Ahmed
Institute of Chemical and Environmental Engineering
Khwaja Fareed University of Engineering and Information Technology
Rahim Yar Khan, Punjab, Pakistan

Saira Ajmal
School of Mechanical Engineering
Dongguan University of Technology
Dongguan, Guangdong, China

Sajjad Ali
Energy, Water, and Environment Lab
College of Humanities and Sciences
Prince Sultan University
Riyadh, Saudi Arabia

Umair Azhar
Institute of Chemical and Environmental Engineering
Khwaja Fareed University of Engineering and Information Technology
Rahim Yar Khan, Punjab, Pakistan

Mohamed Bououdina
Energy, Water, and Environment Lab
College of Humanities and Sciences
Prince Sultan University
Riyadh, Saudi Arabia

Moazzam H. Bhatti
Department of Chemistry
Allama Iqbal Open University
Islamabad, Pakistan

Syeda Maida Batool
Department of Chemistry
Allama Iqbal Open University
Islamabad, Pakistan

Dipak Kumar Das
Department of Chemistry
GLA University
Mathura, India

Abdul Haq
Institute of Chemistry
Khwaja Fareed University of Engineering and Information Technology
Rahim Yar Khan, Punjab, Pakistan

Safana Haqani
Institute of Chemical and Environmental Engineering
Khwaja Fareed University of Engineering and Information Technology
Rahim Yar Khan, Punjab, Pakistan

Muhammad Ikram
Department of Chemistry
Abdul Wali Khan University Mardan
Mardan, Khyber Pakhtunkhwa Pakistan

Anuj Kumar
Department of Chemistry
GLA University
Mathura, India

Shahab Khan
School of chemistry and Chemical engineering
Shaanxi Normal University
Xian, Shaanxi, China

Jai Kumar
College of Materials Science and Engineering
Beijing University of Chemical Technology
Beijing, China

Jasvinder Kaur
Department of Chemistry
School of Sciences
IFTM University
Moradabad, Uttar Pradesh, India

Andleeb Mehmood
College of Physics and Optoelectronic Engineering
Shenzhen University
Guangdong, China

Muhammad Mubeen
College of Materials Science and Engineering
Beijing University of Chemical Technology
Beijing, China

Muhammad Asim Mushtaq
Institute for Advanced Study
Shenzhen University
Guangdong, China

Noor Muhammad
Shenyang National Laboratory for Materials Science
Institute of Metal Research
Chinese Academy of Sciences
Shenyang, China
and
School of Materials Science and Engineering
University of Science and Technology of China
Hefei, China

Muhammad Nadeem
Department of Chemistry
Allama Iqbal Open University
Islamabad, Pakistan

Rana R. Neiber
Beijing Key Laboratory of Ionic Liquids Clean Process
CAS Key Laboratory of Green Process and Engineering
Institute of Process Engineering
Chinese Academy of Sciences
Beijing, China
and
School of Chemistry and Chemical Engineering
University of Chinese Academy of Sciences
Beijing, China

Tuan Anh Nguyen
Institute for Tropical Technology
Vietnam Academy of Science and Technology
Hanoi, Vietnam

Nasira Parveen
Department of Physical Education and Sports Sciences
University of Gujrat
India

Waseem Raza
Institute for Advanced Study
Shenzhen University
Guangdong, China

Nadeem Raza
Chemistry Department
College of Science
Imam Mohammad Ibn Saud Islamic University (IMSIU)
Riyadh, Saudi Arabia

Harimohan Sharma
Department of Chemistry
GLA University
Mathura, India

Monika Singh
Department of Chemistry
GLA University
Mathura, India

Muhammad Sagir
Institute of Chemical and Environmental Engineering
Khwaja Fareed University of Engineering and Information Technology
Rahim Yar Khan, Punjab, Pakistan

Mohammad Tabish
College of Materials Science and Engineering
Beijing University of Chemical Technology
Beijing, China

Unaiza Talib
Institute of Chemistry
Khwaja Fareed University of Engineering and Information Technology
Rahim Yar Khan, Punjab, Pakistan

Ghulam Yasin
School of Environment and Civil Engineering
Dongguan University of Technology
Dongguan, Guangdong, China

Uzma Yunus
Department of Chemistry
Allama Iqbal Open University
Islamabad, Pakistan

Chao Zeng
Institute of Advanced Materials
Key Lab of Fluorine and Silicon for Energy Materials and Chemistry of Ministry of Education
College of Chemistry and Chemical Engineering
Jiangxi Normal University
Nanchang, China

Muhammad Zahoor
Department of Biochemistry
University of Malakand
Chakdara Dir Lower, KPK, Pakistan

Part I

Fundamentals

1 Multifunctional Coordination Materials
An Introduction

Ghulam Yasin

1.1 INTRODUCTION

Coordination materials (CMs), characterized by their extensive molecular structures, exhibit outstanding chemical and physical properties along with thermal conductivity and corrosion resistance. Furthermore, these materials exhibit countless arrangements of coordination units established by coordination bonds between metal entities and organic ligands [1]. In the last two decades, porous coordination materials have garnered increasing attention due to their potential applications in various fields, including gas adsorption and separation, catalysis, sensing, proton conduction, and electrochemistry. Multifunctional CMs were synthesized in the 18th century, although the first one, Prussian blue, was synthesized in the same decade. In contrast to traditional porous inorganic materials (such as zeolites and activated carbon), CMs have a distinct geometry of metal ions/clusters and organic ligands [2]. In addition, CMs exhibit crystallinity, which is useful for describing relationships between structure and property [3]. These materials are commonly referred to as Metal-organic frameworks (MOFs), covalent organic frameworks (COFs), and porous coordination polymers (PCPs).

To enhance our grasp of versatile CMs, it is imperative to write the comprehensive fundamentals and current updates. In this chapter, we summarize the specific characteristics of MOFs, COFs, and PCPs, as well as their attractive structures, and we highlight remarkable performance in various applications. It is believed that by pointing out potential directions for basic research and applications in this emerging field, this chapter will increase awareness about multifunctional CMs.

1.2 MULTIFUNCTIONAL METAL-ORGANIC FRAMEWORKS

Various functionalities, porosity, and dimensions of MOFs are attracting significant interest in structural exploration and innovative applications [4]. MOFs, which were introduced in 1995, are considered vital inventions in chemistry as they combine organic and inorganic materials. It brings a new perspective to this field by emphasizing its framework-like nature, similar to that of zeolite. Due to their diverse characteristics, including conductivity, reactivity, fluorescence, luminescence, and ferromagnetism, they exhibit significant porosity and are widely used in various

DOI: 10.1201/9781003345886-2

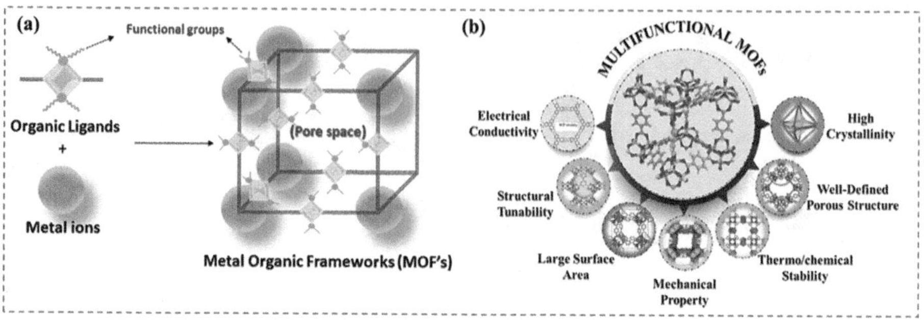

FIGURE 1.1 (a) Schematic illustration of the construction of MOFs. Reproduced with permission from reference [5] copyright from Royal Society of Chemistry. (b) The multifunctional promising properties of MOFs.

fields [5]. In Figure 1.1, metal clusters or ions are connected with organic ligands to create a hybrid structure that has high porosity and functionality. Due to their crystalline and porous properties, MOFs are versatile materials capable of adapting to a variety of applications in various sectors [6].

Recently, MOFs as emerging revolutionary materials have been extensively studied [7]. In order to understand structure-property relationships, MOFs' crystalline nature allows precise characterization through diffraction studies [8]. As a material and structural design, these are valuable materials for many applications because of their unique attributes. Though some MOFs can be designed in a predictable manner, the term "rational design" may be contested [9]. These structures are derived from metal cluster geometries and organic linker shapes. $Zn_4O(COO)_6$ is an example of a MOF with octahedral geometry, which embodies the rational design principle. There have been several realizations of isoreticular MOFs, for example, $Zn_4O[R(COO)_2]_3$ as shown in Figure 1.2a [10]. MOFs with different topologies, pore sizes, and curvatures were formed from $Cu_2(COO)_4$ secondary building units and organic linkers in Figure 1.2b [11]. Systematically tuning the porosity of MOF materials is an essential aspect of developing multifunctional materials as shown in Figure 1.2c. Additionally, MOF versatility is greatly enhanced by additional functional sites on pore surfaces and nodes. Organic linkers allow MOFs to recognize small molecules by incorporating specific sites like pyridyl and amine into their surfaces. Multiple functional sites can be immobilized on pores using multivariate MOFs [12].

In MOFs, metal sites can be systematically immobilized via post-synthesis, and these metal sites within metal cluster nodes can be utilized to develop functional properties in mixed-metal-organic frameworks. Because MOF materials combine metal ions/clusters with organic linkers, they tend to have a variety of electronic, magnetic, and optical properties [13]. MOFs can hold tiny particles, metal mixtures, and colorful dyes because of their special trapping abilities. In addition to traditional organic/inorganic hybrid materials, they have many other multifunctional properties and uses in many applications including gas storage and separation, photonics, electromagnetism, chemical sensing, catalysis, biomedicine, and green energy devices.

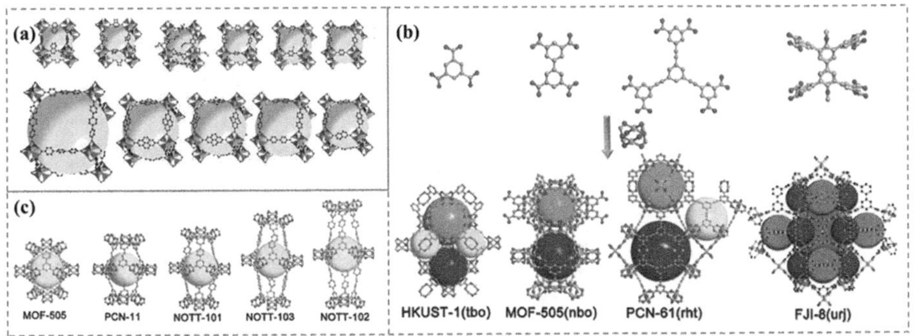

FIGURE 1.2 (a) A crystal structure for IRMOF-n with general formula $Zn_4O[R(COO)_2]_3$. (b) Organic ligands containing $Cu_2(COO)_4$ units serve as secondary building blocks in nbo-type 9 MOFs. (c) Tunable pore structures (pore sizes and curvatures) in the series of nbo-type MOFs. Reproduced with permission from [14] copyright from John Wiley & Sons.

1.3 MULTIFUNCTIONAL COVALENT-ORGANIC FRAMEWORKS

Reticulated chemistry is used in organic synthesis to synthesize COFs, which differ from Metal-Organic Frameworks. Contrary to MOFs, COFs are entirely organic (H, B, C, N, O) and possess a robust, three-dimensional crystalline structure based on covalent bonds; due to this, they are structurally stronger and have a different composition. The scientific community has been drawn to COFs due to their ability to produce high-quality, targeted materials at the nanoscale with atomic precision shown in Figure 1.3a [15]. In polymer chemistry, COFs are directional and strong organic structures with well-defined, permanent porosity rendering these materials rigid, low-density, stable, and resistant. Although COFs are a relatively new class, they show promising results that can be comparable with or even surpass more established materials such as zeolites, MOFs, and various organic polymer classes, such as Hyper Crosslinked Polymers, Polymers with Intrinsic Microporosity, Porous Aromatic Frameworks, and Conjugated Microporous Polymers are significant among others.

COFs, porous organic materials formed by an organic framework with covalent bonds, can be customized for a family of applications (Figure 1.3b) through pore engineering, taking advantage of their stability, porosity, and tunability. As a result of their porous, stable, and crystalline properties, their honeycomb-like structure suggests promise for diverse applications [16]. The structure of materials has been improved over time through innovations to reduce production costs and to improve their environmental friendliness [17]. COFs exhibit high catalytic activity and broad stability by mimicking the structure of natural enzymes. In addition to their biodegradability and porosity, these materials are environmentally friendly and can be used for a diverse range of purposes [18]. In mesoporous materials, such as zeolites and carbon, porosity is an important property. Due to structural randomness introduced by coupling reactions, it is difficult to predict crucial pore characteristics in these materials. So, their applications might be less efficient due to this unpredictability [19].

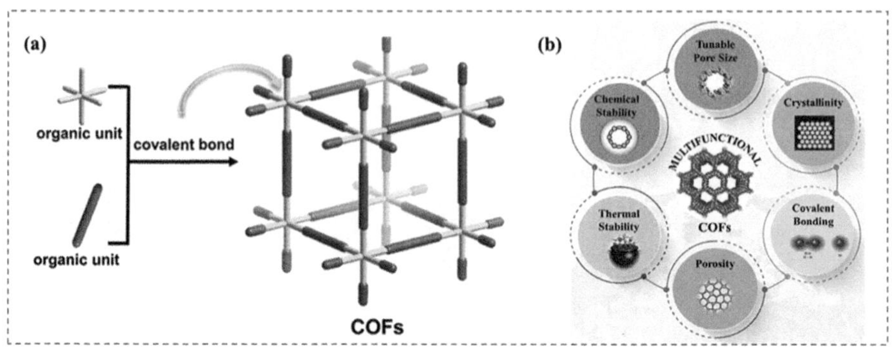

FIGURE 1.3 (a) General COF structure, reproduced with permission from reference [15] copyright from American Association for the Advancement of Science and (b) its most important properties.

COFs, with the ability to spatially and structurally define pores in terms of size and shape, facilitate the storage, release, or separation of molecular materials within their empty yet precisely tailored spaces. During reversible condensation, COFs form covalent bonds that enable them to form crystals [20]. Through rearranging the crystal lattice by breaking and rebuilding bonds, these covalent bonds allow slipup correction and fine-tuning. The literature primarily reports on 2D structures, but 3D structures can also exist [21]. Covalently bonded organic frameworks, commonly forming stacked sheets through Van der Waals forces, often adopt AA stacking ($C^{3+}C^{2+}$ stacking) for maximum pore size [22]. Scientists are addressing the challenge of optimizing the steric orientation in 2D COFs without compromising stability by developing methodologies to control stack positions during crystal growth for multifunctional uses.

1.4 MULTIFUNCTIONAL POROUS COORDINATION POLYMERS

PCPs have a flexible structure, a large surface area, and abundant active sites [23]. Researchers investigated synthesizing CMs with robust frameworks like zeolites, studying MOFs and PCPs to enhance the reversibility of molecule accommodation and successfully reverting the as-synthesized crystals after molecule removal from their pores [24]. PCPs can acquire the necessary properties by rationally selecting building blocks, given their crystalline backbones consisting of organic linkers and inorganic nodes. Moreover, these porous materials can be created in many different ways because there are a lot of different building blocks and ways to combine them [25]. As a result, PCPs have become an important class of porous materials, laying the foundation for future development, and they were predicted to be superior in 1998 based on the existing understanding of layered and interpenetrated frameworks [26].

The variation of PCP crystal structures has been reported due to chemical (adsorption of gases, incorporation of substrates) and physical (temperature, pressure, light, electric field) stimuli. Soft porous crystals are porous solids with a highly ordered network and structurally transformable properties [27]. It has been possible to develop

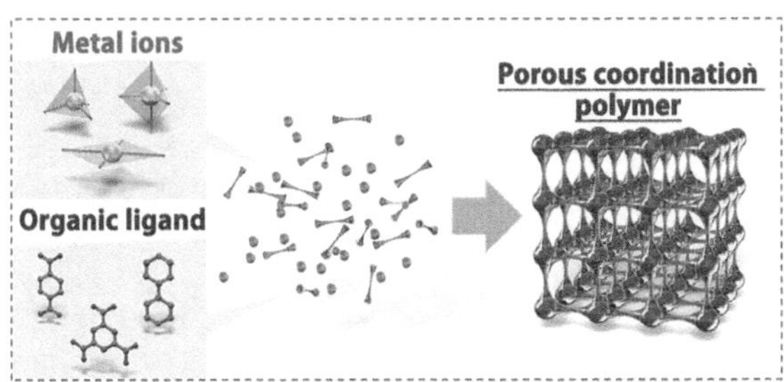

FIGURE 1.4 A schematic illustration of porous coordination polymers. Reproduced with permission from reference [30], copyright from Atomis.

CPCs through supramolecular chemistry and crystal engineering. Supramolecular chemistry and crystal engineering have worked together to understand how solids self-assemble and how they interact with one another. Asymmetric coordination polymers exhibit metal cations of different geometries, such as linear, tetrahedral, square-planar, square-pyramidal, triangular bipyramidal, octahedral, triangular-prismatic, and pentagonal-bipyramidal. These polymers further form infinite networks (1D, 2D, and 3D) based on ligand strength, directionality, and reaction conditions [28]. Figure 1.4 illustrates how multifunctional ligands (e.g., 4,40-bipy) can form zigzag, diamonded, linear, ladder, square-grid, honeycomb, and octahedral frameworks. Mixing ligands or incorporating mixed functionality into single ligands not only enhances the stability of CPCs but also contributes to the construction of porous CPCs using both carboxylic and pyridyl ligands [29].

The synthesis of CPCs has also been carried out using multiple bridge ligands. Different topologies can arise as a result of building blocks interfering with the reaction, so the desired network may not be achieved. In addition, porous coordination materials exhibit diverse applications across various fields. From constructing pillar-like networks with pyridyl and dicarboxylate ligands to creating ladder-like networks with monocarboxylate ligands, these materials offer versatile possibilities. Moreover, in photo-dimerization reactions, acetate and trifluoroacetate serve as effective bridging agents. The broad range of potential applications highlights the significance of porous coordination polymers and/or materials in evolving advancements across different scientific and technological domains.

1.4.1 Applications of Multifunctional Coordination Materials

Multifunctional CMs, including MOFs, COFs, and PCPs, are among the most exciting areas of materials science due to their high surface areas and tunable properties, and their precise engineering makes them ideal candidates for a wide range of applications. As a result of their exceptional porosity, these can be used in many applications; the crystalline structures and their bonds open up new avenues for gas separation, sensing, electronics, catalysis, biomedical, and electrochemical applications. Figure 1.5 illustrates how porous CMs can be applied in various functional applications and next-generation devices. There are a variety of physical

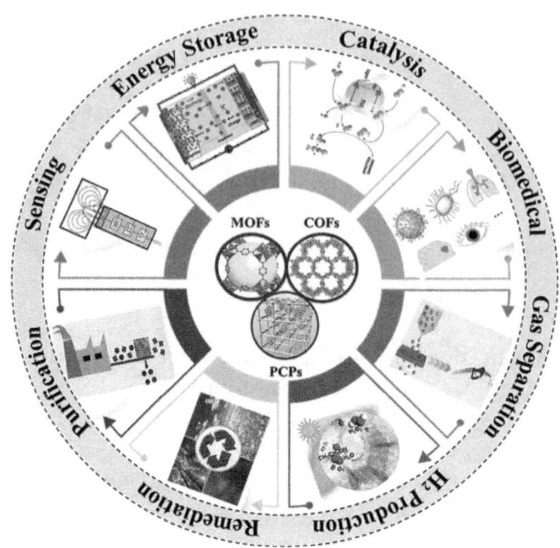

FIGURE 1.5 Promising applications of multifunctional CMs.

and chemical properties associated with this family of materials. Coordination of diverse components with exceptional features broadens the application scope of multifunctional CMs, including crystallinity, high porosity, functional pores, thermal stability, and well-organized frameworks. In addition to their extended conjugation and tunable chemistry, these materials exhibit excellent chemical and thermal stability. Interestingly, multifunctional CMs have been used in probing the storage and separation of gases [31], chemo-encapsulation [32], catalysis [33], photo-redox catalysis [34], light emittance [35], chemo-sensing [36], energy storage and conversion [37], environmental remediation [38], gas storage [39], purification, drug delivery and biomedical [40], sensing [41], gas separation [42], and photocatalytic $_{H2}$ production [43].

1.5 CONCLUSION AND OUTLOOKS

Multifunctional CMs are demonstrating impressive results across various fields of scientific research and technological advancements, enabling competitive solutions for commercial uses. There are a variety of applications for this emerging class of porous materials, owing to their excellent properties such as sensitivity, selectivity, porosity, large surface areas, and tunability. Besides, activated surface areas and multifunctional sites in COFs, MOFs, and PCPs enable rapid electrochemical properties and their ability for swift electron and energy transfer. Similarly, advanced and efficient fabrication and immobilization techniques can improve the reusability and stability of these revolutionary materials. It is also possible to design high porosity by choosing the right linkers and ligands for tunable synthesis. In conclusion, there is a crucial need for continued exploration of porous CMs for various functional uses, especially those involving integrated functional-structural and electrochemical applications, to broaden their usage, and further research is needed on porous CMs with a balanced combination of functional, structural, and physiochemical properties to enable commercial applications.

ACKNOWLEDGMENTS

This work was supported by the Research Fund for International Scientists (RFIS-Grant number: 52150410410) National Natural Science Foundation of China.

REFERENCES

[1] S.R. Batten, N.R. Champness, X.-M. Chen, J. Garcia-Martinez, S. Kitagawa, L. Öhrström, M. O'Keeffe, M. Paik Suh, J. Reedijk, Terminology of Metal–Organic Frameworks and Coordination Polymers (IUPAC Recommendations 2013), 85 (2013) 1715–1724.

[2] L. Chai, J. Song, A. Kumar, R. Miao, Y. Sun, X. Liu, G. Yasin, X. Li, J. Pan, Bimetallic-MOF Derived Carbon with Single Pt Anchored C4 Atomic Group Constructing Super Fuel Cell with Ultrahigh Power Density and Self-Change Ability, *Advanced Materials*, 36 (2024) 2308989.

[3] H.M. Asif, Z. Shakoor, S. Ibraheem, A.M. Salama, M.A. Khan, T.A. Nguyen, G. Yasin, Chapter 12 – Polyoxometalate-Based Metal Organic Frameworks (POMOFs) for Lithium-Ion Batteries, in: R.K. Gupta, T.A. Nguyen, G. Yasin (Eds.) *Metal-Organic Framework-Based Nanomaterials for Energy Conversion and Storage*, Elsevier, 2022, pp. 245–268.

[4] S. Daliran, A.R. Oveisi, Y. Peng, A. López-Magano, M. Khajeh, R. Mas-Ballesté, J. Alemán, R. Luque, H. Garcia, Metal–Organic Framework (MOF)-, Covalent-Organic Framework (COF)-, and Porous-Organic Polymers (POP)-Catalyzed Selective C–H Bond Activation and Functionalization Reactions, *Chemical Society Reviews*, 51 (2022) 7810–7882.

[5] V. Unnikrishnan, O. Zabihi, M. Ahmadi, Q. Li, P. Blanchard, A. Kiziltas, M. Naebe, Metal–Organic Framework Structure–Property Relationships for High-Performance Multifunctional Polymer Nanocomposite Applications, *Journal of Materials Chemistry A*, 9 (2021) 4348–4378.

[6] M. Nadeem, G. Yasin, M. Arif, M.H. Bhatti, K. Sayin, M. Mehmood, U. Yunus, S. Mehboob, I. Ahmed, U. Flörke, Pt-Ni@PC900 Hybrid Derived from Layered-Structure Cd-MOF for Fuel Cell ORR Activity, *ACS Omega*, 5 (2020) 2123–2132.

[7] G. Yasin, S. Ibrahim, S. Ibraheem, A. Kumar, T.A. Nguyen, Chapter 1 – MOF-Based Nanostructures and Nanomaterials for Next-Generation Energy Storage: An Introduction, in: R.K. Gupta, T.A. Nguyen, G. Yasin (Eds.) *Metal-Organic Framework-Based Nanomaterials for Energy Conversion and Storage*, Elsevier, 2022, pp. 3–10.

[8] C. Goyal, A. Kumar, G. Yasin, R.K. Gupta, MOFs-Derived Metal Oxides-Based Compounds for Flexible Supercapacitors, in: *Smart and Flexible Energy Devices*, CRC Press, 2022, pp. 501–517.

[9] S. Ibraheem, S. Ibrahim, G. Yasin, A. Kumar, M. Tabish, T.A. Nguyen, S. Ajmal, Chapter 33 – MOF-Based Advanced Nanomaterials for Electrocatalysis Applications, in: R.K. Gupta, T.A. Nguyen, G. Yasin (Eds.) *Metal-Organic Framework-Based Nanomaterials for Energy Conversion and Storage*, Elsevier, 2022, pp. 749–763.

[10] M. Eddaoudi, J. Kim, N. Rosi, D. Vodak, J. Wachter, M. O'Keeffe, O.M. Yaghi, Systematic Design of Pore Size and Functionality in Isoreticular MOFs and Their Application in Methane Storage, *Science*, 295 (2002) 469–472.

[11] J. Pang, F. Jiang, M. Wu, C. Liu, K. Su, W. Lu, D. Yuan, M. Hong, A Porous Metal-Organic Framework with Ultrahigh Acetylene Uptake Capacity under Ambient Conditions, *Nature Communications*, 6 (2015) 7575.

[12] Z. Ji, H. Wang, S. Canossa, S. Wuttke, O.M. Yaghi, Pore Chemistry of Metal–Organic Frameworks, *Advanced Functional Materials*, 30 (2020) 2000238.
[13] S.H. Mir, L.A. Nagahara, T. Thundat, P. Mokarian-Tabari, H. Furukawa, A. Khosla, Review – Organic-Inorganic Hybrid Functional Materials: An Integrated Platform for Applied Technologies, *Journal of The Electrochemical Society*, 165 (2018) B3137.
[14] B. Li, H.-M. Wen, Y. Cui, W. Zhou, G. Qian, B. Chen, Emerging Multifunctional Metal–Organic Framework Materials, *Advanced Materials*, 28 (2016) 8819–8860.
[15] W.-K. Han, Y. Liu, X. Yan, Z.-G. Gu, Coordination Directed Metal Covalent Organic Frameworks, *Materials Chemistry Frontiers*, 7 (2023) 2995–3010.
[16] S. Zhang, F. Zhao, G. Yasin, Y. Dong, J. Zhao, Y. Guo, P. Tsiakaras, J. Zhao, Efficient Photocatalytic Hydrogen Evolution: Linkage Units Engineering in Triazine-Based Conjugated Porous Polymers, *Journal of Colloid and Interface Science*, 637 (2023) 41–54.
[17] L. Zhou, X. Luo, J. Gao, G. Liu, L. Ma, Y. He, Z. Huang, Y. Jiang, Facile Synthesis of Covalent Organic Framework Derived Fe-COFs Composites as a Peroxidase-Mimicking Artificial Enzyme, *Nanoscale Advances*, 2 (2020) 1036–1039.
[18] M. Nadeem, G. Yasin, M. Arif, H. Tabassum, M.H. Bhatti, M. Mehmood, U. Yunus, R. Iqbal, T.A. Nguyen, Y. Slimani, H. Song, W. Zhao, Highly Active Sites of Pt/Er Dispersed N-Doped Hierarchical Porous Carbon for Trifunctional Electrocatalyst, *Chemical Engineering Journal*, 409 (2021) 128205.
[19] M.S. Lohse, T. Bein, Covalent Organic Frameworks: Structures, Synthesis, and Applications, *Advanced Functional Materials*, 28 (2018) 1705553.
[20] W.Q. Khan, M. Ali, M.J. Anjum, T. Mehtab, G. Yasin, N. Muhammad, M.U. Malik, S.Q. Lodhi, M.A. Yousaf, M.J.S.O.A.M. Noman, Energy Economised Strategy for Synthesis of Silica and Graphene Oxide Modified Porous Barium Magnesium Niobate Ceramic with Enhanced Dielectric Properties, *Science of Advanced Materials*, 11 (2019) 1118–1125.
[21] Q. Guan, G.-B. Wang, L.-L. Zhou, W.-Y. Li, Y.-B. Dong, Nanoscale Covalent Organic Frameworks as Theranostic Platforms for Oncotherapy: Synthesis, Functionalization, and Applications, *Nanoscale Advances*, 2 (2020) 3656–3733.
[22] X. Wu, X. Han, Y. Liu, Y. Liu, Y. Cui, Control Interlayer Stacking and Chemical Stability of Two-Dimensional Covalent Organic Frameworks via Steric Tuning, *Journal of the American Chemical Society*, 140 (2018) 16124–16133.
[23] A.M.P. Peedikakkal, N.N. Adarsh, Porous Coordination Polymers, in: M.A. Jafar Mazumder, H. Sheardown, A. Al-Ahmed (Eds.) *Functional Polymers*, Springer International Publishing, Cham, 2018, pp. 1–44.
[24] H. Li, L. Li, R.-B. Lin, W. Zhou, Z. Zhang, S. Xiang, B. Chen, Porous Metal-Organic Frameworks for Gas Storage and Separation: Status and Challenges, *EnergyChem*, 1 (2019) 100006.
[25] F.O. Boakye, K. Harrath, M. Tabish, G. Yasin, K.A. Owusu, S. Ajmal, W. Zhang, H. Zhang, Y.-G. Wang, W. Zhao, Phosphorus Coordinated Co/Se2 Heterointerface Nanowires: In-Situ Catalyst Reconstruction Toward Efficient Overall Water Splitting in Alkaline and Seawater Media, *Journal of Alloys and Compounds*, 969 (2023) 172240.
[26] A.J. Blake, N.R. Champness, P. Hubberstey, W.-S. Li, M.A. Withersby, M. Schröder, Inorganic Crystal Engineering Using Self-Assembly of Tailored Building-Blocks, *Coordination Chemistry Reviews*, 183 (1999) 117–138.
[27] S. Ibraheem, G. Yasin, A. Kumar, M.A. Mushtaq, S. Ibrahim, R. Iqbal, M. Tabish, S. Ali, A. Saad, Iron-Cation-Coordinated Cobalt-Bridged-Selenides Nanorods for Highly Efficient Photo/Electrochemical Water Splitting, *Applied Catalysis B: Environmental*, 304 (2022) 120987.
[28] S.-I. Noro, J. Mizutani, Y. Hijikata, R. Matsuda, H. Sato, S. Kitagawa, K. Sugimoto, Y. Inubushi, K. Kubo, T. Nakamura, Porous Coordination Polymers with Ubiquitous and Biocompatible Metals and a Neutral Bridging Ligand, *Nature Communications*, 6 (2015) 5851.

[29] A.M.P. Peedikakkal, Y.-M. Song, R.-G. Xiong, S. Gao, J.J. Vittal, Cobalt(II) Coordination Polymers Containing Trans-1,2-Bis(4-pyridyl)ethene and Their Magnetic Properties, *European Journal of Inorganic Chemistry*, 2010 (2010) 3856–3865.
[30] https://www.atomis.co.jp/en/porous-2/
[31] Y. Zhao, K.X. Yao, B. Teng, T. Zhang, Y. Han, A Perfluorinated Covalent Triazine-Based Framework for Highly Selective and Water–Tolerant CO2 Capture, *Energy & Environmental Science*, 6 (2013) 3684–3692.
[32] K. Venkata Rao, R. Haldar, T.K. Maji, S.J. George, Dynamic, Conjugated Microporous Polymers: Visible Light Harvesting via Guest-Responsive Reversible Swelling, *Physical Chemistry Chemical Physics*, 18 (2016) 156–163.
[33] Y. He, D. Gehrig, F. Zhang, C. Lu, C. Zhang, M. Cai, Y. Wang, F. Laquai, X. Zhuang, X. Feng, Highly Efficient Electrocatalysts for Oxygen Reduction Reaction Based on 1D Ternary Doped Porous Carbons Derived from Carbon Nanotube Directed Conjugated Microporous Polymers, *Advanced Functional Materials*, 26 (2016) 8255–8265.
[34] C.-A. Wang, Y.-W. Li, X.-L. Cheng, J.-P. Zhang, Y.-F. Han, Eosin Y Dye-Based Porous Organic Polymers for Highly Efficient Heterogeneous Photocatalytic Dehydrogenative Coupling Reaction, *RSC Advances*, 7 (2017) 408–414.
[35] J.-X. Jiang, A. Trewin, D.J. Adams, A.I. Cooper, Band Gap Engineering in Fluorescent Conjugated Microporous Polymers, *Chemical Science*, 2 (2011) 1777–1781.
[36] T.-M. Geng, H. Zhu, W. Song, F. Zhu, Y. Wang, Conjugated Microporous Polymer-Based Carbazole Derivatives as Fluorescence Chemosensors for Picronitric Acid, *Journal of Materials Science*, 51 (2016) 4104–4114.
[37] C. Zhang, Y. He, P. Mu, X. Wang, Q. He, Y. Chen, J. Zeng, F. Wang, Y. Xu, J.-X. Jiang, Toward High Performance Thiophene-Containing Conjugated Microporous Polymer Anodes for Lithium-Ion Batteries through Structure Design, *Advanced Functional Materials*, 28 (2018) 1705432.
[38] Q. Wang, Q. Gao, A.M. Al-Enizi, A. Nafady, S. Ma, Recent Advances in MOF-Based Photocatalysis: Environmental Remediation under Visible Light, *Inorganic Chemistry Frontiers*, 7 (2020) 300–339.
[39] L. Tang, Q. Xu, Y. Zhang, W. Chen, M. Wu, MOF/PCP-Based Electrocatalysts for the Oxygen Reduction Reaction, *Electrochemical Energy Reviews*, 5 (2022) 32–81.
[40] P. Ghosh, P. Banerjee, Drug Delivery Using Biocompatible Covalent Organic Frameworks (COFs) Towards a Therapeutic Approach, *Chemical Communications*, 59 (2023) 12527–12547.
[41] X. Liu, D. Huang, C. Lai, G. Zeng, L. Qin, H. Wang, H. Yi, B. Li, S. Liu, M. Zhang, R. Deng, Y. Fu, L. Li, W. Xue, S. Chen, Recent Advances in Covalent Organic Frameworks (COFs) as a Smart Sensing Material, *Chemical Society Reviews*, 48 (2019) 5266–5302.
[42] Z. Li, X. Feng, S. Gao, Y. Jin, W. Zhao, H. Liu, X. Yang, S. Hu, K. Cheng, J. Zhang, Porous Organic Polymer-Coated Band-Aids for Phototherapy of Bacteria-Induced Wound Infection, *ACS Applied Bio Materials*, 2 (2019) 613–618.
[43] R.S. Sprick, Y. Bai, A.A.Y. Guilbert, M. Zbiri, C.M. Aitchison, L. Wilbraham, Y. Yan, D.J. Woods, M.A. Zwijnenburg, A.I. Cooper, Photocatalytic Hydrogen Evolution from Water Using Fluorene and Dibenzothiophene Sulfone-Conjugated Microporous and Linear Polymers, *Chemistry of Materials*, 31 (2019) 305–313.

2 Design Principles and Functional Characteristics of Multifunctional Coordination Materials

Syeda Maida Batool, Uzma Yunus, Muhammad Nadeem, Moazzam H. Bhatti, Nasira Parveen, Mohammad Tabish and Ghulam Yasin

Abbreviations

SPh	S (Sulfur) Ph (Phenyl)
Py	Pyridine
H_2tbta	tetrabromoterephthalic acid
CH_3CN	methyl cyanide
Spy	pyridyl sulfide
THF	Tetra Hydro Furan
Et	Ethyl
Ln	Lanthnides
bbta	1H,5Hbenzo(1,2-d:4,5-d′)bistriazolate
en	1,2-ethanediamine
bpy	2,2‘-bipyridine
pp	4,4′-dimethyl-2,2′-bipyridine
MTSO	p-tolyl sulfoxide
N_3	azido
EE	(end to end)
MeOH	methanol
EO	end on
SR_4	tetrathiolate-bound
S_4	3,7-dithianonane-1,9-dithiolate
qnalH	N-(8-quinolyl)-2-hydroxy-1-naphthaldimine
Dmit	4,5-dithiolato-1,3-dithiole-2-thione
Cp	cyclopentadienyl

DOI: 10.1201/9781003345886-3

CN	cyano
cat	catechol
tip	2,4,6-triisopropylbenzene
mim	2-methylimidazolate
tbmpho	tert-butyl(methyl)phosphine oxide
tp	terephthalate
Hophen	2-hydroxy-1,10- phenanthroline
tepa	tetraethylenepentamine
peha	pentaethylenehexamine
atta	4-(2-aminoethyl)-triethylenetetramine
apda	N-(2-aminoethyl)piperazine-1,4-diethylamine
bpda	N,N-bis(2-aminoethyl)piperazine-1,4-diethylamine
NCS	thiocyanate
DMSO	dimethylsulfoxide
DMF	dimethyl formamide
Acacen	shown in figure 10
bpdbe	1,6-bis(2-pyridyl)-2,5-diaza-1,5-hexadiene
pcih	2-pyridinecarbaldehyde isonicotinoylhydrazonate
dmph	2,9-dimethyl-1,10-phenanthroline(neocuproine)}
salten	2,2′- [iminobis(3,1-propanediylnitrilomethylidyne)]bis-phenol)
bipytz	3,6-bis(4-pyridyl)- 1,2,4,5-tetrazine)
Pro	Proline
Trp	Tryptophan
Ala	Alanine
Phe	Phenylalanine
Val	Valine
Leu	Leucine
Arg	Arginine

2.1 INTRODUCTION

Coordination compounds are made up of central metal atoms or ion-surrounding molecules known as ligands. The ligands cling to the metal with their electron pairs. Bonding in coordination compounds is explained by Werner's theory. Alfred Werner proposed this theory in 1893. Some naturally occurring coordination compounds are vitamin B_{12}, hemoglobin, chlorophyll, etc. Coordination compounds are known to be synthesized by an enormous number of reactions. However, the vast majority of these synthesis methods can be classified into a rather small number of general types. The synthesis of coordination compounds presented in this chapter can describe some important types of reactions. The field of coordination chemistry is so vast that no single volume can provide complete coverage of the body of knowledge that exists concerning the reactions involved in the synthesis of a wide variety of complexes. Different types of synesthetic schemes and methods can be adopted; here these methods are discussed in detail.

2.2 SYNTHETIC METHODS

2.2.1 One Pot Solution Method

By adding ligand and metal salt at the same time, a coordination compound can be synthesized in a single pot. Single-pot solution reaction is mostly used for the synthesis of coordination compounds. for example, the formation of $[Ni(py)_4(NO_2)_2]$ results when $Ni(py)_4(SO_4)$ (whereas py stands for pyridine) with $NaNO_2$ solution in water results in replacement of only one ligand but all ligands can be removed like $[Co(en)_3]Cl_3$ is formed by the reaction of $[CoCl_3(NH_3)_3]$ with 1,2-ethanediamine (en) in aqueous solution. Water is commonly used in such types of reactions. The solubility of ligands in water affects the reaction so, in case of low solubility, other organic solvents are used. For complexes that are not stable in water, anhydrous conditions are used, for example, a reaction of $CoCl_2$ with pyridine resulted in the stable tetrahedral $[CoCl_2(py)_2]$ complex whereas in the presence of water results in dissociation [1].

2.2.2 Diffusion Reactions

Coordination compounds are prepared by diffusion reactions; it is a slow reaction and forms a single-size crystal of high thermodynamic stability. In diffusion reaction, an H-shaped vessel is used in which a solution of reactants diffuses to form single crystals of coordination compounds. In literature, the diffusion reaction $Cd(OAc)_2.2H_2O$ with H-Me-trz-pba (shown in Figure. 2.1a) has been carried out at room temperature for twenty-one days, which resulted in the formation of $[Cd(H\text{-}Me\text{-}trz\text{-}pba)_2(H_2O)_4]$ complex. Similarly, ligand Me-3py-trz-pba (shown in Figure. 2.1b) reacted with cadmium salt by diffusion reaction and resulted in the formation of $Cd(Me\text{-}3py\text{-}trz\text{-}pba)_2(H_2O)_2].H_2O$. A four-fold interpenetrated network structure $[Zn(H\text{-}Me\text{-}trz\text{-}pba)_2].4H_2O$ formed by diffusion method reaction of zinc nitrate and H-Me-trz-pba resulted in a four-fold structure with diamond pattern [2].

Synthesis of two different types of co-ordination polymers is possible by reaction of zinc nitrate or chloride with an organic ligand by diffusion reaction. The versatile co-coordination polymers reported in literature are {[Zn(btp)(1,4-bdc)][Zn(btp)

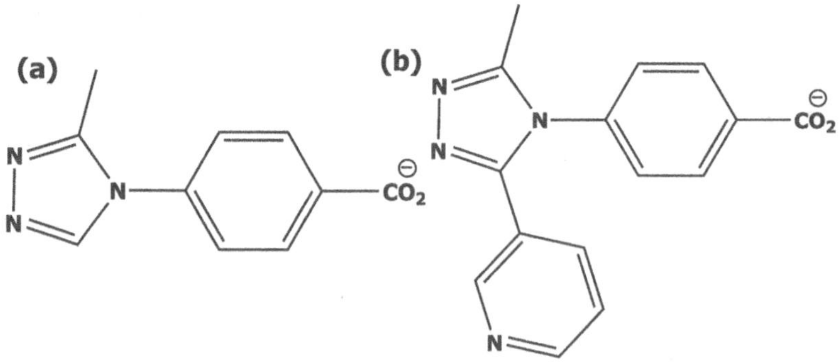

FIGURE 2.1 Structure of ligands (a) H-Me-trz-pba and (b) Me-3py-trz-pba.

$(1,4\text{-bdc})_{0.5}Cl]H_2O\}_n$ and $\{[Zn(btp)(1,4\text{-bdc})].\ CH_3OH.2H_2O\}_n$. Here btp stands for c along with 1,4-bdc which is 1,4-benzenedicarboxylate [3].

Similarly different metals nitrate including Pd (II), Cd(II), Ni(II), Cu (II), Zn (II) or Ag (II) form coordination compounds by reacting with ligands such as H_2tbta (tetrabromoterephthalic acid) through diffusion reaction [4].

2.2.3 Mechanical Grinding

Grinding of reactants to produce coordination compounds is an environmentally friendly method also known as mechanochemical grinding for the synthesis of coordination compounds. Though, grinding is a less popular method, as mechanical grinding takes place either in the presence or absence of solvent, so structural changes occur due to the loss or uptake of solvent, which in turn leads to the self-assembly of coordination compounds. Further, mechanical grinding results in a change of structure, which alters reactivity. For example, coordination complex $[Zn(bpe)_2(H_2O)_4](NO_3)_2.3H_2O.2/3bpe$ where bpe stands for 4,4'-bipyridylethylene results in increased photo chemical [2+2] cycloaddition reaction as a result of grinding [5]. In the case of formation of coordination polymers $[Zn(m\text{-bpe})(Br)_2]$ and $[Cd(m\text{-bpe})(m\text{-Br})_2]$, bromine in KBr replaces the carboxylate ligand in CP $[Zn_2(m\text{-bpe})_2\text{-}(m\text{-}CH_3CO_2)_2(CF_3CO_2)_2]$ and CP (coordination polymer) $[Cd(CH_3CO_2)_2(m\text{-bpe})(H_2O)]$ during mechanical grinding as bromine is capable of bridging [6, 7]. Further, $Pt(en)(NO_3)_2$ forms coordination product $[Pt(en)(4,4'\text{-bipy})]_4(NO_3)_8$ with 4,4'-bipyridine by mechanical grinding in the presence of solvent while oligo(bipyridine) powder on grinding with $[(CH_3CN)_4Cu]PF_6$ results in the formation of Copper (I) Bipyridine Based Helicates [8].

The formation of covalent bonds through mechanical grinding includes the formation of *cis*-$(Ph_3P)_2PtCl_2$ and *cis*-$(Ph_3P)_2PtCO_3$, which are platinum complexes. Similarly, the formation of ferrocenyl complexes such as $[Fe(\eta^5\text{-}C_5H_4\text{-}1\text{-}C_5H_4N)_2]$ (shown in Figure 2.2) takes place faster in the absence of solvent in mechanical grinding as compared to the solution approach. This mechanochemically formed complex reacts further with salts of Ag, Cd, Cu, and Zn to form macrocyclic complexes. Formation of coordination polymers such as $Ag[N(CH_2CH_2)_3N]_2[CH_3COO]\cdot 5H_2O$ also takes place through grinding silver salt with ligand and by replacing silver salt with Zn and on prolonged grinding with the same ligand $[N(CH_2CH_2)_3N]$ forms $Zn[N(CH_2CH_2)_3N]Cl_2$ [9].

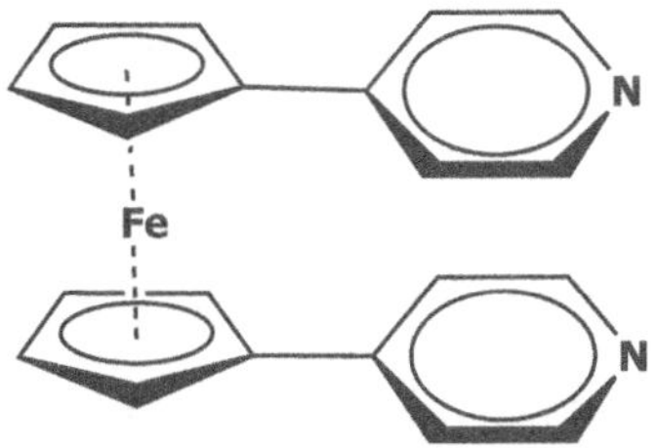

FIGURE 2.2 Structure of $[Fe(\eta^5\text{-}C_5H_4\text{-}1\text{-}C_5H_4N)_2]$.

2.2.4 Microwave Heating

Electronic factors that slow down the synthesis of coordination compounds can be increased by microwave irradiation, although this method is commonly used for the synthesis of organic compounds, whereas enantiomeric-rich compounds are required but in the case of inorganic coordination compounds asymmetric compounds can be obtained by microwave heating. Ease of formation is more in the case of third-row transition metals and decreases for the fourth and fifth rows. Racemic mixture of cis-$[Ru(bpy)_2(Cl)_2].2H_2O$ (where as bpy = 2,2'-bipyridine) and racemic cis- $[Ru(phen)_2(Cl)_2]$ (phen stands for phenanthroline) on reaction with enantiomerically pure methyl p-tolyl sulfoxide (MTSO) in the presence of microwave radiations resulted in the formation of ruthenium complexes with a high level of asymmetry [10]. 2,2'-bipyridine (bpy) and reacts with metals such as Fe, Ru, and Os to form metal (bpy) complexes in the presence of microwave irradiation under more rigorous conditions [11].

Examining the effect of microwave radiation on the formation of coordination complexes, molybdenum is the most-studied element; for example, the formation of $Mo(py)_n(CO)_n$ takes place by reacting molybdenum hexacarbonyl with pyridine in the presence of microwave radiation. The number of n in $Mo(py)_n(CO)_n$ increases from 1 to 3 on increasing the time of microwave radiation. This observation serves as good evidence of the fact that microwave heating can accelerate the reaction to completion within seconds, which is otherwise difficult to achieve with conventional heat [11]. Similarly, $Mo(CO)_6$ forms an N-coordinated complex by reaction with ethyl 2-(5-(pyridin-2-yl)-1H-pyrazol-1-yl)acetate in the presence of microwave radiation for a few minutes [12].

2.2.5 Ligand Substitution Method

The ligand substitution as in equation (2.1) is one of the most important methods for the synthesis of coordination compounds. In the case of ligand substitution, the reaction stereochemistry of the product remains unaltered.

$$ML_3X + Y \rightarrow ML_3Y + X \tag{2.1}$$

An example is the reaction of $Ni(CO)_4$ with PCl_3 to substitute CO with chlorine. The different parameters that can affect the ligand substitution reaction are the coordination number of metal, nature of metal, the charge on metal, replacement ligand, leaving group, the position of a group, and geometries [13]. The ethyl group has replaced the chloride ion in the substitution reaction as follows in equation (2.2).

$$K_2PtCl_4 + 2(C_2H_5)2S \rightarrow \{Pt[(C_2H_5)_2S]Cl_2\} + 2KCl \tag{2.2}$$

In the case of $Rh(PCP)(PPh_3)$, triphenyl phosphine is replaced by other ligands like carbon monoxide, isocyanate, and ethene. The reaction with ethene is as shown in the following figure [14].

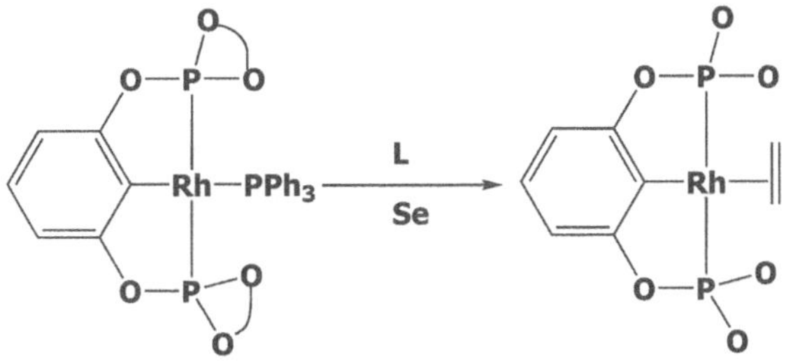

In ligand substitution reaction main ligands are added before metal important auxiliary ligands. The metal-important auxiliary ligands are pyridine, bi-pyridine, 1,10-Phenanthroline, hexamethyl phosphoryl triamide, triphenylphosphine, and triphenylphosphine oxide. The purpose of auxiliary ligands is added in order to increase the stability as well as the solubility of coordination compounds in solvents [15].

Substitution reaction of trinuclear cluster $[Pd_3(GaCp^{*Ph})(\mu_2\text{-}GaCp^{*Ph})(\mu_3\text{-}GaCp^{*Ph})_2\ (dvds)]$ and dinuclear cluster $[Pd_2(GaCp^{*Ph})_2(\mu_2\text{-}GaCp^{*Ph})_3]$ whereas $(Cp^{*Ph}=C_5Me_4Ph)$ reacts with GaCp*, AlCp* or phosphines and forms substitution products. Ligand substitution with Cp*Ga leads to the formation of the trinuclear species $[Pd_3(GaCp^*)_4(\mu_2\text{-}GaCp^*)_4]$ shown in equation [2.16].

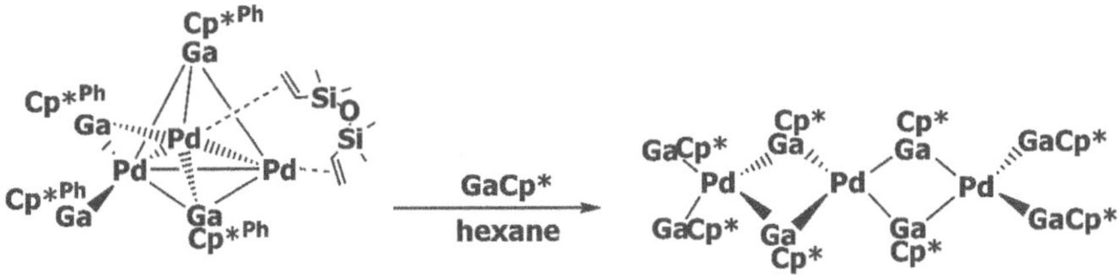

2.2.6 Metal Exchange Reactions

Replacement of one metal with another is difficult because metal connects the ligands together, so, in this case, stability is disrupted as well as the problem of metal selectivity. Therefore, a complex containing a single metal can be used for the preparation of heterometallic complexes with selectivity by metal on exchange. So, it is important to maintain connectivity and crystallinity in case of metal exchange. Pb ions replace Cd ions in $Cd_{1.5}(H_3O)_3[(Cd_4O)_3(hett)_8].6H_2O$ (hett structure is shown in scheme 1) by adding aqueous $Pb(NO_3)_2$ without disturbing structural geometry and crystallinity as shown in Scheme 2.1 [17].

Removal of Mn(II) from a complex containing two metals takes place by reaction of a bimetallic complex like $[FeMnL_2Cl]$ (where as L_2 is shown in Figure. 2.3) with $FeCl_2$ resulting in the formation of $[Fe_2Cl\ L_2]$, but the same reaction if carried out with $CoCl_2$ $[Co_2ClL_2]$ does not form $[Fe_2Cl\ L_2]$. Instead $[FeCoL_2Cl]$ is formed with an irreversible conversion but on using $[CoMnL_2Cl]$ with $CoCl_2$ results in the replacement of Mn(II) and formation of $[Co_2ClL_2]$, irreversible conversion due to stability of product [18].

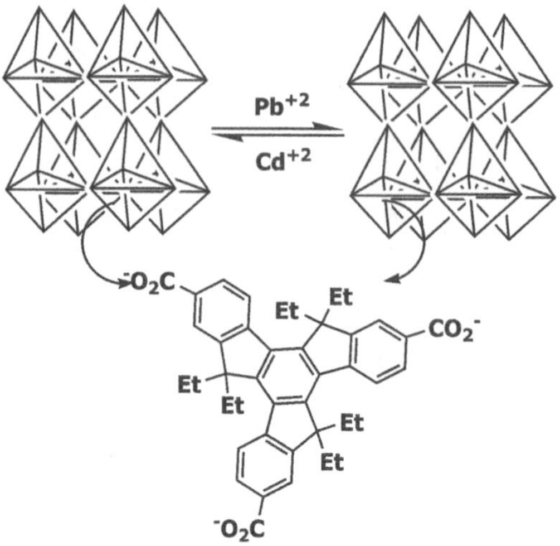

SCHEME 2.1 Structure of hett with the exchange of Pb^{+2} ions by Cd^{+2} ions.

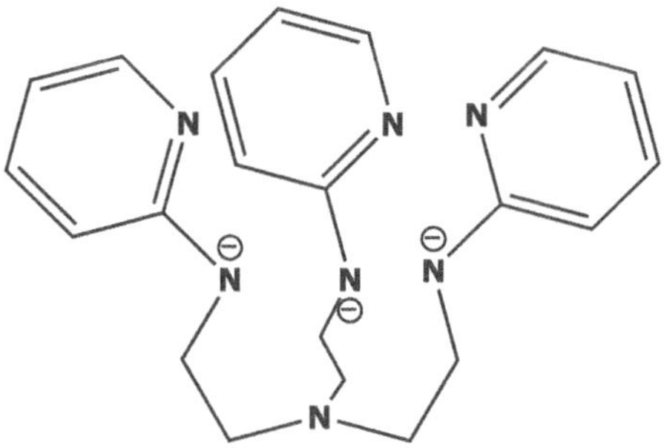

FIGURE 2.3 The structure of L_2 as ligand.

Similar metal-ion exchange occurs in the methanolic solution of $MnCl_2$ to create a single crystal of Cd(II) coordination polymer, resulting in the replacement of cadmium with manganese without altering crystallinity through single-crystal to single-crystal fashion. Similarly, $\{[Zn(L)_2(H_2O)_2(PF_6)pyrene_2(H_2O)\}_n$, where L is benzene-1,3,5-triyltriisonicotinate, converted to $[Cu(L)_2(H_2O)_2(PF_6)pyrene_2(H_2O)\}_n$ on addition of copper nitrate methanolic solution and resulted in the substitution of zinc metal with copper. Complete conversion occurred in 6 hours; a change in color was observed in 15 minutes. Some conversions of metals take a lot of time to complete and are not completely converted, for example, $\{[Cu(L)_2(H_2O)_2(PF_6)$ $pyrene_2(H_2O)\}_n$. In this case replacement of Cu(II) with cadmium Cd(II) is only half converted even after 10 days [19].

2.2.7 Hydrothermal/Solvothermal Method

Teflon autoclave is used for hydrothermal/solvothermal reactions using aqueous or organic solvent at a temperature of 100–250°C under self-generated pressure.

Such a method favors the formation of crystals. For example, manganese chloride reacted with $Na[N(CN)_2]$ and sodium azide to form colorless crystals of $[Mn\{Mn_3(\mu_3\text{-}OH)(bta)_3(H_2O)_6\}_2]$ and by reaction of $CdCl_2$ with $Na[N(CN)_2]$ and sodium azide to form $[Cd\{Cd_3(\mu_3\text{-}OH)(bta)_3(H_2O)_6\}_2]$ similarly by reaction of KF with $Na[N(CN)_2]$ and sodium azide to form $[Cd\{Cd_3(\mu_3\text{-}F)(bta)_3(H_2O)_6\}_2]$ [20]. Similarly, $[Mn_3O(Me_3CCO_2)_6(py)_3]$ reacts with $H_2O_3PPh(Me)_2$ and $[Ph_2CHPO_3H_2]$ resulting in formation of $[Mn_{19}O_8(O_2CCMe_3)_{10}(OCH_3)_{16}(O_3PPh(Me)_2)_6]$ and $[Mn_{16}(Ph_2CHPO_3)_8(O)_8(OMe)_4(HOMe)_2(Me_3CCO_2)_6(OH_2)(Py)_2]$ respectively by the solvothermal process at 125°C for 12 hours [21]. Other examples include metal–tetrazole coordination complexes. The complexes involve formation of [Zn(OH)(3-PTZ)] whereas PTZ=5-(3-pyridyl)tetrazolate). Changing the temperature of the reaction from 160°C to 105°C results in the formation of $[(3\text{-}PTZ)_2Zn]$, which is also a complex of tetrazole. Another tetrazole complex formed is $[CdN_3(3\text{-}PTZ)]$ by using $CdCl_2$ along with sodium azide and 3-cyanopyridine. [ZnCl(4-PTZ)] is also formed under hydrothermal conditions as shown in Scheme 2.2 [22].

Hydrothermal reaction for the synthesis of $[Zn_3(btrc)_2(1,10\text{-}phen)_2(H_2O)_2]_n$ and $[Co_3(btrc)_2(1,10\text{- }phen)_2(H_2O)_2]_n$ whereas btrc stands for 2,4-benzenetricarboxylate on reaction with N-heteroaromatic compound with Zn and Co acetate [23]. Hydrothermal synthesis of $C_{48}H_{32}Cd_2N_8O_{18}$ and $C_{17}H_{13}CdN_3O_5$ by reaction of pydcy which stands for 4,4′-bis(3,3′-dicyano)pyridine with cadmium nitrate hexahydrate.

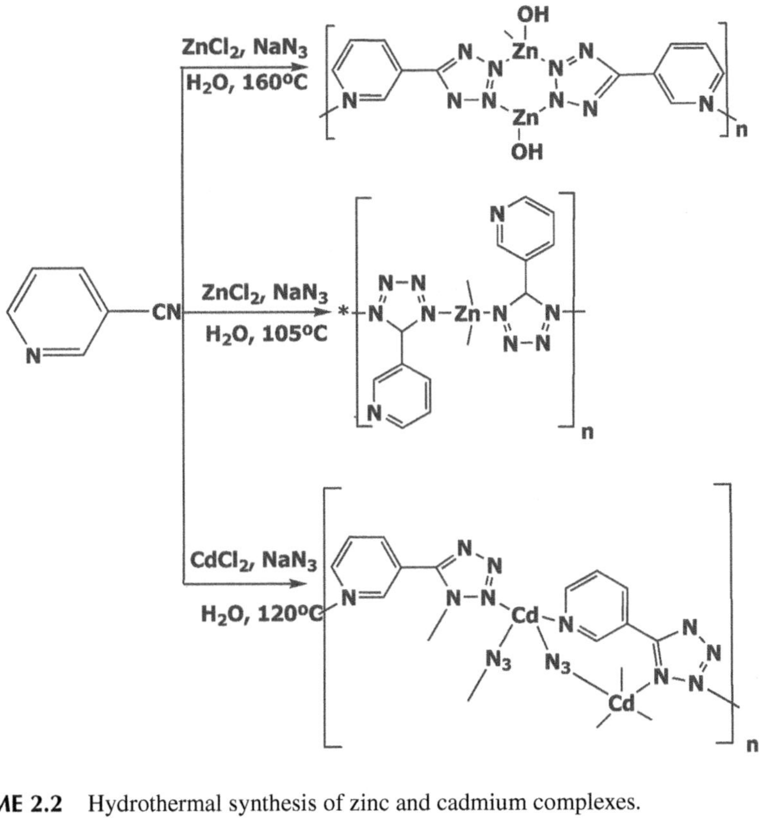

SCHEME 2.2 Hydrothermal synthesis of zinc and cadmium complexes.

In this hydrothermal reaction, the solvent affects the structure of formed cadmium compounds [24]. Hydrothermal synthesis of $[Ln_7(\mu_3\text{-}OH)_8]^{13+}$, which is a lanthanide cluster, is synthesized by hydrothermal method [25].

2.2.8 Redox Interaction Methods

2.2.8.1 Oxidation of Metal

In this method metal is reacted with a ligand in the presence of an oxidizing agent, for example, compound $[Cr_2(2gb)_4(\mu\text{-}OH)_2](ClO_4)_4{\cdot}5H_2O$ obtained from which is a Cr(III) complex formed from an acid aqueous solution of Cr(II) in the presence of $HClO_4$ followed by air oxidation, resulting in the formation of chromium complex (where 2gb stands for 2-guanidinobenzimidazole) [26]. The oxidation of uranium metal by dichalcogenides resulted in the formation of molecular uranium chalcogenolates and a small amount of iodine was used as an oxidizing agent. $[U(SPh)_4(py)_3]$, $[U(SePh)_4(py)_3]$, $[U(SPh)_4(CH_3CN)_2]_2$, $[U(SePh)_4(CH_3CN)_2]_2$, $[U(Spy)_4(thf)]$ and $[U(py)_2(SePh)(\mu_3\text{-}Se)(\mu_2\text{-}SePh)]_4.4py$ resulted from oxidation of uranium metal. Dichalcogenide in the reaction mixtures was intended to result in the oxidation of uranium metal to U(III) chalcogenolate complexes. Uranium metal can be oxidized by 1.5 equivalent of iodine in pyridine to form $UI_3(py)_4$ [27]. Similarly, Ln(III) pyridyl thiolates have been prepared by the oxidation of elemental lanthanide with dipyridyl disulfide to form both anionic $[PEt_4][Ln(Spy)_4]$ and neutral $[Ln(Spy)_3(py)_2]$ complexes. Similarly, synthesis of $[(THF)_3Ce(SC_6F_5)_3]$, $[Er(THF)_3\ (SC_6F_5)_3]$ and $[Ho(THF)_3(SC_6F_5)]$ have been performed by the reaction of Ce, Er, and Ho metal with $Hg(SC_6F_5)_2$ in the presence of THF, respectively [28].

2.2.8.2 Oxidation of Low-Valent Metal

Fe (II) can be easily oxidized to Fe (III) by O_2 in the air under the solvo(hydro)thermal conditions. Mn (II) MOF (MAF-X25) with monomer unit results in 1H,5Hbenzo(1,2-d:4,5-d′)bistriazolate (bbta), which exhibited a 3D framework with honeycomb-like structure. About half of the Mn(II) ions in MAF-X25 can be oxidized by H_2O_2 into Mn(III) ions, resulting in the first Mn(III) MOF with retention of the framework and crystallinity as shown below in Figure. 2.4 [29].

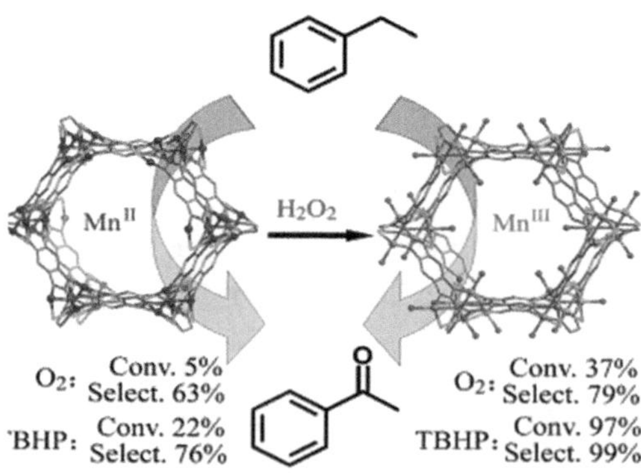

FIGURE 2.4 Oxidation of Mn(II) to Mn(III) in MAF-X25 MOF.

2.2.9 Chiral Resolution Method

Chiral coordination compounds, due to their distinct properties, possess tremendous applications in the fields of material sciences, life sciences, and catalysis. Several synthetic methods have been recognized for the synthesis of enantio-rich transition metal complexes. Werner in 1911 was the first who performed resolution of the octahedral coordination compounds $[Co(en)_2X(NH_3)]^{+2}$ in its enantiomers [30]. Different chiral auxiliaries used for asymmetric synthesis are tartrate, chiral monodentate sulfoxide, Chiral Salicycloxazolines, cleavable chiral linkers, boronic acid-saccharide, and chiral anions. Ruthenium complex $[Ru(bpy)_3]^{2+}$, which is configurationally inert, reacted with (*R*,*R*)-(+)-tartrate and resulted in the formation of ruthenium complex with enantiomeric ratio of 63:37. Racemic mixture of cis-$[Ru(pp)_2(Cl)_2]$ on reaction with enantiomerically pure methyl p-tolyl sulfoxide (MTSO) in the presence of microwave radiations resulted in the formation of ruthenium complexes with a high level of asymmetry. The chiral salicyloxazolinate ligand induces chirality in ruthenium complexes. At the end, replacement of the salicyloxazolinate by bpy resulted in complete retention of configuration and a high level of enantiopurity. Cleavable chiral auxiliaries are not linked directly with ligands; they linked with metal and then the connection was made with the ligand. An octahedral chiral two-bladed iron complex is formed by using a chiral tartrate linker as the chiral auxiliary. D-glucose is used as a chiral auxiliary for stereoselective synthesis of $[Co(bpy)_3]^{3+}$ [31].

The conglomerates of $[Mn_3(L_1)_2(N_3)_6]_n$ and $[Mn_2(L_2)_2(N_3)_3]_n(ClO_4)_n$,$nH_2O$ were synthesized by resolution of L1 and L2 induces spontaneous resolution whereas L1 named (1,4-bis(2-pyridyl)-1-amino-2,3-diaza-1,3- butadiene) and L_2 was (1,4-bis(2-pyridyl)-1, 4-diamino-2,3-diaza-1,3-butadiene), where single EE (end-to-end) azido bridges the tri- or binuclear chiral units produced by the diazine ligands to form homochiral one dimensional (1D) chains. Chiral crystals are produced when the chains are stacked in a homochiral manner via hydrogen bonding [32]. In another example, $[Mn_2(L_3)(N_3)_4]_n$,nMeOH and $[Mn_2(L_3)(N_3)_4]_n$, here L3 named as (2,5-bis(2-pyridyl)-3,4-diaza-2, 4-hexadiene) and resulted incomplete resolution, but occurred spontaneously, where single EE azido bridges connect binuclear chiral units to form homochiral two-dimensional (2D) layers. The polymer is layered either homochirally or heterochirally to produce a racemic compound at the same time. Another ligand 1,4-bis(2-pyridyl)-2, 3-diaza-1,3-butadiene) (referred to as L4) did not produce resolution spontaneously. The complex formed in this case is $[Mn_2(L_4)(N_3)_4]$. By alternating single diazine, single EE azido, and double EO (end on) azido bridges, a 2D centrosymmetric layered molecule resulted. These substances have remarkably diverse magnetic properties. Specifically, $[Mn_2(L_2)_2(N_3)_3]_n(ClO_4)_n$,$nH_2O$ acted as a 1D antiferromagnet material with alternating antiferromagnetic interactions, while $[Mn_2(L_3)(N_3)_4]_n$,nMeOH and $[Mn_2(L_3)(N_3)_4]_n$ acted as spin-canted weak ferromagnets with distinct critical temperatures, and $[Mn_2(L_4)(N_3)_4]$ likewise acted as a spin-canted weak ferromagnet but displayed two-step magnetic transitions at two different temperatures [32].

2.2.10 Electrochemical Oxidation

Coordination compounds can be selectively prepared by electrochemical oxidation as it is highly economical, produces good yield, and there are fewer chances of impurities. As an example, when catechol undergoes electrochemical oxidation, it builds a complex with tin (II); such types of compounds possess an unusual oxidation state. $Sn(O_2Ar)$ with Sn(II) oxidation state on electrochemical reaction with catechol derivates produces Sn(IV) complex as given in equation (2.4). Several cells in series are used to yield products in large quantities [33].

$$\left[Sn^{II}\left(O_2Ar\right)\right]^0 \rightarrow \left[Sn^{IV}\left(O_2Ar\right)\right]^{+2} \tag{2.4}$$

Electrochemical reactions are irreversible. Other examples of electrochemical oxidation include $[Fe_4S_4(SR)_4]^{-1}$ from $[Fe_4S_4(SR)_4]^{-2}$ (R is aryl and alkyl substituent groups); the electrochemical oxidation of hydroxoiron(III) porphyrins resulted in the formation of Fe(IV)==O(phor) complexes; electrochemical oxidation of Cu (I) to Cu (II) takes place by three-coordinate copper(I) center [34].

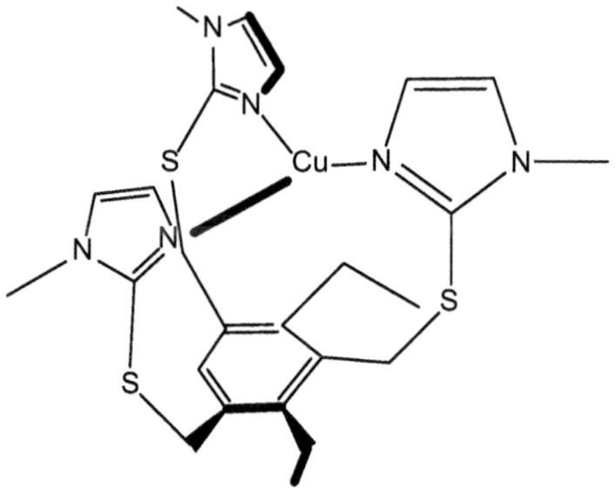

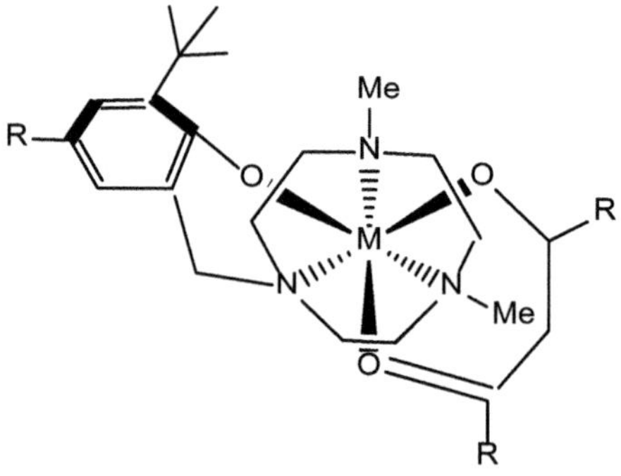

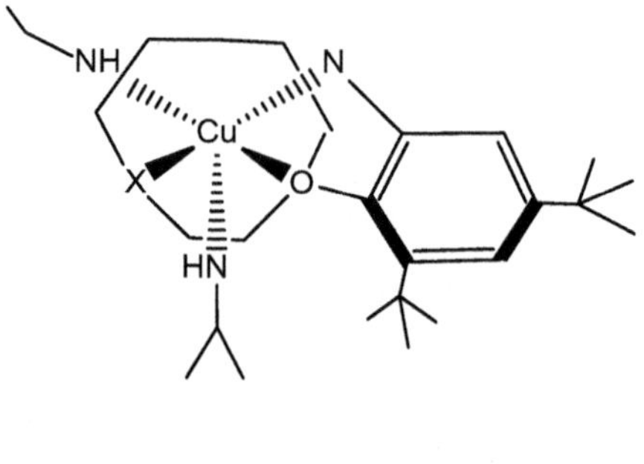

(1) (2) (3)
(M=Cu(II), $n = 1$) R'=Ph or Bu (X=Cl–, $n = 1$)
(X=CH_3CN, $n = 2$)

The mononuclear Cu(II)-phenoxyl radical species such as (1), (2), and (3) formed from Cu(II) phenolate complexes on electrochemical oxidation [35].

Conducting crystals of $[Fe(qnal)_2][Pd(dmit)_2]_5$ acetone are formed by electrochemical oxidation of $[Fe(qnal)_2][Pd(dmit)_2]$ by applying a constant voltage in an H-type glass cell [36].

The stability of complexes formed by electrochemical oxidation depends on the metal and solvent used. So, in the case of metallocene containing cyclopentadienyl CP_2M^{+2} moiety, one electron transfer is a reversible change.

Conversion of Fe(III) to Fe(II) in ferricyanide, $[Fe(III)(CN)_6]$, takes place electrochemically. Here the ECE mechanism is suggested (first electron transfer step (E) chemical step (C) then second electron transfer step E step) for the oxidation of $[Fe(III)(CN)_6]^{3-}$, depicting the presence of a Fe(IV) intermediate and the reductive removal of two cyanide ligands to generate a $[Fe(III)(CN)_4(CH_3CN)_2]^{1-}$ complex. DFT simulations helped with the speciation of oxidation of ferricyanide. The present investigation broadens the scope of the redox chemistry of $[Fe(III)(CN)_6]^{3-}$ beyond the traditional reversible $[Fe(III/II)(CN)_6]^{3-/4-}$ pair. This implies the possibility of novel redox reactions in closely related cyanometallates, such as Prussian blue and its derivatives [37].

2.2.11 Metalloligand Method

Metalloligand is such type of method in which metal complexes contain an electron-rich ligand that can act as a donor to other electrophilic metal ions. It is a strategy for the synthesis of extended coordination in complexes. Metalloligands may be divided into two types. Type-1 includes metal-containing ligand systems in which metal

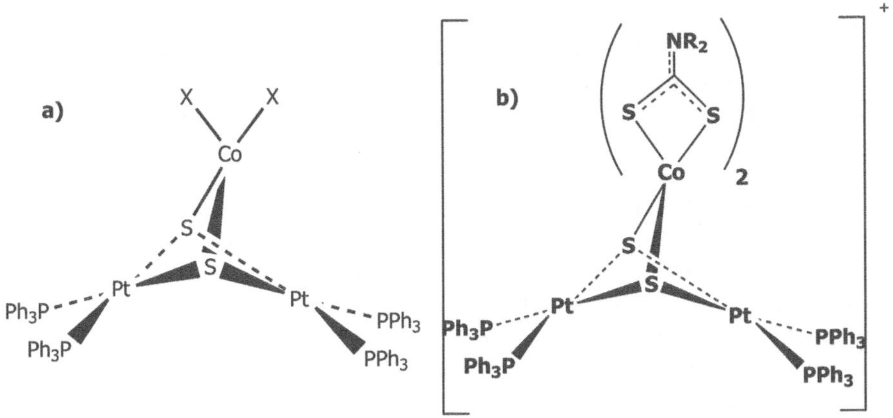

FIGURE 2.5 (a) and (b) metalloligand synthesis of heterometallic complexes.

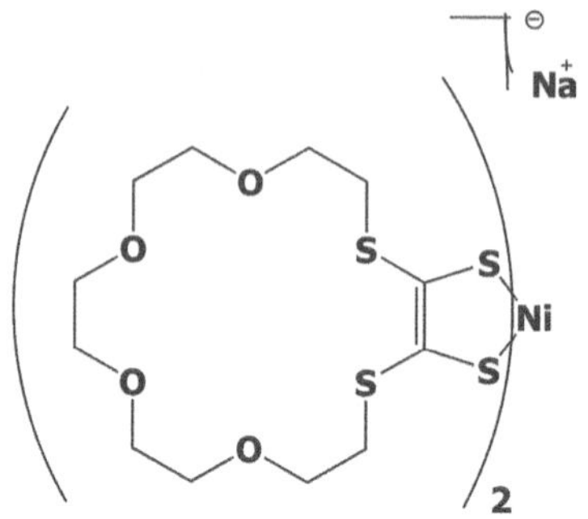

FIGURE 2.6 Structure of crown ether complex.

ions are not involved in coordination. Type-2 of metalloligands includes metal ions that result in further coordination by directing an unbound ligand. For example, the metalloligand complex $[Pt_2(\mu\text{-}S)_2(PPh_3)_4]$ contains sulfide as a donor ligand to form cobalt (II) and cobalt (III) complexes. Cobalt complexes 5a and 5b are produced by metalloligand reaction with cobalt chloride and $NaS_2CNEt_2.3H_2O$, respectively [38].

Similarly, $[\{Ni(EtN_2S_2)\}2Fe_4S_4(Stip)_2]$ was synthesized using the metalloligand approach by reaction of nickel complex containing donor thiolates with $[Fe_4S_4(SR)_4]^{2-}$ [39]. In addition to sulfur, oxygen can also act as a bridging ligand. For example, $[M(cat)_3]^{-3}$ (cat mean catechol) metalloligands can bind to guest metal ions via oxygen atoms. Other complexes used as metalloligands include nitronyl nitroxide metal complexes [40]. Metallocrowns are the type of metallo-ditopic ligands with 3D metal ions, featuring a repeating unit having a cyclic arrangement. Green first described complexes bearing such crown ethers in 1990. An example of a crown-ether substituted paramagnetic complex isolated as its Na^+ salt is shown in Figure. 2.6 [41].

2.2.12 Self-Assemblies

Self-assembly is an excellent method for the construction of interconnected complicated structures such as nanocages, helicates, polyhedra, catenanes, polygons, and knots. These self-assembled species have well-defined cavities. These cavities result in the enhancement of magnetic, optical, and other properties of coordination compounds. Among nanocages heterometallic nanocaged compounds like $[Fe_2Co_3L_6Cl_6]$, $[Fe_2Zn_3L_6Br_6]$ and $[Cr_2Zn_3L_6Br_6]$ formed by coordination-driven self-assembly of FeL and CrL (where L was 1-(4-pyridyl)butane-1,3-dione) metalloligand) with cobalt(II) chloride and zinc(II) chloride, respectively [42]. Similarly, a metallic cage was produced when pyridyl [2] rotaxane ligand self-assembled in the presence of palladium (II) ions to form a Pd_2L_4, a metallo supramolecular cage a metallo-rotaxane [43].

Formation of helicates such as dinuclear helicates $[Mn_2\mathbf{L}_3](ClO_4)_4$, $[Co_2\mathbf{L}_3](BF_4)_4$, $[Ni_2\mathbf{L}_3](BF_4)_4$, $[Cu_2\mathbf{L}_3](BF_4)_4$, and $[Zn_2\mathbf{L}_3](BF_4)_4$ where L is referred as two α-diimine primary donor groups and two secondary 4-pyridyl donor groups uses metalloligand self-assembly [44].

Self-assembly procedure among complexes resulted in the formation of an angle of a polygon; e.g. in the case of a reaction of Cu(I) with ligand 2-methylimidazolate (mim), different molecular polygons have been produced, which includes molecular octa/decagons [45].

Cages can be interlocked to produce catenane, for example, 3D metallosupramolecular assembly catenane was fabricated, when symmetric cages interlock themselves through π-π interactions between ligands [46]. Owing to this procedure, octahedral coordination template metal/anion knots can be synthesized.

The supramolecular structures can be generated through the self-assembly of anion-coordination. Here anion-templated synthesis of some typical macrocyclic molecules is very important; here anions are used as templates to create different types of macromolecules. Examples include the formation of pseudopeptidic macrocycles, oligopyrrolic macrocycles, cyclo[8]pyrrole, hexaurea macrocycle, cyclic tetraureas, and pentagonal cyanostar macrocycles. Among anion-assisted metal-coordination self-assemblies, anion affect the formation of supramolecules as shown later.

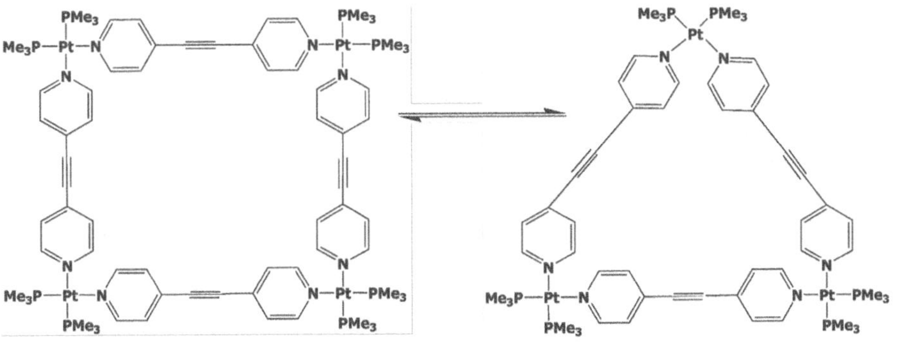

By changing the concentration of anion equilibrium, position also changes. Other examples of anion-assisted self-assembly include the use of oxoanions like AO_4^{-n} whereas A = (Cr, P, W, Mo, Se and S), and *n* ranges from 2–3, resulted in the formation of M_4L_6 ($M = Ni^{2+}$, Zn^{2+}) tetrahedral cage receptors, which showed selective encapsulation of tetrahedral AO_4^{n-} oxoanions from the aqueous solution.

Self-assembly involves the coordination of anions in supramolecules as anions also act as coordination centers. In addition to the synthesis of cages and helicates, anionic coordination is also effective in the synthesis of foldamers. For example, indolocarbazolepyridine hybrid ligand in addition of a sulphate anion resulted in the formation of a helical structure [47].

2.2.13 Schlenk Line Techniques

Wilhelm Schlenk was the first to design an apparatus that was actually a modification of the procedure developed by Boudjouk. Depending on the sensitivity of the reactants and products, changes in apparatus are required. When coordination compounds are sensitive to air and/or moisture, Schlenk line techniques are often required for such manipulations. For example, prior drying and deoxygenation are essentially required for the synthesis of anhydrous tetrahydrofuran complexes from metal chloride hydrates [48].

$[HN(SPPh_2)(SePPh_2)]$ was produced only by Schlenk line technique and this compound was used to form metal-chelated complexes with metals including Bi, Sn, Zn, and Co [49]. The synthesis of many versatile coordination compounds like [IrCl(COD)(κP-(R)-tbmpho)], [AuCl(κP-(S)-tbmpho)], $[RuCl_2(\eta^6$-p-cymene) (κP-(R)-tbmpho)], $[Pd(\mu$-Cl)(κP-(S)-tbmpho$)_2]_2$, and $[Pd(\mu$-OAc)(κP-(S)-tbmpho)2], where tbmpho is tert-butyl(methyl)phosphine oxide, was performed by the Schlenk line technique [50] and synthesis of (Phenolato)-(Carboxylato)iridium hydride complexes was also done by the use of the Schlenk line technique [51].

2.2.14 In-Situ Metal Ligand Reaction

Hydro(solvo) thermal in-situ metal/ligand reaction has rapidly developed as a powerful approach in coordination chemistry and organic synthesis due to its effectiveness, simplicity, and environmental friendliness. The novel coordination complexes and organic ligands produced by hydro-/solvo-thermal in-situ reactions are inaccessible or not easily obtainable by routine organic synthetic methods. In fact, several in-situ ligand reactions are important in organic chemistry, including cycloaddition of organic nitriles with azide and ammonia, carbon-carbon bond formation, and hydroxylation of aromatic rings [52], for example, a mixed-valence Cu(I)/Cu (II) complex $[Cu_4(ophen)_4(tp)]$ (Hophen=2-hydroxy-1,10- phenanthroline, tp=terephthalate) with in-situ generated Hophen ligand, by heating $Cu(NO_3)_2$, 1,10- phenanthroline, H_2tp, NaOH, and H_2O at 160°C for 144 hours [52]. The in-situ metal-ligand approach provides a new perspective to generate unconventional molecules such as a ligand that might not be easily possible like tepa (tetraethylenepentamine) and peha (pentaethylenehexamine), which undergo interesting in-situ ligand rearrangements in the presence of the halide/pseudohalide and metal ions and resulted in the formation of 4-(2-aminoethyl)-triethylenetetramine (atta), N-(2-aminoethyl)piperazine-1,4-diethylamine (apda), and N,N-bis(2-aminoethyl)piperazine-1,4-diethylamine (bpda) ligands. Mercury (II) complexes synthesized were $[Hg(atta)][HgCl_4]$ and cadmium (II) complexes formed were $[Cd_5(apda)_2Cl_{10}]_n$, $[Cd_3(apda)_2(N_3)_4]_n(ClO_4)_{2n}$ and $[Cd_2(bpda)(NCS)_4]_n$. Through strong metal-ligand covalent bonds combined with weak non-covalent forces like hydrogen bonds and S . . . S interactions, such in-situ metal-ligand processes produce novel MOFs [53].

2.2.15 Direct Synthesis from Zerovalent Metals

Zerovalent metals are used for the synthesis of all types of coordination compounds. Zerovalent metals produce π-complexes by gas-phase reactions while developing molecular adducts in liquid-phase reactions. Diverse organic ligand systems can be used in the direct synthesis method. In addition, metals of all groups can be employed in direct synthesis. The reaction of metal vapors with alkenes and haloderivatives result in the formation of π-complexes in the gas phase. For example, the reaction of florosubstituted alkene with palladium produced π-complexes at 77K as shown below [54].

$$Pd + F_3C{-}\underset{F}{C}{=}\underset{F}{C}{-}CF_3 \xrightarrow{77K} Pd{-}\left[\,(F)(CF_3)C{=}C(F)(CF_3)\,\right]$$

Other metals such as Fe, Cr, Mn, and Co can also generate π-complexes on reaction with an unsaturated source in the gas phase. The metal chelates such as metal acetylacetonates upon reaction with metal vapors result in the formation of β-ketonane [54].

The synthesis of metal complexes from zerovalent metal in the liquid phase can be performed either through direct reaction or breaking up metal in water-free solvent or through the application of electricity. The direct reaction of metal with ligand involves the reaction of 1,2-dicyanobenzene with any of zerovalent metals such as Cd, Sn, Zn, Cu, Ni, and Fe resulting in the formation of phthalocyanine complexes [54]. The structure of the phthalocyanine complex is shown in Figure. 2.7.

Metal in a zero oxidation state reacts with an oxidant in the presence of a ligand and causes it to produce coordination compounds. Molecular complexes formed in this way include $CrCl_{(1-3)}$ $DMSO_{(1-3)}$, which is neutral while $[Fe(III)(DMF)_6]$ $[Fe(III)Cl_4]_4$ is an example of an ionic complex; here dinuclear and polynuclear molecular complexes can be formed in this way. Similarly, the tetra nuclear cluster structure of $[Cu_4(OC1)_6(DMF)_2]$ (shown in Figure. 2.8) was obtained by the reaction of DMF and CCl_4 with zerovalent copper [54].

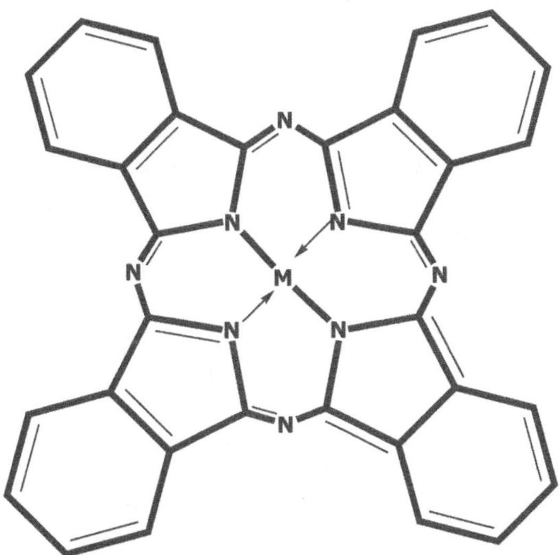

FIGURE 2.7 Structure of phthalocyanine complexes.

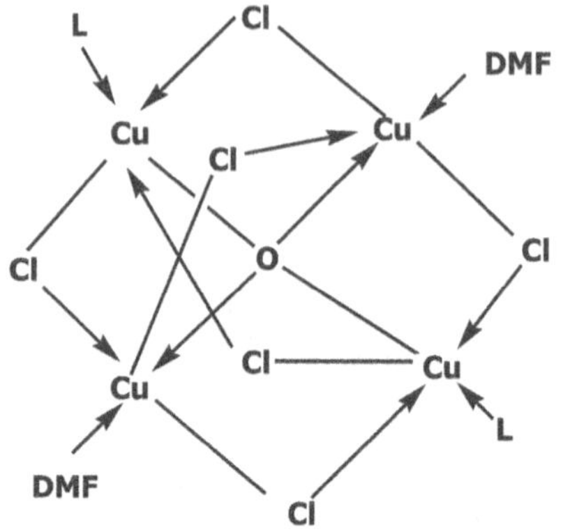

FIGURE 2.8 Structure of [$Cu_4OC1_6(DMF)_2$].

The other way for the synthesis of coordination compounds is electrosynthesis of complexes, which is the process carried out by dissolving cathodes and anodes with zerovalent metal in a liquid phase. For example, [$Cd(OH)_2(Phen)_2$], which is a molecular complex, was obtained during electrosynthesis, where the dissolution of cadmium along with acetonitrile solution containing N-(hydroxyethyl)salicylideneimine and 1,10-phenanthroline had taken place [54]. Among π-complexes bis(η^5-cyclopentadienyl), metal derivatives were produced by electrosynthesic dissolution [54]. The preparation of trans-[$PdCl(PPh_3)_2$(NH,NH-NHC)] (NHC is called imidazolylidene) was achieved by the reaction of zerovalent metal and complex [$Pd(PPh_3)_4$] following oxidative addition and protonation [55].

2.2.16 Langmuir-Blodgett Deposition

For the formation of layered structures, the Langmuir-Blodgett (LB) technique is utilized as it renders fidelity on a molecular level. In this technique, molecules self-assemble into a monolayer arrangement and then this monolayer forms Langmuir-blodgett (LB) film. Langmuir-Blodgett films serve as a medium for supramolecular chemistry and for nanofabrication [56]. By using the Langmuir-Blodgett deposition process, the ability of the solution is possible when the complex is immobilized, and a long alkyl chain is used in the ligand. Different complexes were synthesized using this method including [$Co(L_2)BF_4$] where L is 4-florophenyl [57]. Similarly, fullerene-Ru-porphyrin conjugates were deposited on Langmuir-Blodgett films [58]. Ln(III) amphiphiles can be prepared by three methods. In the first method, the surface of a sub-phase containing Ln(III) (produced in situ) is treated with a free ligand. The ligands then coordinate with Ln(III) in the sub-phase as the barriers are squeezed, forming a complex. In the second method, pre-formed complexes with an amphiphilic ligand and Ln(III) are first complexed, and the resultant amphiphilic complex is subsequently applied to the Langmuir-blodgett (LB) trough surface. In the third method Ln(III) forms complexes with amphiphilic counter ions; in these systems, the ion-pair produced is applied to the surface of the LB trough and the counter ion (anion or cation) has an amphiphilic quality [59].

2.2.17 Layer-by-Layer Self-Assembly

Layer-by-layer self-assembly procedure is receiving attention due to the advancement in nanotechnology. Metal complexes formed are self-assembled on a gold surface and have been widely studied. A number of methods are used for the layer-by-layer self-assembly of coordination compounds on the solid surface such as coordination, chemisorption, electrostatic, and supramolecular interactions, etc. Bis(dithiocarbamato)copper(II) complex was deposited through layer-by-layer formation on a gold surface using small building blocks [60].

The system of polyelectrolytes is involved in layer-by-layer self-assembly through electrostatic methods. This method produces multi-layers that are formed by alternate deposition of BiPE (a complex electrolyte) that has permanent positive charges due to metal-ion receptors and an oppositely charged polyelectrolyte system such as poly(styrene sulfonate). Adsorption of a metal ion from the solution results in multiple layers. Microcontact stamping is another approach that results in spatially confined regions of metallo-units in layer-by-layer assembly. A binding motive to assembled multi-layers is itself metal-ion coordination. AFM is used to study the binding force between multi-layers that arises due to metal-ion coordination [61].

2.2.18 Combinatorial Synthesis of Coordination Compounds

For the synthesis of new coordination complexes, the combinatorial system helps in creating changes in structure and is also helpful in finding synergistic effects of ligand binding, coordination geometry, and other factors such as steric and electronic effects, stability, as well as activity of metal [62]. For example, acacen-type (Figure. 2.9) coordination compounds are synthesized by this method utilizing a transamination reaction [63].

In addition to homometallic complexes, hetero bimetallic complexes are also synthesized by the combinatorial method. Examples include [(cymene)Ru(μ-Cl)$_3$ Mo(CO)$_2$(η^3-C$_3$H$_5$)] shown in Figure 2.10, which utilizes {(cymene)RuCl$_2$} and {MoCl (C$_3$H$_5$)(CO)$_2$} as two fragment structures [64].

Synthesis of manganese tricarbonyl (Figure. 2.11) is expected to show excellent antibacterial activity through combinatorial synthesis among 420 proposed metal complexes [65].

2.2.19 Sonochemical Synthesis of Coordination Compounds

It is a very powerful tool to prepare coordination compounds in a very short time, and without the use of hard reaction conditions of conventional synthesis, the

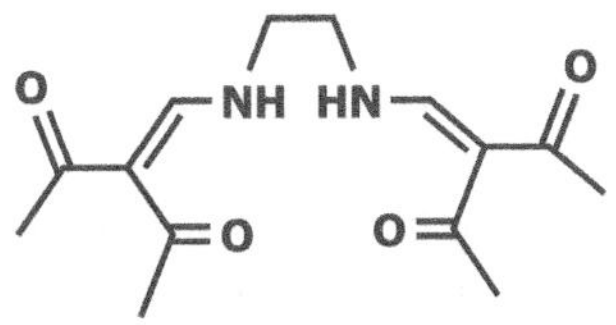

FIGURE 2.9 Structure of acacen.

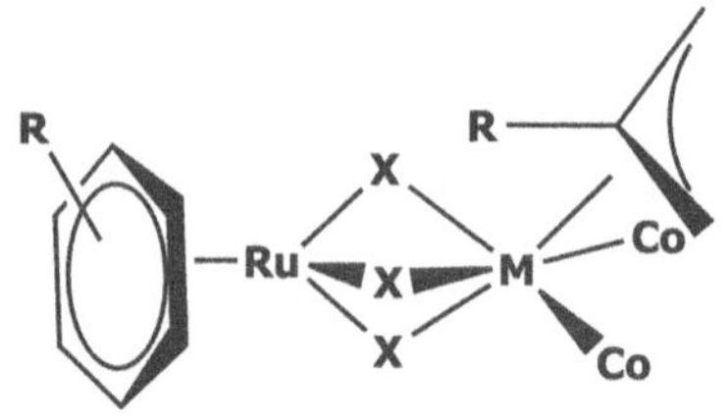

FIGURE 2.10 Manganese tricarbonyl complex (M=Mo).

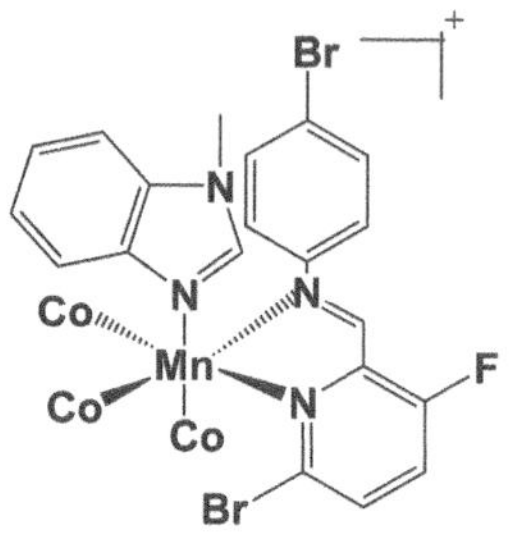

FIGURE 2.11 Structural formula of manganese tricarbonyl complex.

high-intensity ultrasonic radiations are required. Examples include sonochemical formation of metal-azido complexes containing 2,9-dimethyl-1,10-phenanthroline as a ligand with lead(II) by the use of a high-density ultrasonic probe operating at 20 kHz with a maximum power output of 600 W [66].

$$Pb(CH_3COO)_2 + NaN_3 + dmp \rightarrow \text{nanohexagonal rods of compounds}$$

Other examples include sonochemical formation of $[Pb_2(\mu\text{-2-bpdbe})_2(NO_3)_4]_n$ where bpdbe is regarded as 1,6-bis(2-pyridyl)-2,5-diaza-1,5-hexadiene} [67], $[Co(en)_3]Cl_3{\cdot}3H_2O$, $[Co(dien)_2]Cl_3{\cdot}3.5H_2O$, and $[Co(NH_3)_6]Cl_3{\cdot}2H_2O$, and $[Pb(pcih)_2]$ (pcih stands for 2-pyridinecarbaldehyde isonicotinoylhydrazonate) were synthesized by the sonochemical method [68, 69].

The reaction of neocuproine (ligand) along with Zn(II) acetate and potassium iodide in the presence of sonochemical rays resulted in the formation of nanoparticles of coordination complex $[Zn(dmph)I_2]$. Here dmph denotes 2,9-dimethyl-1,10-phenanthroline(neocuproine) [70].

2.2.20 Post Synthetic Modifications of Coordination Compounds

Novel coordination complexes have been explored by using the post synthetic strategy based on organic coupling reactions. As coordination compounds are less soluble in organic solvents these post-synthetic modifications are done in a heterogeneous way. Different methods used for post-synthetic modifications include covalent, coordination, and noncovalent methods. An example of a covalent modification is the conversion of [$Cu_2(OOCC_6H_4N_3)_4$·(quinoline)$_2$], (which is a dinuclear copper(II) paddlewheel structure with four azide groups at the peripheral positions and two quinoline molecules at the axial positions) to Cu_2(OOC-C_6H_4-C_2N_3H-$COOCH_3$)$_4$·(quinoline)$_2$ by click reaction in the presence of methyl propiolate [71]. The post-synthetic modification of [Fe(NH_2-trz)$_3$](NO_3)$_2$ (NH_2-trz is referred as 4-amino-1,2,4-triazole), which is a spin crossover compound, reacts with p-anisaldehyde, resulting in formation of imine by reaction of NH_2 and carbonyl of p-anisaldehyde. By adjusting the reaction period, varied transformation degrees, ranging from 23% to 100%, were obtained, [72] Post synthetic modifications also result in the formation of complexes that can't be prepared by using precursors, for example, [Fe(salten)$_2$(bipydz)](BPh_4)$_2$•EtOH, whereas salten stands for 2,2′- [iminobis(3,1-propanediylnitrilomethylidyne)]bisphenol) and bipydz is for 3,6-bis(4-pyridyl)- 1,2,4,5-tetrazine. Formation also takes place by post-synthetic modification in synthesized complex [Fe(salten)$_2$(bipytz)] (BPh_4)$_2$•EtOH [73]. Coordination strategies for post-synthetic modification involve metal nodes or linkers, metal nodes involve like copper-copper coordination cages and one of copper coordinates with a silicon nanowire to form a complex that has a similar structure but enhanced charge that results in better transistor coordination linking strategy in post-synthetic modification and can be combined with covalent [74].

Post synthetic modifications of Cd-MOF to form new MOFs by using amino acid result in formation of [Cd(1,4-BDC) (Pro) (H2O)]∞, [Cd(1,4-BDC) (L-Trp) (H2O)]∞, [Cd(1,4-BDC) (L-Ala) (H2O)]∞, [Cd(1,4-BDC) (L-Phe) (H2O)]∞, [Cd(1,4-BDC) (L-Val) (H2O)]∞, [Cd(1,4-BDC) (L-Leu) (H2O)]∞, [Cd(1,4-BDC) (L-Arg) (H2O)]∞ [75]. PSM are also done in order to introduce chirality in achiral MOF such as MIL-101 (Matérial Institut Lavoisier) by use of chiral ligands such as chiral natural amino acid to form CMOF-1, CMOF-2 and CMOF-3 [76] shown in Figure 2.12.

2.3 SUMMARY

Synthesis of coordination compounds is enchanting as small changes in reaction conditions result in completely different complexes with new structures and different applications. Similarly, there is a broad option for choice of metal and ligand. The choice of metal is affected by factors such as its oxidation state, size, and electronic configuration. These factors, in turn, dictate the subsequent physical and chemical properties of the coordination compounds. After the selection of metal, the choice of ligand is also important as its purpose is to stabilize the metal atom. Synthesis of coordination compounds can take place by a lot of methods. Simple methods involve the synthesis of coordination compounds in a single step just by mixing precursors to form the complex, whereas other methods include a lot of steps, controlled reaction conditions, and innovative techniques. So, depending upon the method used, the arrangement of ligands around the metal also changes, which in turn results in new applications. Advanced techniques also help in characterizing the structural properties of coordination compounds.

Some of the methods, including one pot solution method, diffusion reactions, mechanical grinding, microwave heating, ligand substitution method, metal exchange reactions, hydrothermal/solvothermal method, redox interaction methods, chiral

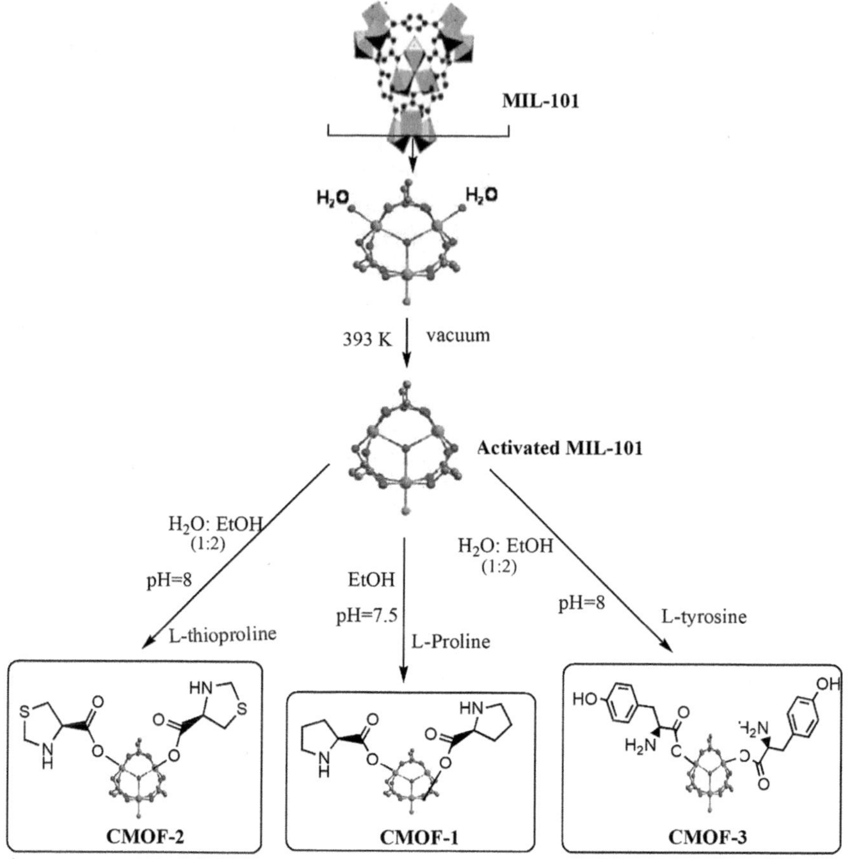

FIGURE 2.12 Post synthetic modifications of achiral MOFs to form chiral MOFs.

resolution, electrochemical oxidation, metalloligand method, self-assemblies, schlenk line techniques, in-situ metal-ligand reaction, direct synthesis from zero-valent metals, langmuir-blodgett deposition, layer-by-layer self-assembly, Sono chemical method computational methods, and post-synthetic modifications are discussed here.

In inorganic chemistry, the synthesis of coordination materials is a complex and thought-provoking subject of study. It allows chemists to build and create compounds with unique features for specific purposes by combining creativity and precision. Coordination compound synthesis has a broad impact and opens doors for breakthroughs in many scientific fields, including medical chemistry and materials research.

ACKNOWLEDGMENTS

This work was supported by the Research Fund for International Scientists (RFIS-Grant number: 52150410410) National Natural Science Foundation of China.

REFERENCES

[1] Lawrance, G.A., *Introduction to coordination chemistry*. 2013: John Wiley & Sons.
[2] Lässig, D., et al., Solid-state syntheses of coordination polymers by thermal conversion of molecular building blocks and polymeric precursors. *Inorganic Chemistry*, 2012. **51**(11): p. 6180–6189.
[3] Wang, J., et al., A polythreading coordination array formed from 2D grid networks and 1D chains. *CrystEngComm*, 2011. **13**(10): p. 3342–3344.
[4] Li, C.-P., J. Chen, and M. Du, Structural diversification and metal-directed assembly of coordination architectures based on tetrabromoterephthalic acid and a bent dipyridyl tecton 2,5-bis(4-pyridyl)-1,3,4-oxadiazole. *CrystEngComm*, 2010. **12**(12): p. 4392–4402.
[5] Peedikakkal, A.M.P. and J.J. Vittal, Solid-state photochemical [2+ 2] cycloaddition in a hydrogen-bonded metal complex containing several parallel and crisscross C=C bonds. *Chemistry–A European Journal*, 2008. **14**(17): p. 5329–5334.
[6] Toh, N.L., M. Nagarathinam, and J.J. Vittal, Topochemical photodimerization in the coordination polymer [{(CF3CO2)(μ-O2CCH3) Zn} 2 (μ-bpe) 2] n through Single-crystal to single-crystal transformation. *Angewandte Chemie*, 2005. **117**(15): p. 2277–2281.
[7] Nagarathinam, M., et al., Mechanochemical reactions of coordination polymers by grinding with KBr. *Chemical Communications*, 2012. **48**(20): p. 2585–2587.
[8] Orita, A., et al., Solventless reaction dramatically accelerates supramolecular self-assembly. *Chemical Communications*, 2002. (13): p. 1362–1363.
[9] Braga, D., et al., Mechanochemical preparation of molecular and supramolecular organometallic materials and coordination networks. *Dalton Transactions*, 2006. (10): p. 1249–1263.
[10] Pezet, F., et al., Highly diastereoselective preparation of ruthenium bis (diimine) sulfoxide complexes: New concept in the preparation of optically active octahedral ruthenium complexes. *Organometallics*, 2000. **19**(20): p. 4008–4015.
[11] Abe, T., et al., Microwave-assisted synthesis of metal complexes. *Mini-Reviews in Organic Chemistry*, 2011. **8**(3): p. 315–333.
[12] Coelho, A.C., et al., Microwave assisted synthesis of molybdenum and tungsten tetracarbonyl complexes with a pyrazolylpyridine ligand. Crystal structure of cis-[Mo (CO) 4 {ethyl [3-(2-pyridyl)-1-pyrazolyl] acetate}]. *Molecules*, 2006. **11**(12): p. 940–952.
[13] Richens, D.T., Ligand substitution reactions at inorganic centers. *Chemical Reviews*, 2005. **105**(6): p. 1961–2002.
[14] Rubio, M., et al., Rhodium complexes with pincer diphosphite ligands. Unusual olefin in-plane coordination in square-planar compounds. *Organometallics*, 2009. **28**(2): p. 547–560.
[15] Zhang, L., et al., Synthesis and X-ray crystal structures of biscyclopentadienylthulium benzothiazole-2-thiolate Cp2Tm (SBT)(THF). *Chinese Journal of Structural Chemistry*, 2001. **20**(1): p. 40–43.
[16] Buchin, B., et al., Coordination chemistry of Ga (C5Me4Ph): Novel homoleptic d10 cluster complexes of palladium. *Inorganic Chemistry*, 2006. **45**(4): p. 1789–1794.
[17] Das, S., H. Kim, and K. Kim, Metathesis in single crystal: Complete and reversible exchange of metal ions constituting the frameworks of metal– organic frameworks. *Journal of the American Chemical Society*, 2009. **131**(11): p. 3814–3815.
[18] Eisenhart, R.J., L.J. Clouston, and C.C. Lu, Configuring bonds between first-row transition metals. *Accounts of Chemical Research*, 2015. **48**(11): p. 2885–2894.
[19] Mukherjee, G. and K. Biradha, Post-synthetic modification of isomorphic coordination layers: Exchange dynamics of metal ions in a single crystal to single crystal fashion. *Chemical Communications*, 2012. **48**(36): p. 4293–4295.
[20] Liu, N., et al., Coordination compounds of bis (5-tetrazolyl) amine with manganese (II), zinc (II) and cadmium (II): Synthesis, structure and magnetic properties. *Dalton Transactions*, 2008. (34): p. 4621–4629.

[21] Konar, S. and A. Clearfield, Solvothermal synthesis and characterization of two high-nuclearity mixed-valent manganese phosphonate clusters. *Inorganic Chemistry*, 2008. **47**(9): p. 3489–3491.

[22] Zhao, H., et al., In situ hydrothermal synthesis of tetrazole coordination polymers with interesting physical properties. *Chemical Society Reviews*, 2008. **37**(1): p. 84–100.

[23] Yan, Y., C.D. Wu, and C.Z. Lu, Hydrothermal synthesis of two new transition metal coordination polymers with mixed ligands. *Zeitschrift für anorganische und allgemeine Chemie*, 2003. **629**(11): p. 1991–1995.

[24] Yin, G.-J., B.-M. Ji, and C.-X. Du, In situ hydrothermal synthesis of two novel Cd (II) coordination compounds. *Inorganic Chemistry Communications*, 2012. **15**: p. 21–24.

[25] Guo, Y., et al., Chiral silver–lanthanide metal–organic frameworks comprised of one-dimensional triple right-handed helical chains based on [Ln7 (μ3-OH) 8] 13+ clusters. *Inorganic Chemistry*, 2018. **57**(3): p. 995–1003.

[26] Ceniceros-Gómez, A.E., et al., Synthesis, X-ray and spectroscopic characterisation of chromium (III) coordination compounds with benzimidazolic ligands. *Polyhedron*, 2000. **19**(15): p. 1821–1827.

[27] Gaunt, A.J., B.L. Scott, and M.P. Neu, U (IV) chalcogenolates synthesized via oxidation of uranium metal by dichalcogenides. *Inorganic Chemistry*, 2006. **45**(18): p. 7401–7407.

[28] Melman, J.H., et al., Trivalent lanthanide compounds with fluorinated thiolate ligands: Ln– F dative interactions vary with Ln and solvent. *Inorganic Chemistry*, 2002. **41**(1): p. 28–33.

[29] Liao, P.Q., et al., Drastic enhancement of catalytic activity via post-oxidation of a porous MnII triazolate framework. *Chemistry–A European Journal*, 2014. **20**(36): p. 11303–11307.

[30] Meggers, E., Asymmetric synthesis of octahedral coordination complexes. *European Journal of Inorganic Chemistry*, 2011. **2011**(19): p. 2911–2926.

[31] Meggers, E., Chiral auxiliaries as emerging tools for the asymmetric synthesis of octahedral metal complexes. *Chemistry–A European Journal*, 2010. **16**(3): p. 752–758.

[32] Gao, E.-Q., et al., From achiral ligands to chiral coordination polymers: Spontaneous resolution, weak ferromagnetism, and topological ferrimagnetism. *Journal of the American Chemical Society*, 2004. **126**(5): p. 1419–1429.

[33] Hesari, M., D. Nematollahi, and L. Fotouhi, Electrochemical synthesis and study of coordination compounds part 1: Tin (II) catechol complexes. *Journal of Coordination Chemistry*, 2008. **61**(11): p. 1744–1750.

[34] Hagen, K.S., J.G. Reynolds, and R. Holm, Definition of reaction sequences resulting in self-assembly of [Fe4S4 (SR) 4] 2-clusters from simple reactants. *Journal of the American Chemical Society*, 1981. **103**(14): p. 4054–4063.

[35] Constable, E.C., G. Parkin, and L. Que, *Comprehensive coordination chemistry III*. 2021: Elsevier.

[36] Takahashi, K., et al., Evidence of the chemical uniaxial strain effect on electrical conductivity in the spin-crossover conducting molecular system:[FeIII (qnal) 2][Pd (dmit) 2] 5 acetone. *Journal of the American Chemical Society*, 2008. **130**(21): p. 6688–6689.

[37] Cheah, M.H. and P. Chernev, Electrochemical oxidation of ferricyanide. *Scientific Reports*, 2021. **11**(1): p. 23058.

[38] Clarke, H.M., W. Henderson, and B.K. Nicholson, Extending the coordination chemistry of cobalt with the metalloligand [Pt2 (μ-S) 2 (PPh3) 4]: Synthesis of the first cobalt (III) derivatives. *Inorganica Chimica Acta*, 2011. **376**(1): p. 446–455.

[39] Osterloh, F., W. Saak, and S. Pohl, Unidentate and bidentate binding of nickel (II) complexes to an Fe4S4 cluster via bridging thiolates: Synthesis, crystal structures, and electrochemical properties of model compounds for the active sites of nickel containing CO dehydrogenase/acetyl-CoA synthase. *Journal of the American Chemical Society*, 1997. **119**(24): p. 5648–5656.

[40] Zhu, M., et al., Nitronyl nitroxide–metal complexes as metallo-ligands for the construction of hetero-tri-spin (2p–3d–4f) chains. *Chemical Communications*, 2014. **50**(15): p. 1906–1908.
[41] Famengo, A., et al., Dithiolene complexes as metallo-ligands: A crown-ether approach. *New Journal of Chemistry*, 2012. **36**(3): p. 638–643.
[42] Li, F. and L.F. Lindoy, Metalloligand strategies for assembling heteronuclear nanocages–recent developments. *Australian Journal of Chemistry*, 2019. **72**(10): p. 731–741.
[43] Wong, K.K.G., et al., Self-assembly of a porous metallo-[5] rotaxane. *Chemical Communications*, 2020. **56**(72): p. 10453–10456.
[44] Wallis, M.J., et al., Investigating the conformations of a family of [M2L3] 4+ helicates using single crystal X-ray diffraction. *Molecules*, 2023. **28**(3): p. 1404.
[45] Huang, X.-C., J.-P. Zhang, and X.-M. Chen, A new route to supramolecular isomers via molecular templating: Nanosized molecular polygons of copper (I) 2-methylimidazolates. *Journal of the American Chemical Society*, 2004. **126**(41): p. 13218–13219.
[46] Westcott, A., et al., Self-assembly of a 3-D triply interlocked chiral [2] catenane. *Journal of the American Chemical Society*, 2008. **130**(10): p. 2950–2951.
[47] Yang, D., et al., Anion-coordination-directed self-assemblies. *Organic Chemistry Frontiers*, 2018. **5**(4): p. 662–690.
[48] Davis, C.M. and K.A. Curran, Manipulation of a Schlenk line: Preparation of tetrahydrofuran complexes of transition-metal chlorides. *Journal of Chemical Education*, 2007. **84**(11): p. 1822.
[49] Sekar, P. and J.A. Ibers, Improved synthesis of HN (SPPh2)(SePPh2) and some coordination chemistry of [N (SPPh2)(SePPh2)]–. *Inorganica Chimica Acta*, 2001. **319**(1–2): p. 117–122.
[50] Gallen, A., et al., Synthesis and coordination chemistry of enantiopure t-BuMeP (O) H. *Dalton Transactions*, 2018. **47**(15): p. 5366–5379.
[51] Ladipo, F.T., M. Kooti, and J.S. Merola, Oxidative addition of oxygen-hydrogen bonds to iridium (I): Synthesis and characterization of (phenolato)-and (carboxylato) iridium (III) hydride complexes. *Inorganic Chemistry*, 1993. **32**(9): p. 1681–1688.
[52] Chen, X.-M. and M.-L. Tong, Solvothermal in situ metal/ligand reactions: A new bridge between coordination chemistry and organic synthetic chemistry. *Accounts of Chemical Research*, 2007. **40**(2): p. 162–170.
[53] Satapathi, S., et al., A set of new coordination compounds of cadmium (II)/mercury (II) halides/pseudohalides containing polyamines: Syntheses involving in situ metal–ligand reactions, crystal structures and molecular properties. *Inorganica Chimica Acta*, 2012. **384**: p. 37–46.
[54] Garnovskii, A.D., et al., Direct synthesis of coordination compounds from zerovalent metals and organic ligands. *Russian Chemical Reviews*, 1995. **64**(3): p. 201.
[55] Das, R., et al., Synthesis of complexes with protic NH, NH-NHC ligands via oxidative addition of 2-Halogenoazoles to zero-valent transition metals. *Organometallics*, 2014. **33**(23): p. 6975–6987.
[56] Ariga, K., et al., 25th Anniversary article: What can be done with the Langmuir-Blodgett method? Recent developments and its critical role in materials science. *Advanced Materials*, 2013. **25**(45): p. 6477–6512.
[57] Laverick, R.J., et al., Solution processible Co (III) quinoline-thiosemicarbazone complexes: Synthesis, structure extension, and Langmuir-Blodgett deposition studies. *Journal of Coordination Chemistry*, 2021. **74**(1–3): p. 321–340.
[58] Da Ros, T., et al., Synthesis, electrochemistry, Langmuir–Blodgett deposition and photophysics of metal-coordinated fullerene–porphyrin dyads. *Journal of Organometallic Chemistry*, 2000. **599**(1): p. 62–68.
[59] Wales, D.J. and J.A. Kitchen, Surface-based molecular self-assembly: Langmuir-Blodgett films of amphiphilic Ln (III) complexes. *Chemistry Central Journal*, 2016. **10**: p. 1–8.

[60] Cao, R., et al., Building layer-by-layer a bis (dithiocarbamato) copper (II) complex on Au {111} surfaces. *Journal of the American Chemical Society*, 2007. **129**(21): p. 6927–6930.

[61] Krass, H., G. Papastavrou, and D.G. Kurth, Layer-by-layer self-assembly of a polyelectrolyte bearing metal ion coordination and electrostatic functionality. *Chemistry of Materials*, 2003. **15**(1): p. 196–203.

[62] Francis, M.B., N.S. Finney, and E.N. Jacobsen, Combinatorial approach to the discovery of novel coordination complexes. *Journal of the American Chemical Society*, 1996. **118**(37): p. 8983–8984.

[63] Tomažin, U.A., et al., Combinatorial synthesis of acacen-type ligands and their coordination compounds. *ACS Combinatorial Science*, 2017. **19**(6): p. 386–396.

[64] Gauthier, S., et al., Combinatorial synthesis of bimetallic complexes with three halogeno bridges. *Chemistry–A European Journal*, 2004. **10**(11): p. 2811–2821.

[65] Scaccaglia, M., et al., *Discovery of antibacterial manganese (I) tricarbonyl complexes through combinatorial chemistry*. 2023. Chemical Science, 15 (2024) 3907–3919.

[66] Hanifehpour, Y., B. Mirtamizdoust, and S.W. Joo, Sonochemical synthesis and characterization of the new micro-hexagonal-rod lead (II)-azido coordination compound. *Journal of Inorganic and Organometallic Polymers and Materials*, 2012. **22**: p. 916–922.

[67] Hashemi, L. and A. Morsali, Sonochemical synthesis of nano-structured lead (II) complex: Precursor for the preparation of PbO nano-structures. *Journal of Coordination Chemistry*, 2011. **64**(23): p. 4088–4097.

[68] Bala, R., et al., Sonochemical synthesis, characterization, antimicrobial activity and textile dyeing behavior of nano-sized cobalt (III) complexes. *Ultrasonics Sonochemistry*, 2017. **35**: p. 294–303.

[69] Chung, J.-H., et al., From discrete to infinite 3D coordination polymer: Sonochemical syntheses and structural characterization of a new nano flower lead (II) coordination compound. *Journal of Molecular Structure*, 2014. **1076**: p. 698–703.

[70] Ranjbar, M., et al., Sonochemical synthesis and characterization of nano-sized zinc (II) coordination complex as a precursor for the preparation of pure-phase zinc (II) oxide nanoparticles. *Nanochemistry Research*, 2017. **2**(1): p. 120–131.

[71] Han, S., et al., Postsynthetic modification of a coordination compound with a paddlewheel motif via click reaction: DOSY and ESR studies. *Inorganic Chemistry Communications*, 2012. **15**: p. 78–83.

[72] Enríquez-Cabrera, A., et al., Complete post-synthetic modification of a spin crossover complex. *Dalton Transactions*, 2019. **48**(45): p. 16853–16856.

[73] Komatsumaru, Y., et al., Post-synthetic modification of a Dinuclear spin crossover iron (III) complex. *Zeitschrift für anorganische und allgemeine Chemie*, 2018. **644**(14): p. 729–734.

[74] Liu, J., et al., Post-synthetic modifications of metal–organic cages. *Nature Reviews Chemistry*, 2022. **6**(5): p. 339–356.

[75] Ahmed, I., et al., Post synthetically modified compounds of Cd-MOF by l-amino acids for luminescent applications. *Journal of Solid State Chemistry*, 2020. **287**: p. 121320.

[76] Yasmeen, F., et al., The development of chiral metal–organic frameworks for enantioseparation of racemates. *RSC Advances*, 2023. **13**(24): p. 16651–16662.

3 Synthesis Strategies of Coordination Materials

Muhammad Mubeen, Mohammad Tabish, Noor Muhammad, Muhammad Asim Mushtaq and Ghulam Yasin

3.1 INTRODUCTION

The area of coordination nanomaterials has seen an unparalleled upsurge in interest, propelled by their distinctive and adaptable characteristics. Coordination materials refer to a wide range of compounds, particularly porous coordination polymers (PCPs), metal-organic frameworks (MOFs), and covalent organic frameworks (COFs). These materials exhibit structured metal ions or clusters arranged with organic ligands, leading to the creation of one-dimensional (1D) elongated chains, two-dimensional (2D) layers or sheets, or three-dimensional (3D) frameworks. In this categorization, PCPs denote structures that include pores. MOFs, porous coordination polymers, are coordination networks that are characterized by voids. COFs emphasize the existence of organic units that are covalently connected in their frameworks [1]. These coordination materials possess a diverse array of adjustable characteristics and are used in several domains, including gas storage, separation, catalysis, and sensing, highlighting their adaptability and importance in the area of materials research. The fast growth of coordination nanomaterial research has been driven by the increasing need for materials that provide highly specific functionality. These materials possess adjustable structures, large surface areas, adjustable pore sizes, and unique electrical and optical characteristics [2–5].

The field of coordination nanomaterials synthesis is undergoing constant evolution as of 2024. The discipline has seen significant advancements, with a growing number of papers and patents focused on innovative synthesis methodologies. Recent research indicates a significant increase in papers focused on the synthesis of MOFs for sustainable energy applications. This highlights the rising significance of these materials in tackling global concerns.

The purpose of this chapter is to provide a recent review of the most recent approaches used to create coordination nanomaterials, as categorized in Figure 3.1. We will explore the complex compositions and characteristics of these materials, analyzing recent advancements and novel developments. In addition, the chapter will emphasize recent significant advancements and accomplishments in the synthesis of PCPs, MOFs, and COFs. This will include the most recent scientific discoveries, practical uses in industry, and emerging patterns, highlighting the ever-changing and evolving character of the topic. Throughout this study, our objective is to provide a thorough comprehension of existing synthesis methodologies as well as to provide

DOI: 10.1201/9781003345886-4

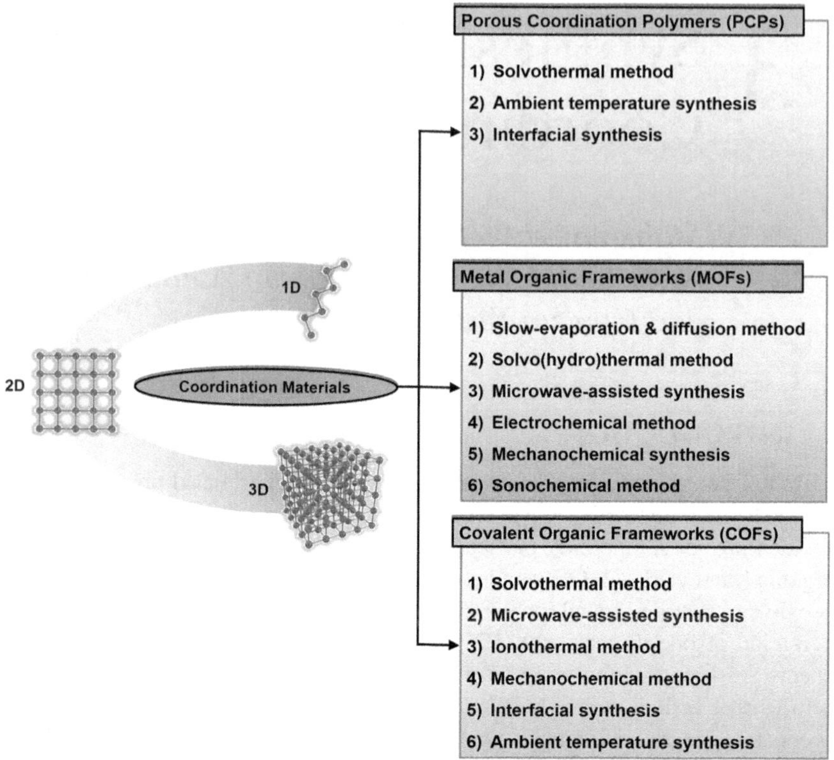

FIGURE 3.1 Different synthetic strategies of coordination materials.

insights into cutting-edge research in this dynamic and continuously progressing field. Our purpose is to inspire future discoveries in this sector.

3.2 OVERVIEW AND CHARACTERISTICS OF COORDINATION NANOMATERIALS

Coordination nanomaterials is a fascinating group that is characterized by their complex structures and distinctive behaviors at the nanoscale. The importance of materials resides in their ubiquitous influence on human existence and their capacity to mold the trajectory of technological advancement. The key aspect of these materials is the organization of metal ions or clusters and organic ligands, which directs their arrangement into well-defined structures, as seen in Figure 3.2. The production of these materials exhibits nanoscale accuracy, which provides a wide range of distinct features. These features are listed in Table 3.1, highlighting their uniqueness in the field of materials research.

3.2.1 Porous Coordination Polymers (PCPs)

Porous Coordination Polymers (PCPs) are a kind of polymeric solid that have structures with one-dimensional (1D), two-dimensional (2D), or three-dimensional (3D)

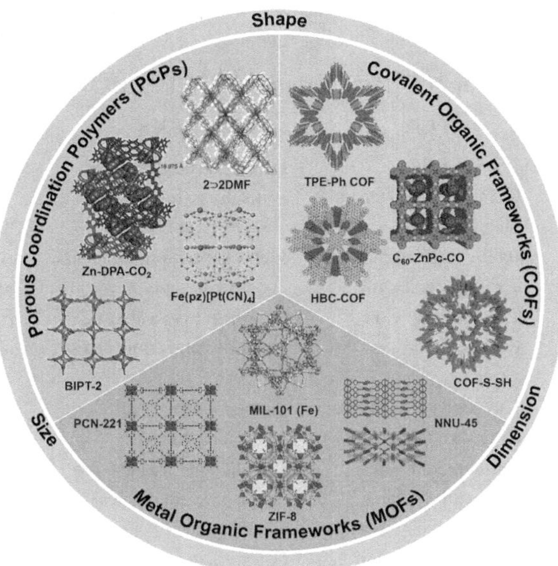

FIGURE 3.2 Structural diversity of coordination materials (PCPs; 2⊃2DMF [6], Zn-DPA-CO_2 [7], Fe(pz)[Pt$(CN)_4$] [8], BIPT-2 [9], MOFs; MIL-101(Fe) [10], NNU-45 [11], ZIF-8 [12], PCN-221 [13], COFs; TPE-Ph COF [14], HBC-COF [15], C_{60}-ZnPc-CO [16], COF-S-SH [17]).

porosity. These materials are created by using the nodes and linkers of metal ions or clusters and organic ligands respectively, which form coordination bonds. PCPs, in contrast to typical zeolites and other inorganic porous materials, possess distinctive features like exceptional crystallinity, customizable and adjustable porosity, and structural adaptability. PCPs provide unique characteristics that allow them to be used in a wide range of applications, including gas storage, separation, catalysis, luminescence, sensing, drug administration, and ionic/electronic conductivity. The study of PCPs, including both their chemistry and physics, has been an extremely active field of research since the 1990s, characterized by significant and quick advancements. Moreover, in nanoscale porous materials, the decrease in diffusion is of utmost importance, notably in the fields of catalysis and sorption, particularly in liquid-phase applications [18]. The significance of reducing the size and manipulating the crystal morphology of porous coordination frameworks is highlighted by recent advancements in porous membranes, thin film devices, and carrier particles for drug administration. These breakthroughs greatly enhance the role of these frameworks in the area of nanotechnology [19].

3.2.2 Metal-Organic Frameworks (MOFs)

Tomic's groundbreaking publication in 1965 on porous materials and metal-organic frameworks (MOFs) sparked significant attention within the scientific community, representing a significant milestone in the study of these materials. MOFs are a kind of porous coordination polymers that were first proposed by Yaghi et al. in the late 1990s. They have gained considerable interest in the last twenty years [20]. These

TABLE 3.1
Structural and Functional Characteristics in PCPs, MOFs, and COFs

Features	PCPs	MOFs	COFs
Structural Type	Varied structures including 1D, 2D, and 3D networks	Diverse structures encompassing 1D, 2D, and 3D frameworks	Typically, 2D structures with layered arrangements
Composition	Metal-containing units and organic ligands	Metal-containing units and organic ligands	Covalently bonded organic units
Porosity	High porosity with well-defined void spaces	Significant porosity due to metal-ligand coordination	Tunable porosity with ordered pore structures
Metal-Ion Precursors	Metal salts (e.g. sulfates, chlorides, nitrates)	Metal salts (e.g., sulfates, chlorides, nitrates)	N/A (organic units are covalently bonded)
Organic Ligands	Multidentate ligands (e.g., carboxylates, nitriles)	Multidentate ligands (e.g., carboxylates, nitriles)	Various organic linkages (e.g., imines, boronate esters)
Applications	Gas storage, separation, catalysis, sensing	Gas storage, separation, catalysis, sensing	Gas storage, separation, catalysis, sensing
Functionalization	Functional materials can be integrated for enhanced properties	Diverse functionalization possibilities through metal doping, combining with conductive polymers, etc.	Incorporation of functional groups to tailor properties

materials have a distinct composition consisting of metal-containing units – often known as secondary building units (SBUs) – and organic ligands. Over twenty-thousand unique variants of the MOF family have been created, and their growth has been accomplished by modifying both the SBUs and the functionalities of the organic ligands. [21–25]. It is worth mentioning that MOFs can be constructed by using affordable precursors – usually inorganic salts (such as sulfates, chlorides, and nitrates) as sources of metal ions – and multidentate organic ligands (such as carboxylates, nitriles, or azoles) [26–29]. Metal-organic frameworks (MOFs) possess exceptional characteristics, such as elevated specific surface areas, enhanced porosity, varied topologies, and functions, in comparison to traditional inorganic porous materials. In addition, a wide range of functional materials have been effectively integrated into MOFs, therefore improving their characteristics via the creation of diverse composites. These composites include nickel oxalate, conducting polymers, metal doping, hydroxide metal salts, carbon materials, and other similar substances [30–32].

3.2.3 Covalent Organic Frameworks (COFs)

The main objective in material sciences is to customize the porosity structure using basic molecular components. Crystalline extended organic frameworks, known as

COFs, are formed by assembling organic building units in order to create polymeric architectures in 2D or 3D. The COFs possess a backbone structure consisting of lightweight components, including C, O, N, and B, which are interconnected by strong covalent bonds [33]. The concept of reticular synthesis for the creation of COFs was introduced by Yaghi et al. in 2005 [34]. The COFs are equipped with a crystalline conjugated network including numerous pores, functional groups, and channels. As previously mentioned, COFs are part of the porous category, where building units intricately combine to create organized pores. The fabrication of COFs for specific purposes is a novel category of porous materials that is now gaining prominence [34]. COFs have numerous notable characteristics in comparison to MOFs, including a vast surface area, wide pores, low display densities resulting from their composition of light elements, and exceptional stability [35]. By applying reticular chemistry principles, researchers have achieved improved control over the adjustment of pore metrics and rationalized the design of COFs. As a result, they have developed frameworks that exhibit a high charge carrier mobility of 8.1 cm2 V s, a high surface area of 4,210 m2/g, an ultralow density of 0.17 g/cm3, and large pore sizes of up to 4.7 nm [36]. The strategically engineered porous and chemically durable nanostructured materials have significant potential for substantially improving the efficiency of energy storage devices, chemical sensors, catalysis, and electrical devices [37]. By precisely assembling molecular components using strong chemical bonds, it is possible to create extended crystalline formations. A new discipline of chemistry, known as reticular chemistry, has emerged. This field is used to synthesis MOFs and, more recently, COFs.

3.3 SYNTHESIS STRATEGIES OF COORDINATION NANOMATERIALS

The fabrication of coordination nanomaterials encompasses a range of techniques, such as solvothermal, hydrothermal, and mechanochemical processes. These methods enable researchers to precisely adjust the dimensions, configuration, and characteristics of the produced nanomaterials, so promoting progress in both basic comprehension and practical use. The wet chemical technique has become more famous in the production of porous materials, such as PCPs, MOFs, and COFs. This strategy has more recently been used to create coordination nanomaterials, where a high level of purity is required for three essential chemical components: a metal salt, an appropriate organic linker, and a selected solvent. Usually, the selective solvent is mixed with the metal salt and organic ligand in a suitable container, or the two are dissolved in the solvent separately and then combined in a single pot [38].

Numerous forces, including hydrogen bonding, ion-dipole, dipole-dipole, dipole-induced dipole, solvo-phobic, and dispersion or London interactions, are involved in solute and solvent interactions. Solubility, reactivity, stability, donor number, dielectric constant, polarity, and boiling point temperature are some of the variables that often determine how well solvents function in liquid-phase synthesis [39, 40]. Hence, solvents or co-solvents have a vital function in effectively regulating the structure of porous materials [41]. Various parameters, including the reagent concentration, temperature and length of heating, solution pH, and precursor characteristics, may affect the morphologies, crystal sizes, topologies, and phase purity of the final material [42]. The sections that follow include descriptions of many trustworthy techniques for the liquid-phase synthesis of different coordination nanomaterials.

3.3.1 Synthesis Strategies of PCPs

The synthesis of Porous Coordination Polymers (PCPs) entails applying several ways to engineer their structure, morphology, and characteristics. Solvothermal synthesis, which involves combining metal ions and organic ligands in an appropriate solvent and heating them at high temperatures in a sealed reaction vessel is a widely utilized method. This technique allows for the creation of well-specified PCP structures with clear and distinct forms and sizes. Mechanochemical synthesis is a method that promotes the creation of covalent bonds by grinding the precursor components together. This technology offers an alternate approach for fabricating PCPs. Moreover, the process of interfacial synthesis, which takes place at the boundary between two solvents that do not mix, enables the creation of thin films of PCP with accurate manipulation of their thickness. Each technique has distinct benefits, enhancing the adaptability of PCP synthesis for customized applications in diverse sectors such as gas storage, catalysis, and sensing.

3.3.1.1 Solvothermal Method

Solvothermal synthesis is a very adaptable technique for creating PCPs, resulting in precisely defined structures that possess substantial porosity and surface area. The solvothermal synthesis of PCPs yielded micro-sized rods with a hexagonal face, specifically represented by CPP-3. This was achieved using a simple and direct approach, as seen in Figure 3.3a. The synthesis included the dissolution of 1,4-benzenedicarboxylic acid (H2BDC) in N,N-dimethylformamide (DMF). Afterward, a solution of DMF containing In(NO3)3·xH2O was added to the mixture, resulting in a homogenous mixture. The final blend received a solvothermal treatment, with the whole reaction system heated to 100°C for 10 min. The solvothermal process, which was meticulously regulated, as shown by Won Cho et al. [43], produced a variety of unique CPP forms, including hexagonal disks, hexagonal lumps, and hexagonal rods. The solvothermal method, which involves the use of DMF and high temperature in a sealed environment, enables the specific creation of well-defined PCP structures with distinct morphologies.

3.3.1.2 Ambient Temperature Synthesis

PCPs represent a notable breakthrough in the domain of porous materials. This novel method enables the precise formation of PCPs with unique structures and dimensions using mild reaction conditions. Qing Liu et al. showcased a universal approach to do this by adjusting the acid/base conditions of the reaction system at ambient temperature. An inherent benefit of ambient temperature synthesis is the circumvention of elevated pressure, excessive heat, and the use of potentially hazardous solvents like DMF. This eco-friendly technique provides a simple but efficient approach to produce PCPs with well-defined structures, high levels of crystallinity, and outstanding gas adsorption characteristics. Liu et al. investigated porous coordination-polymer MOF-14 microcrystals, as shown in Figure 3.3b. By including sodium acetate in the process, they were able to manipulate the shapes of the crystals, including rhombic dodecahedron, truncated rhombic dodecahedron, cube with truncated sides, and cube. Additionally, they were able to affect the size of the particles by increasing the pace at which nucleation occurs. The method's flexibility is shown by successfully applying it to the synthesis of HKUST-1, highlighting its potential for synthesizing other PCP structures [44]. The synthesis of PCPs at ambient temperature is a promising approach that offers a sustainable and regulated method for creating porous materials with specific features and uses.

3.3.1.3 Interfacial Synthesis

The use of interface-assisted synthesis methods is crucial in the early phases of creating 2D Porous Coordination Polymers (PCPs). These approaches include the combination of metal ions and organic ligands at different interfaces, including solid–gas, liquid-liquid, gas-liquid, or solid-liquid interfaces. The efficacy of these techniques hinges on the creation of an appropriate interface with a sufficient reaction area and

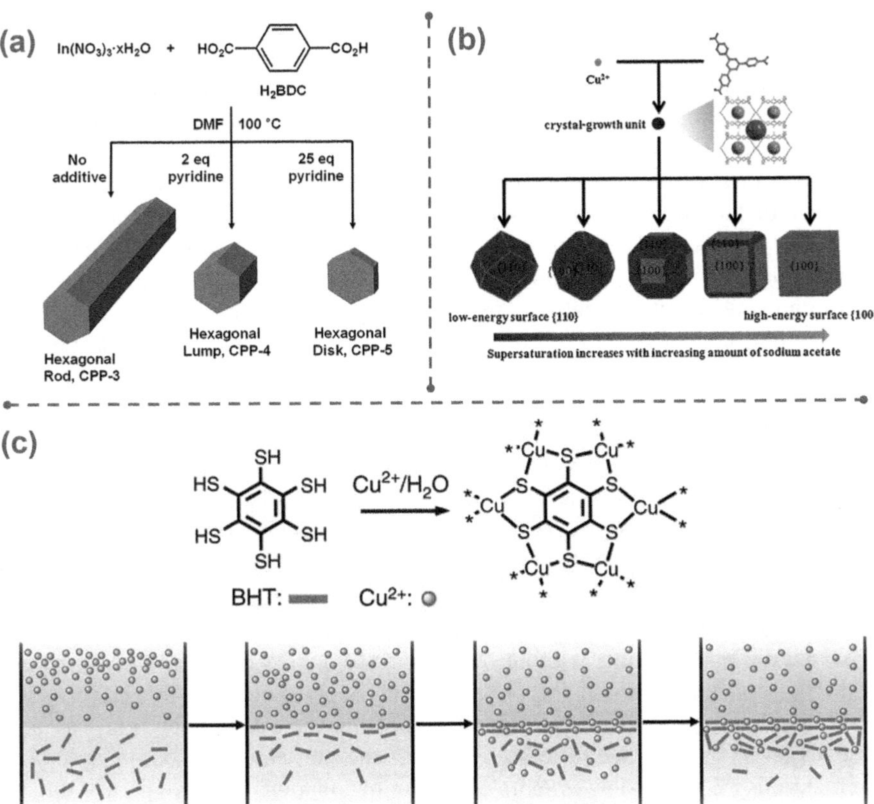

FIGURE 3.3 Synthesis schemes of various PCPs. (a) CPP-3 by solvothermal method [43], (b) MOF-14 synthesis at ambient temperature [44], and (c) interface-assisted synthesis of Cu-BHT [46].

assuring the compatibility between metal ions or ligands to enable the production of elongated nanosheets. A significant advancement in this particular area was reported by Alexandre et al. in 2003 [45], demonstrating the creation of 2D Fe-carboxylate grids (trimellitic acid) nanosheets by depositing organic polytopic ligands and metal atoms onto the surface of Cu(100) via vacuum deposition. Nevertheless, the difficulties linked to high vacuum, level metal surfaces, and significant energy input have inspired researchers to investigate mild conditions for synthesizing interfaces, such as liquid-liquid, gas-liquid, and solid-liquid interfaces. Zhu et al. in 2015 [46] conducted an experiment where they successfully synthesized a two-dimensional p-d conjugated coordination polymer termed Cu-BHT (BHT = benzenehexathiol). This synthesis took place at the interface of BHT/dichloromethane and aqueous copper(II) nitrate solutions, as seen in Figure 3.3c. This demonstrates the adaptability of interfacial techniques in producing customized PCPs with regulated architectures and characteristics.

3.3.2 Synthesis Strategies of MOFs

Metal-organic frameworks (MOFs) are highly regarded as an important group of porous compounds due to their unique functional and structural characteristics. These frameworks are intricately constructed by connecting organic linkers, metal-ion clusters, and metal ions. Without a doubt, the careful selection of main building blocks (PBUs) plays a crucial role in determining the ultimate structure and characteristics of MOFs. Nevertheless, the synthesis process encompasses a multitude of factors, such as temperature, duration of reaction, pressure, pH, and selection of solvent. Different synthetic techniques, including as solvo(hydro) thermal, slow diffusion, electrochemical, mechanochemical, microwave-assisted, ultrasound, and electric heating, provide various strategies for creating MOFs. Each method contributes to the distinct structures and functions of the resultant materials.

3.3.2.1 Slow-Evaporation and Diffusion Method

Both the slow evaporation and diffusion processes are performed under ambient conditions and do not need an additional energy supply. The standard technique of crystal formation, known as slow evaporation, has been recently used to synthesize certain MOFs such as MOF-5, MOF-177, and HKUST-1, as shown in Figure 3.4a [47]. Although it is a low-temperature method, MOF synthesis is selected preferentially, although with the drawback of needing longer reaction periods [38]. In order to accelerate the reaction time without sacrificing the quality of the crystal, one might use higher concentrations of the precursor and apply agitation [48]. Nevertheless, the product created under these circumstances may not completely correspond to those achieved by alternate processes used for the fabrication of extremely crystalline materials [49]. The slow evaporation technique involves the progressive removal of the solvent at a constant low temperature, resulting in the concentration of the original components in the solution. At times, co-solvents are added to improve the solubility of reagents and speed up the process by causing low-boiling solvents to evaporate more quickly [38].

The diffusion strategy seeks to progressively facilitate interactions between different species. Solvent liquid diffusion is a method that involves the formation of two layers with different densities. One layer contains the precipitant solvent, while the other layer has the product dissolved in a solvent. The layers are divided by a solvent layer, and crystal formation takes place at the boundary after the progressive diffusion of the precipitant solvent into the distinct layer. Another form of diffusion includes the slow movement of reactants via physical barriers, such as two vials of

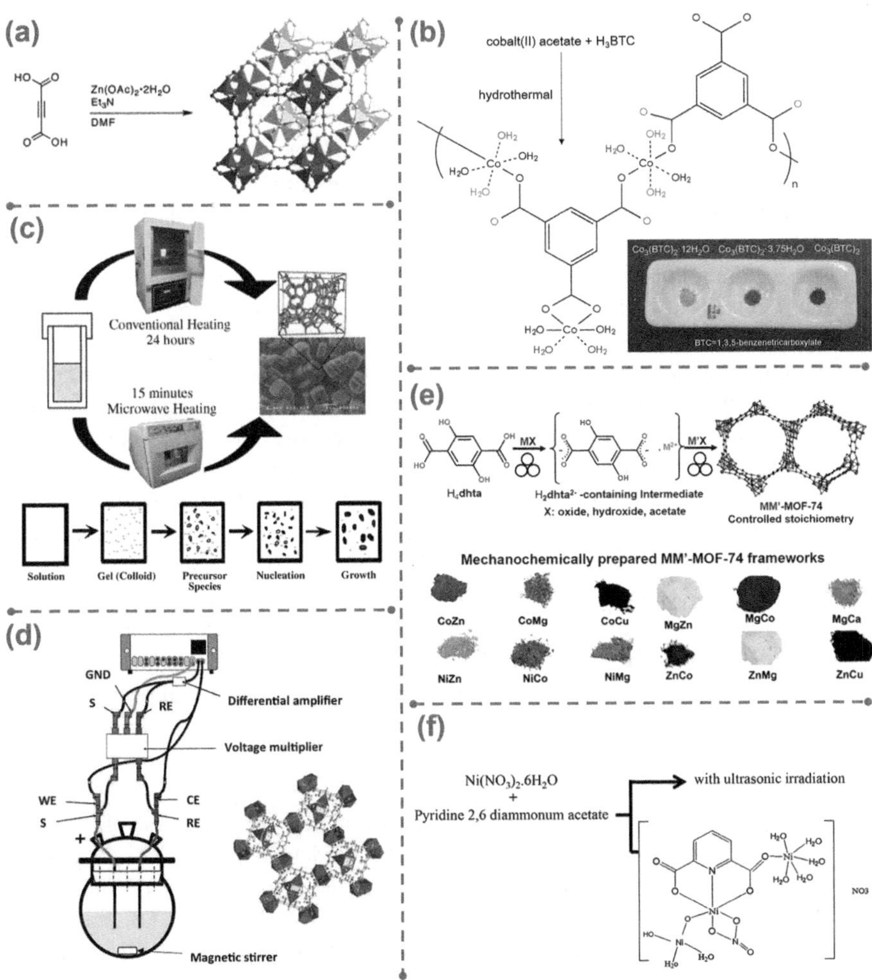

FIGURE 3.4 Different synthetic routes for MOF preparation. (a) MOF-5, MOF-177, HKUST-1 by slow-evaporation and diffusion method [47], (b) $Co_3(BTC)_2{\cdot}12H_2O$ by solvo(hydro)thermal method [54], (c) IR-MOF-1, 2, and 3 by microwave-assisted method [59], (d) HKUST-1, ZIF-8, MIL-100(Al), MIL-53(Al), and NH2-MIL-53(Al) by electrochemical method [65], (e) MM'-MOF-74 by mechanochemical method [73], and (f) Ni-MOF was prepared by ultrasonically [78].

varying sizes. Gels are sometimes used as mediums for crystallization and diffusion in order to decelerate diffusion and hinder the precipitation of the bulk material. The diffusion approach is very beneficial for acquiring single crystals that are acceptable for x-ray diffraction investigation, particularly in cases when the resulting products have restricted solubility [50, 51].

3.3.2.2 Solvo(Hydro)Thermal Method

Complex porous materials, such as metal oxides, zeolites, and metal-organic frameworks (MOFs), have been developed for many years using the solvo(hydro)thermal synthesis process. The term "hydrothermal" is used to describe chemical reactions that take place in either a homogeneous or heterogeneous state that includes both water or organic solvents. In the presence of water, it is considered a dipolar protic condition at a higher pressure of more than 1 atm with increased boiling temperature. The existence of organic solvent signifies protic or aprotic conditions depending upon the polar groups and called as solvothermal technique [52, 53].

Owing to the capability of the organic solvents to self-assemble the products from soluble precursors, solvothermal methods were utilized. This technique is primarily devised for the preparation of zeolite and is considered appropriate for the synthesis of MOF, $Co_3(BTC)_2{\cdot}12H_2O$ synthesis. The operational temperature may vary between 80 to 260°C followed by the adaptive cooling rate at the end, operating in the sealed vessel under autogenous pressure as shown in Figure 3.4b [38, 54]. Different materials may demand prolonged reaction time, ranging to days or weeks for solvothermal or hydrothermal approaches for diffusion processes [55]. Numerous MOFs have been successfully synthesized using this method, incorporating different metals and linkers, including Mn-MOF, Co-MOF, Zn-MOF-5, Cu-HKUST-1, Zr-PCN-111, etc.

For optimal regulation of autogenous pressure, the volume of the organic solvent must not surpass 75% of the vessel's overall capacity, while the quantity of the volume must not surpass 30% of the entire volume. Based on their chemical characteristics, solvents can be either aprotic (dimethyl sulfoxide, acetonitrile) or protic-hydrogen donors (water, ethanol). On the other hand, difficulties consist of the production of waste, dangers posed by anions such as nitrates and chlorides, and problems arising from oxidation and corrosion [56, 57].

3.3.2.3 Microwave-Assisted Synthesis

Microwave-assisted synthesis is deemed an instant, cost-effective, and clean process, originally used for developing nanosized metal oxides, as presented in Figure 3.4c. Although this approach is not widely used for the synthesis of crystalline MOF but useful for the rapid preparation of MOFs such as IR-MOF-1, 2, and 3, enabling size control over the resulting particles in the 10–15 nm range [58, 59]. Microwave irradiation generates the energy that is utilized for the formation of the controlled-sized MOFs in a shorter duration, suggesting high productivity, small particle size, selectivity, and morphology control [60, 61]. Like solvo(hydro)thermal synthesis, the microwave method efficiently produces MOFs with desirable morphology, textural properties, and adsorption affinity; however, there are a number of shortcomings to consider, such as possible high reaction temperatures, health and safety concerns, and the requirement for solvents with lower boiling points under pressure [62].

3.3.2.4 Electrochemical Method

The electrochemical method has appeared as an appealing and eco-friendly technique for the synthesis of MOFs, due to its facile fabrication conditions, rapid crystal growth, and scalability [63, 64]. This approach offers the advantage of controlling reaction conditions during synthesis, making it particularly beneficial for industrial applications where continuous production of metal salts is feasible [65, 66]. The three primary phases of the electrochemical synthesis are as follows: (a) combining the organic linker and electrolyte in a selective solvent; (b) supplying metal ions by anodic dissolution; and (c) increasing the pH value by increasing the water content in the solution (Figure 3.4d).

Martinez Joaristi, Alberto, et al. [65] synthesized ZIF-8, MIL-53(Al), HKUST-1 MIL-100(Al), and NH2-MIL-53(Al) using this approach. The technique utilizes an electrode to produce metal ions, which subsequently undergo a reaction with deprotonated linkers to create MOF molecules in close proximity to the electrode surface [67]. The quality of the product is influenced by several characteristics, including specific current, applied voltage, synthesis duration, electrode gap, linker chemistry, solvent type, nitrate availability, electrolyte concentration, and water content [66]. However, there are certain challenges involving morphology particle size and MOF film thickness, for which electrochemical methods have been implemented to control the concentration of the reaction [68, 69]. The industrial-scale manufacturing of MOF powders has effectively utilized this technology, which has advantages such as the exclusion of anions like nitrates, reduced reaction temperatures, and quick synthesis in comparison to solvothermal methods [70, 71]. The electrochemical synthesis allows for further refinement by adjusting voltage or specific signals, hence increasing the method's adaptability.

3.3.2.5 Mechanochemical Method

Mechanochemical synthesis, which entails the mechanical disruption of intramolecular bonds, has played an important role in the field of synthetic chemistry and is consumed in diverse areas, including the creation of pharmaceutically active co-crystals through multicomponent reactions and the study of inorganic solid-state chemistry [72]. Environmental concerns are propelling research into the mechanically activated synthesis of metal-organic frameworks (MOFs). This approach eliminates the need for organic solvents during reactions, which allows for ambient temperature reactions to produce small components with high quantitative yields in a relatively short amount of time. According to Ayoub et al. [73], the MM'-MOF-74 was produced by mechanical means, with M = Mg, Zn, Co, Ni, Cu; M′ = Ca, Mg, Co, Mg, Cu and Zn being a 1:1 stoichiometric concentration of divalent transition metals and as illustrated in Figure 3.4e. A strategy that is beneficial to the environmental benign involves replacing metal salts with metal oxides, with water being the only resultant of this process. By utilizing organic reactants with low melting temperatures and hydrated metal salts, such as metal acetates or carbonates, it is possible to overcome the limited solubility of metal oxides in reactions that are based on solvents. Liquid-assisted grinding (LAG) is a technique that involves the use of very small volumes of solvents to speed up mechanochemical reactions. This technique has the potential to improve reactant mobility and exhibit structure-directing characteristics. This technique has been extended to include ion- and liquid-assisted grinding (ILAG), which is effective for the selective synthesis of pillared-layered MOFs and demonstrates the influence that ions and solvents have on the structure-directing qualities of the MOFs [74, 75].

3.3.2.6 Sonochemical Method

The sonochemical method, employing ultrasonic radiation within the frequency range of 20–1,000 kHz, represents a fast, cost-effective, and eco-friendly technique for producing MOFs without requiring high pressure or temperature conditions [76].

When ultrasound is applied to liquids, it creates bubbles by using sound waves. These bubbles then develop and quickly burst, causing localized changes in temperature and pressure. Acoustic cavitation, a phenomenon in which sound waves cause changes in a solution of MOF precursors, is known to increase chemical or physical reactions [77]. Ultrasound induces severe circumstances, including increased heat and the generation of free radicals, which facilitate chemical processes by rapidly forming crystallization nuclei of Ni-MOF (as illustrated in Figure 3.4f) [78].

The properties of the resultant Zn-HKUST-1 are affected by the duration, temperature, and power of ultrasonication. The morphology, crystal size, and yield of the material rely on the reaction time and concentrations of the precursors [79]. Sonochemical approaches, which can provide uniform nucleation centers and decrease the time required for crystallization compared to traditional hydrothermal processes, are used both independently as synthesis techniques and as pre-primary mixing tools in solvothermal/hydrothermal systems [38]. Sonochemistry, which utilizes high-energy ultrasound, is utilized in diverse chemical reactions, such as the production of organic and nanomaterials. This makes it a versatile instrument in MOF science, where the objective is to develop rapid, environmentally friendly, energy-efficient, and ambient temperature-compatible synthesis techniques to facilitate the expansion of MOF production [80, 81].

3.3.3 Synthesis Strategies of COFs

Selecting an effective synthetic approach plays a pivotal role in advancing the development of Covalent Organic Frameworks (COFs). Traditional synthesis methods encompass solvothermal, ionothermal, microwave, and room-temperature techniques. Achieving highly ordered and crystalline COFs requires a deep understanding of construction units, synthetic pathways, and the thermodynamic equilibria of covalent bonding. The solvothermal method, initially successful under the guidance of Yaghi et. al., paved the way for subsequent research groups in laboratory-scale exploration [34]. However, the challenge lies in scaling up these methodologies for the mass production of COFs. The scientific community has explored alternative, simpler, and more productive synthesis methods, including ionothermal, microwave, mechanochemical, room-temperature, and interface methods [82]. Each method offers unique advantages and contributes to the diverse toolkit available for COF synthesis.

3.3.3.1 Solvothermal Method

Most Covalent Organic Frameworks (COFs) have historically been synthesized through solvothermal methods, a widely employed approach dependent on the solubility and reactivity of building blocks and the reversibility of reactions. The crucial factors for preparing crystalline porous COFs via this method include reaction time, temperature, solvent conditions, and catalyst concentration. In a typical procedure, a mixture of suitable monomers, catalysts, and solvents is sonicated, degassed, sealed in a Pyrex tube, and maintained at a specific temperature. The resulting precipitate is collected, washed, and dried, with some COFs amenable to large-scale preparation. Despite its popularity, solvothermal synthesis presents challenges for industrial scalability and slow reaction rates. Significantly, this approach enables the large-scale synthesis of certain COFs including TPT-COF-1, derived from 2,4,6-tris(4-aminophenoxy)-1,3,5-triazine (TPT-NH2) and 2,4,6-tris(4-formylphenoxy)-1,3,5-triazine (TPT-CHO), as shown in Figure 3.5a. The resulting TPT-COF-1 exhibits an impressive BET surface area of 1589 $m^2 g^{-1}$, coupled with high crystallinity. Recent innovations explore solution-based synthesis under ambient pressure, offering potential advantages in terms of scalability and reaction efficiency [83].

3.3.3.2 Microwave-Assisted Synthesis

In response to the prolonged reaction times associated with solvothermal methods, the microwave technique has been investigated by Campbell et al. for the swift fabrication of crystalline porous Covalent Organic Frameworks (COFs) [84]. COF-5, COF-102, and imine-linked TpPa-COF, as well as crystalline covalent triazine frameworks (CTFs), have been successfully synthesized using the microwave approach (represented in Figure 3.5b for COF-102) [84–86]. The general microwave method involves sealing a mixture of monomers in a suitable solvent in a microwave tube, heating under nitrogen or vacuum, and stirring for a designated period, such as 60 min at 100°C. Post-reaction, the resulting COFs undergo various purification steps, including solvent extraction, filtration, and drying under vacuum [86]. This method not only efficiently removes oligomers but also enhances the resulting COFs' porosity. The application of microwave irradiation for COF synthesis has demonstrated faster reactions and improved properties, such as higher surface areas, compared to solvothermal synthesis.

3.3.3.3 Ionothermal Method

Ionothermal methods, employing ionic liquids as solvents, catalysts, or structural directing agents, have played a crucial role in the production of crystalline porous COFs [87]. Among them, CTF-1 and CTF-2 stand out as crystalline porous materials synthesized under ionothermal conditions [82, 88]. A typical ionothermal synthesis involves evacuating and sealing a Pyrex ampule for 40 hours containing monomers and $ZnCl_2$, followed by controlled heating to 400°C. The resulting mixture is cooled, grounded, washed, and subjected to additional purification steps to yield the CTF-2, as depicted in Figure 3.5c. Additionally, microwave-assisted ionothermal synthesis has been successfully applied for CTF-1 [89]. Ionic liquids, due to their unique features such as low vapor pressure and extensive liquid range, offer a green and versatile solvent for COF synthesis. Guan et al. devised a novel ionothermal synthetic route for the 3D-IL-COFs (where IL stands for ionic liquid) under ambient reaction conditions. Notably, these 3D-IL-COFs demonstrated significant surface areas (517–870 $m^2 g^{-1}$) [90]. The ionothermal approach has also been employed for the growth of triazine-based COFs with enhanced surface area and porosity. Furthermore,

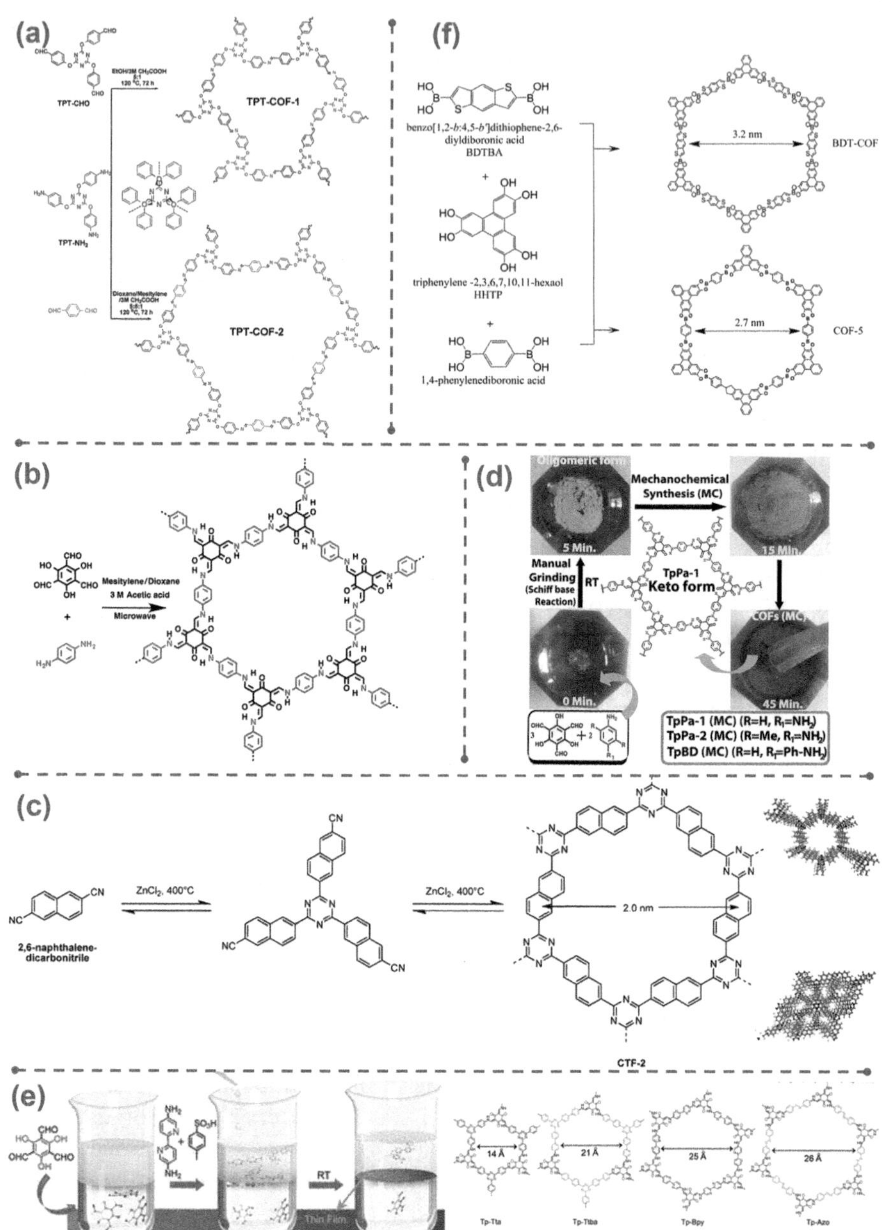

FIGURE 3.5 Various synthetic routes to COFs. (a) TPT-COF-1 by solvothermal method [83], (b) COF-102 by microwave-assisted method [86], (c) CTF-2 by ionothermal method [82], (d) TpPa-1, TpPa-2, and TpBD by mechanochemical method [93], (e) Tp-Tta, Tp-Ttba, Tp-Bpy, and Tp-Azo by interfacial method [96], and (f) BDT-COF, and COF-5 were prepared at ambient temperature [100].

challenging syntheses, like polyimide-linked COFs (PI-COFs) and perylene-based COFs, have been accomplished using ionothermal conditions, addressing difficulties encountered in conventional solvothermal methods [91].

3.3.3.4 Mechanochemical Method

Exploring a straightforward synthetic method as an alternative to solvothermal and microwave reactions, mechanochemical synthesis offers a promising strategy for COF materials. This method involves grinding of monomers at ambient conditions in a mortar as shown in Figure 3.5d, resulting in COFs like TpPa-X (where X refers to1, 2, NO_2, F_4), TpBD, TpBD- Y (Y=NO_2)$_2$, Me_2, OMe)$_2$) [92, 93]. To optimize mechanochemical conditions, the liquid-assisted mechanochemical/grinding method is implemented, incorporating a small amount of catalyst solution during grinding to enhance reaction rates and crystallinity. This approach proves to be an economical, environmentally benign, and scalable method for COF synthesis [94]. However, despite the advantages of being easy and fast, original mechanical grinding of COF precursor mixtures often results in amorphous structures (having poor crystallinity). To overcome this limitation, researchers have expanded the technique by adding solvents and molecular organizers like p-toluenesulfonic acid and applying additional heating and/or recrystallization steps, resulting in ordered COFs (having crystalline and porous structures). Additionally, studies have shown that pre-synthesized unoriented 2D frameworks can be aligned into anisotropic packed layers that are perpendicular to the axis of the mechanical force by applying mechanical stress [95].

3.3.3.5 Interfacial Synthesis

Interfacial synthesis stands out as a promising technique for the precise fabrication of COF thin films, offering control over thickness from 4–150 nm and structural integrity. This innovative approach involves the interaction of monomers at various interfaces, such as liquid-liquid, liquid-air, solid-liquid, and solid-vapor. Noteworthy COFs, including Tp-Z (Z=Bpy, Azo, Ttba, Tta), and COFTTA-DHTA, also 2DCCOF-1, and 2DCCOF-2 through Suzuki-coupling (a dilute toluene solution with monomers and the $Pd(PPh_3)_4$ catalyst is carefully deposited atop an aqueous K_2CO_3 solution). After that, reactions occur at the toluene-water interface in argon at 2°C. Over a month, using this rigorous technique stable, extensive AB-staggered COF sheets [96] have been successfully produced using this method, schematic, and analog structures for certain COFs are represented in Figure 3.5e. The solvothermal strategy is a key synthetic approach in solid-liquid interface synthesis, while techniques like ultra-high vacuum and drop-casting are employed for interfaces between solid and vapor. Liquid-liquid and liquid-air interfaces yield free-standing films with the added advantage of transferability to other substrates. Despite substantial progress, challenges persist, particularly in terms of production capacity [97, 98]. The interfacial synthesis approach not only demonstrates versatility but also presents opportunities for further exploration and optimization in the realm of COF thin-film fabrication.

3.3.3.6 Ambient Temperature Synthesis

Recently, Peng and colleagues invented a room-temperature batch synthetic method for imine, azine, and enamine-linked COFs [99]. The monomer mixture was homogenized by sonication and left for 3 days, resulting in crystalline and porous COFs when utilizing linkers with strong π-stacking abilities. In cases where building blocks exhibited lower π-stacking ability, such as hydrazine and TFPB (1,3,5-tris(4-formylphenyl)benzene), a COF was still synthesized with a reduced BET surface area. While solvothermal conditions are common for COF synthesis, the appeal of room-temperature synthesis is evident, particularly for delicate building blocks and/or substrates. Vapor-assisted synthesis at ambient temperature proved effective in producing COF films including BDTCOF and COF-5, achieved by drop-casting a mixture of acetone and ethanol solution of COF precursors onto a clean glass substrate, followed by incubation in a desiccator containing mesitylene and dioxane for 72 hours to prepare these boroxine-based thin films, as shown in Figure 3.5f [100]. Additionally, a constant-flow approach for the LZU-1 demonstrated efficient room-temperature synthesis, producing high-quality COF, a strategy also employed by Dichtel et al. for synthesizing thin films [101]. Furthermore, Lin et al. achieved the rapid formation of a COF shell of imine around Fe_3O_4 nanoparticles at ambient temperature in just 5 min [102].

3.4 CONCLUSIONS

The synthesis strategies employed for the fabrication of PCPs, MOFs, and COFs are diverse and have a significant impact in tailoring the properties and applicability of these materials. Solvothermal methods, including traditional solvothermal and ambient temperature approaches, provide controlled conditions for the formation of crystalline structures, while mechanochemical methods offer a simpler, environmentally benign route for the synthesis of COFs. Interfacial methods contribute to the synthesis of PCPs. Each of these synthesis strategies brings unique advantages and challenges, influencing the morphology, crystallinity, and properties of the resulting materials. The exploration of these diverse approaches showcases the dynamic nature of the field and provides avenues for tailoring porous coordination materials to specific applications in areas such as gas storage, catalysis, separation, sensing, luminescence, drug delivery, and ionic/electronic conductivity. As the research progresses, a deeper understanding of the synthesis-structure-property relationships will undoubtedly lead to the development of advanced materials with enhanced performance and functionality.

ACKNOWLEDGMENTS

This work was supported by the Research Fund for International Scientists (RFIS-Grant number: 52150410410) National Natural Science Foundation of China.

REFERENCES

[1] S.R. Batten, N.R. Champness, X.-M. Chen, J. Garcia-Martinez, S. Kitagawa, L. Öhrström, M. O'Keeffe, M. Paik Suh, J. Reedijk, Terminology of Metal–Organic Frameworks and Coordination Polymers (IUPAC Recommendations 2013), *Pure and Applied Chemistry* 85(8) (2013) 1715–1724.

[2] H.-C. Zhou, J.R. Long, O.M. Yaghi, Introduction to Metal–Organic Frameworks, *Chemical Reviews* 112(2) (2012) 673–674. https://doi.org/10.1021/cr300014x

[3] A. Morozan, F. Jaouen, Metal Organic Frameworks for Electrochemical Applications, *Energy & Environmental Science* 5(11) (2012) 9269–9290.
[4] T. Islamoglu, S. Goswami, Z. Li, A.J. Howarth, O.K. Farha, J.T. Hupp, Postsynthetic Tuning of Metal–Organic Frameworks for Targeted Applications, *Accounts of Chemical Research* 50(4) (2017) 805–813. https://doi.org/10.1021/acs.accounts.6b00577
[5] H. Liu, F.-X. Ma, C.-Y. Xu, L. Yang, Y. Du, P.-P. Wang, S. Yang, L. Zhen, Sulfurizing-Induced Hollowing of Co9S8 Microplates with Nanosheet Units for Highly Efficient Water Oxidation, *ACS Applied Materials & Interfaces* 9(13) (2017) 11634–11641. https://doi.org/10.1021/acsami.7b00899
[6] S.-I. Noro, J. Mizutani, Y. Hijikata, R. Matsuda, H. Sato, S. Kitagawa, K. Sugimoto, Y. Inubushi, K. Kubo, T. Nakamura, Porous Coordination Polymers with Ubiquitous and Biocompatible Metals and a Neutral Bridging Ligand, *Nature Communications* 6(1) (2015) 5851. https://doi.org/10.1038/ncomms6851
[7] P. Wu, Y. Li, J.-J. Zheng, N. Hosono, K.-I. Otake, J. Wang, Y. Liu, L. Xia, M. Jiang, S. Sakaki, S. Kitagawa, Carbon Dioxide Capture and Efficient Fixation in a Dynamic Porous Coordination Polymer, *Nature Communications* 10(1) (2019) 4362. https://doi.org/10.1038/s41467-019-12414-z
[8] D. Alvarado-Alvarado, J.H. González-Estefan, J.G. Flores, J.R. Álvarez, J. Aguilar-Pliego, A. Islas-Jácome, G. Chastanet, E. González-Zamora, H.A. Lara-García, B. Alcántar-Vázquez, M. Gonidec, I.A. Ibarra, Water Adsorption Properties of Fe(pz)[Pt(CN)4] and the Capture of CO2 and CO, *Organometallics* 39(7) (2020) 949–955. https://doi.org/10.1021/acs.organomet.9b00711
[9] X.-M. Hao, T.-G. Qu, H. Wang, W.-L. Guo, F. Chen, Y.-B. Wu, D. Yang, Z.-L. Xu, A 3D Porous Coordination Polymer Transformed from a 1D Nonporous Coordination Polymer for Selectively Sensing of Diiodomethane, *Journal of Solid State Chemistry* 268 (2018) 62–66. https://doi.org/10.1016/j.jssc.2018.08.034
[10] H. Hu, H. Zhang, Y. Chen, Y. Chen, L. Zhuang, H. Ou, Enhanced Photocatalysis Degradation of Organophosphorus Flame Retardant using MIL-101(Fe)/persulfate: Effect of Irradiation Wavelength and Real Water Matrixes, *Chemical Engineering Journal* 368 (2019) 273–284. https://doi.org/10.1016/j.cej.2019.02.190
[11] H. Wei, Z. Guo, X. Liang, P. Chen, H. Liu, H. Xing, Selective Photooxidation of Amines and Sulfides Triggered by a Superoxide Radical Using a Novel Visible-Light-Responsive Metal–Organic Framework, *ACS Applied Materials & Interfaces* 11(3) (2019) 3016–3023. https://doi.org/10.1021/acsami.8b18206
[12] M.-R. Ahmadian-Yazdi, N. Gholampour, M. Eslamian, Interface Engineering by Employing Zeolitic Imidazolate Framework-8 (ZIF-8) as the Only Scaffold in the Architecture of Perovskite Solar Cells, *ACS Applied Energy Materials* 3(4) (2020) 3134–3143. https://doi.org/10.1021/acsaem.9b02115
[13] W. Lin, Q. Hu, K. Jiang, Y. Yang, Y. Yang, Y. Cui, G. Qian, A Porphyrin-Based Metal–Organic Framework as a pH-Responsive Drug Carrier, *Journal of Solid State Chemistry* 237 (2016) 307–312. https://doi.org/10.1016/j.jssc.2016.02.040
[14] L. Deng, Z. Ding, X. Ye, D. Jiang, Covalent Organic Frameworks: Chemistry of Pore Interface and Wall Surface Perturbation and Impact on Functions, *Accounts of Materials Research* 3(8) (2022) 879–893. https://doi.org/10.1021/accountsmr.2c00108
[15] S. Dalapati, M. Addicoat, S. Jin, T. Sakurai, J. Gao, H. Xu, S. Irle, S. Seki, D. Jiang, Rational Design of Crystalline Supermicroporous Covalent Organic Frameworks with Triangular Topologies, *Nature Communications* 6(1) (2015) 7786. https://doi.org/10.1038/ncomms8786
[16] L. Chen, K. Furukawa, J. Gao, A. Nagai, T. Nakamura, Y. Dong, D. Jiang, Photoelectric Covalent Organic Frameworks: Converting Open Lattices into Ordered Donor–Acceptor Heterojunctions, *Journal of the American Chemical Society* 136(28) (2014) 9806–9809. https://doi.org/10.1021/ja502692w

[17] Q. Sun, B. Aguila, J. Perman, L.D. Earl, C.W. Abney, Y. Cheng, H. Wei, N. Nguyen, L. Wojtas, S. Ma, Postsynthetically Modified Covalent Organic Frameworks for Efficient and Effective Mercury Removal, *Journal of the American Chemical Society* 139(7) (2017) 2786–2793. https://doi.org/10.1021/jacs.6b12885

[18] S. Diring, S. Furukawa, Y. Takashima, T. Tsuruoka, S. Kitagawa, Controlled Multiscale Synthesis of Porous Coordination Polymer in Nano/Micro Regimes, *Chemistry of Materials* 22(16) (2010) 4531–4538. https://doi.org/10.1021/cm101778g

[19] W. Lin, W.J. Rieter, K.M.L. Taylor, Modular Synthesis of Functional Nanoscale Coordination Polymers, *Angewandte Chemie International Edition* 48(4) (2009) 650–658. https://doi.org/10.1002/anie.200803387

[20] O.M. Yaghi, H. Li, Hydrothermal Synthesis of a Metal-Organic Framework Containing Large Rectangular Channels, *Journal of the American Chemical Society* 117(41) (1995) 10401–10402. https://doi.org/10.1021/ja00146a033

[21] S. Wang, Q. Wang, P. Shao, Y. Han, X. Gao, L. Ma, S. Yuan, X. Ma, J. Zhou, X. Feng, B. Wang, Exfoliation of Covalent Organic Frameworks into Few-Layer Redox-Active Nanosheets as Cathode Materials for Lithium-Ion Batteries, *Journal of the American Chemical Society* 139(12) (2017) 4258–4261. https://doi.org/10.1021/jacs.7b02648

[22] L. Wang, Y. Han, X. Feng, J. Zhou, P. Qi, B. Wang, Metal–Organic Frameworks for Energy Storage: Batteries and Supercapacitors, *Coordination Chemistry Reviews* 307 (2016) 361–381. https://doi.org/10.1016/j.ccr.2015.09.002

[23] H. Furukawa, K.E. Cordova, M. O'Keeffe, O.M. Yaghi, The Chemistry and Applications of Metal-Organic Frameworks, *Science* 341(6149) (2013) 1230444. https://doi.org/10.1126/science.1230444

[24] G. Férey, Hybrid Porous Solids: Past, Present, Future, *Chemical Society Reviews* 37(1) (2008) 191–214.

[25] Y. Wang, G. Ye, H. Chen, X. Hu, Z. Niu, S. Ma, Functionalized Metal–Organic Framework as a New Platform for Efficient and Selective Removal of Cadmium (II) from Aqueous Solution, *Journal of Materials Chemistry A* 3(29) (2015) 15292–15298.

[26] W. Liu, J. Wang, X. Ke, S. Li, Large Piezoelectric Performance of Sn Doped BaTiO3 Ceramics Deviating from Quadruple Point, *Journal of Alloys and Compounds* 712 (2017) 1–6. https://doi.org/10.1016/j.jallcom.2017.04.013

[27] O.M. Yaghi, M. O'Keeffe, N.W. Ockwig, H.K. Chae, M. Eddaoudi, J. Kim, Reticular Synthesis and the Design of New Materials, *Nature* 423(6941) (2003) 705–714. https://doi.org/10.1038/nature01650

[28] S. Qiu, G. Zhu, Molecular Engineering for Synthesizing Novel Structures of Metal–Organic Frameworks with Multifunctional Properties, *Coordination Chemistry Reviews* 253(23) (2009) 2891–2911. https://doi.org/10.1016/j.ccr.2009.07.020

[29] C. Wiktor, M. Meledina, S. Turner, O.I. Lebedev, R.A. Fischer, Transmission Electron Microscopy on Metal–Organic Frameworks–a Review, *Journal of Materials Chemistry A* 5(29) (2017) 14969–14989.

[30] Q.-L. Zhu, Q. Xu, Metal–Organic Framework Composites, *Chemical Society Reviews* 43(16) (2014) 5468–5512.

[31] J. Xu, Y. Li, L. Wang, Q. Cai, Q. Li, B. Gao, X. Zhang, K. Huo, P.K. Chu, High-Energy Lithium-Ion Hybrid Supercapacitors Composed of Hierarchical Urchin-Like WO3/C Anodes and MOF-Derived Polyhedral Hollow Carbon Cathodes, *Nanoscale* 8(37) (2016) 16761–16768. https://doi.org/10.1039/C6NR05480C

[32] B. Wang, J. Park, D. Su, C. Wang, H. Ahn, G. Wang, Solvothermal Synthesis of CoS2–Graphene Nanocomposite Material for High-Performance Supercapacitors, *Journal of Materials Chemistry* 22(31) (2012) 15750–15756. https://doi.org/10.1039/C2JM31214J

[33] O.M. Yaghi, Reticular Chemistry – Construction, Properties, and Precision Reactions of Frameworks, *Journal of the American Chemical Society* 138(48) (2016) 15507–15509. https://doi.org/10.1021/jacs.6b11821

[34] A.P. Côté, A.I. Benin, N.W. Ockwig, M. O'Keeffe, A.J. Matzger, O.M. Yaghi, Porous, Crystalline, Covalent Organic Frameworks, *Science* 310(5751) (2005) 1166–1170. https://doi.org/10.1126/science.1120411

[35] X. Guan, F. Chen, Q. Fang, S. Qiu, Design and Applications of Three Dimensional Covalent Organic Frameworks, *Chemical Society Reviews* 49(5) (2020) 1357–1384.

[36] C.S. Diercks, O.M. Yaghi, The Atom, the Molecule, and the Covalent Organic Framework, *Science* 355(6328) (2017) eaal1585. https://doi.org/10.1126/science.aal1585

[37] A. Altaf, N. Baig, M. Sohail, M. Sher, A. Ul-Hamid, M. Altaf, Covalent Organic Frameworks: Advances in Synthesis and Applications, *Materials Today Communications* 28 (2021) 102612. https://doi.org/10.1016/j.mtcomm.2021.102612

[38] C. Dey, T. Kundu, B.P. Biswal, A. Mallick, R. Banerjee, Crystalline Metal-Organic Frameworks (MOFs): Synthesis, Structure and Function, Acta Crystallographica Section B: Structural Science, *Crystal Engineering and Materials* 70(1) (2014) 3–10.

[39] I. Persson, Solvation and Complex Formation in Strongly Solvating Solvents, *Pure and Applied Chemistry* 58(8) (1986) 1153–1161.

[40] O.A. El Seoud, Understanding Solvation, *Pure and Applied Chemistry* 81(4) (2009) 697–707.

[41] X. Yang, A.E. Clark, Preferential Solvation of Metastable Phases Relevant to Topological Control Within the Synthesis of Metal–Organic Frameworks, *Inorganic Chemistry* 53(17) (2014) 8930–8940. https://doi.org/10.1021/ic5006659

[42] A.J. Howarth, A.W. Peters, N.A. Vermeulen, T.C. Wang, J.T. Hupp, O.K. Farha, Best Practices for the Synthesis, Activation, and Characterization of Metal–Organic Frameworks, *Chemistry of Materials* 29(1) (2017) 26–39. https://doi.org/10.1021/acs.chemmater.6b02626

[43] W. Cho, H.J. Lee, M. Oh, Growth-Controlled Formation of Porous Coordination Polymer Particles, *Journal of the American Chemical Society* 130(50) (2008) 16943–16946. https://doi.org/10.1021/ja8039794

[44] Q. Liu, J.-M. Yang, L.-N. Jin, W.-Y. Sun, Controlled Synthesis of Porous Coordination-Polymer Microcrystals with Definite Morphologies and Sizes under Mild Conditions, *Chemistry – A European Journal* 20(45) (2014) 14783–14789. https://doi.org/10.1002/chem.201402923

[45] A. Dmitriev, H. Spillmann, N. Lin, J.V. Barth, K. Kern, Modular Assembly of Two-Dimensional Metal–Organic Coordination Networks at a Metal Surface, *Angewandte Chemie* 115(23) (2003) 2774–2777.

[46] X. Huang, P. Sheng, Z. Tu, F. Zhang, J. Wang, H. Geng, Y. Zou, C.-A. Di, Y. Yi, Y. Sun, A Two-Dimensional π–d Conjugated Coordination Polymer with Extremely High Electrical Conductivity and Ambipolar Transport Behaviour, *Nature Communications* 6(1) (2015) 7408.

[47] D.J. Tranchemontagne, J.R. Hunt, O.M. Yaghi, Room Temperature Synthesis of Metal-Organic Frameworks: MOF-5, MOF-74, MOF-177, MOF-199, and IRMOF-0, *Tetrahedron* 64(36) (2008) 8553–8557. https://doi.org/10.1016/j.tet.2008.06.036

[48] J.L.C. Rowsell, O.M. Yaghi, Metal–Organic Frameworks: A New Class of Porous Materials, *Microporous and Mesoporous Materials* 73(1) (2004) 3–14. https://doi.org/10.1016/j.micromeso.2004.03.034

[49] J. Hafizovic, M. Bjørgen, U. Olsbye, P.D.C. Dietzel, S. Bordiga, C. Prestipino, C. Lamberti, K.P. Lillerud, The Inconsistency in Adsorption Properties and Powder XRD Data of MOF-5 Is Rationalized by Framework Interpenetration and the Presence of Organic and Inorganic Species in the Nanocavities, *Journal of the American Chemical Society* 129(12) (2007) 3612–3620. https://doi.org/10.1021/ja0675447

[50] R. Abazari, A.R. Mahjoub, J. Shariati, Synthesis of a Nanostructured Pillar MOF with High Adsorption Capacity Towards Antibiotics Pollutants from Aqueous Solution, *Journal of Hazardous Materials* 366 (2019) 439–451. https://doi.org/10.1016/j.jhazmat.2018.12.030

[51] T. Gao, B.-X. Dong, Y.-M. Pan, W.-L. Liu, Y.-L. Teng, Highly Sensitive and Recyclable Sensing of Fe3+ Ions Based on a Luminescent Anionic [Cd(DMIPA)]2- Framework with Exposed Thioether Group in the Snowflake-Like Channels, *Journal of Solid State Chemistry* 270 (2019) 493–499. https://doi.org/10.1016/j.jssc.2018.12.008
[52] M. Yoshimura, K. Byrappa, Hydrothermal Processing of Materials: Past, Present and Future, *Journal of Materials Science* 43 (2008) 2085–2103.
[53] G. Férey, Metal-Organic Frameworks: The Young Child of the Porous Solids Family, in: R. Xu, Z. Gao, J. Chen, W. Yan (Eds.), *Studies in Surface Science and Catalysis*, Elsevier, 2007, pp. 66–84.
[54] J.L. Crane, K.E. Anderson, S.G. Conway, Hydrothermal Synthesis and Characterization of a Metal–Organic Framework by Thermogravimetric Analysis, Powder X-ray Diffraction, and Infrared Spectroscopy: An Integrative Inorganic Chemistry Experiment, *Journal of Chemical Education* 92(2) (2015) 373–377. https://doi.org/10.1021/ed5000839
[55] X. Shi, G. Zhu, S. Qiu, K. Huang, J. Yu, R. Xu, Zn2[(S)-O3PCH2NHC4H7CO2]2: A Homochiral 3D Zinc Phosphonate with Helical Channels, *Angewandte Chemie International Edition* 43(47) (2004) 6482–6485. https://doi.org/10.1002/anie.200460724
[56] A.J. Parker, The Effects of Solvation on the Properties of Anions in Dipolar Aprotic Solvents, *Quarterly Reviews, Chemical Society* 16(2) (1962) 163–187.
[57] H.M. Yang, X.L. Song, T.L. Yang, Z.H. Liang, C.M. Fan, X.G. Hao, Electrochemical Synthesis of Flower Shaped Morphology MOFs in an Ionic Liquid System and Their Electrocatalytic Application to the Hydrogen Evolution Reaction, *RSC Advances* 4(30) (2014) 15720–15726. https://doi.org/10.1039/C3RA47744D
[58] Z. Ni, R.I. Masel, Rapid Production of Metal–Organic Frameworks via Microwave-Assisted Solvothermal Synthesis, *Journal of the American Chemical Society* 128(38) (2006) 12394–12395. https://doi.org/10.1021/ja0635231
[59] G.A. Tompsett, W.C. Conner, K.S. Yngvesson, Microwave Synthesis of Nanoporous Materials, *ChemPhysChem* 7(2) (2006) 296–319. https://doi.org/10.1002/cphc.200500449
[60] P. Horcajada, T. Chalati, C. Serre, B. Gillet, C. Sebrie, T. Baati, J.F. Eubank, D. Heurtaux, P. Clayette, C. Kreuz, J.-S. Chang, Y.K. Hwang, V. Marsaud, P.-N. Bories, L. Cynober, S. Gil, G. Férey, P. Couvreur, R. Gref, Porous Metal–Organic-Framework Nanoscale Carriers as a Potential Platform for Drug Delivery and Imaging, *Nature Materials* 9(2) (2010) 172–178. https://doi.org/10.1038/nmat2608
[61] S.H. Jhung, J.-H. Lee, P.M. Forster, G. Férey, A.K. Cheetham, J.-S. Chang, Microwave Synthesis of Hybrid Inorganic–Organic Porous Materials: Phase-Selective and Rapid Crystallization, *Chemistry – A European Journal* 12(30) (2006) 7899–7905. https://doi.org/10.1002/chem.200600270
[62] R. Vakili, S. Xu, N. Al-Janabi, P. Gorgojo, S.M. Holmes, X. Fan, Microwave-Assisted Synthesis of Zirconium-Based Metal Organic Frameworks (MOFs): Optimization and Gas Adsorption, *Microporous and Mesoporous Materials* 260 (2018) 45–53. https://doi.org/10.1016/j.micromeso.2017.10.028
[63] K. Pirzadeh, A.A. Ghoreyshi, M. Rahimnejad, M. Mohammadi, Electrochemical Synthesis, Characterization and Application of a Microstructure Cu3(BTC)2 Metal Organic Framework for CO2 and CH4 Separation, *Korean Journal of Chemical Engineering* 35(4) (2018) 974–983. https://doi.org/10.1007/s11814-017-0340-6
[64] S. Zhao, Y. Wang, J. Dong, C.-T. He, H. Yin, P. An, K. Zhao, X. Zhang, C. Gao, L. Zhang, J. Lv, J. Wang, J. Zhang, A.M. Khattak, N.A. Khan, Z. Wei, J. Zhang, S. Liu, H. Zhao, Z. Tang, Ultrathin Metal–Organic Framework Nanosheets for Electrocatalytic Oxygen Evolution, *Nature Energy* 1(12) (2016) 16184. https://doi.org/10.1038/nenergy.2016.184
[65] A. Martinez Joaristi, J. Juan-Alcañiz, P. Serra-Crespo, F. Kapteijn, J. Gascon, Electrochemical Synthesis of Some Archetypical Zn2+, Cu2+, and Al3+ Metal Organic Frameworks, *Crystal Growth & Design* 12(7) (2012) 3489–3498. https://doi.org/10.1021/cg300552w

[66] M. Li, M. Dincă, On the Mechanism of MOF-5 Formation under Cathodic Bias, *Chemistry of Materials* 27(9) (2015) 3203–3206. https://doi.org/10.1021/acs.chemmater.5b00899
[67] W.-J. Li, J. Lü, S.-Y. Gao, Q.-H. Li, R. Cao, Electrochemical Preparation of Metal–Organic Framework Films for Fast Detection of Nitro Explosives, *Journal of Materials Chemistry A* 2(45) (2014) 19473–19478.
[68] S. Sachdeva, A. Pustovarenko, E.J.R. Sudhölter, F. Kapteijn, L.C.P.M. De Smet, J. Gascon, Control of Interpenetration of Copper-Based MOFs on Supported Surfaces by Electrochemical Synthesis, *CrystEngComm* 18(22) (2016) 4018–4022.
[69] N. Stock, S. Biswas, Synthesis of Metal-Organic Frameworks (MOFs): Routes to Various MOF Topologies, Morphologies, and Composites, *Chemical Reviews* 112(2) (2012) 933–969. https://doi.org/10.1021/cr200304e
[70] U. Mueller, M. Schubert, F. Teich, H. Puetter, K. Schierle-Arndt, J. Pastré, Metal–Organic Frameworks – Prospective Industrial Applications, *Journal of Materials Chemistry* 16(7) (2006) 626–636. https://doi.org/10.1039/B511962F
[71] Y. Wu, A. Kobayashi, G.J. Halder, V.K. Peterson, K.W. Chapman, N. Lock, P.D. Southon, C.J. Kepert, Negative Thermal Expansion in the Metal–Organic Framework Material Cu3(1,3,5-benzenetricarboxylate)2, *Angewandte Chemie International Edition* 47(46) (2008) 8929–8932. https://doi.org/10.1002/anie.200803925
[72] V.V. Boldyrev, K. Tkáčová, Mechanochemistry of Solids: Past, Present, and Prospects, *Journal of Materials Synthesis and Processing* 8(3) (2000) 121–132. https://doi.org/10.1023/A:1011347706721
[73] M.K. Beyer, H. Clausen-Schaumann, Mechanochemistry: The Mechanical Activation of Covalent Bonds, *Chemical Reviews* 105(8) (2005) 2921–2948. https://doi.org/10.1021/cr030697h
[74] K. Fujii, A.L. Garay, J. Hill, E. Sbircea, Z. Pan, M. Xu, D.C. Apperley, S.L. James, K.D.M. Harris, Direct Structure Elucidation by Powder X-ray Diffraction of a Metal–Organic Framework Material Prepared by Solvent-Free Grinding, *Chemical Communications* 46(40) (2010) 7572–7574. https://doi.org/10.1039/C0CC02635B
[75] V. Štrukil, L. Fábián, D.G. Reid, M.J. Duer, G.J. Jackson, M. Eckert-Maksić, T. Friščić, Towards an Environmentally-Friendly Laboratory: Dimensionality and Reactivity in the Mechanosynthesis of Metal–Organic Compounds, *Chemical Communications* 46(48) (2010) 9191–9193. https://doi.org/10.1039/C0CC03822A
[76] G. Chatel, J.C. Colmenares, Sonochemistry: From Basic Principles to Innovative Applications, *Topics in Current Chemistry* 375(1) (2017) 8. https://doi.org/10.1007/s41061-016-0096-1
[77] K.S. Suslick, The Sonochemical Hot Spot, *The Journal of the Acoustical Society of America* 89(4B_Supplement) (2005) 1885–1886. https://doi.org/10.1121/1.2029381
[78] G. Sargazi, D. Afzali, N. Daldosso, H. Kazemian, N.P.S. Chauhan, Z. Sadeghian, T. Tajerian, A. Ghafarinazari, M. Mozafari, A Systematic Study on the Use of Ultrasound Energy for the Synthesis of Nickel–Metal Organic Framework Compounds, *Ultrasonics Sonochemistry* 27 (2015) 395–402. https://doi.org/10.1016/j.ultsonch.2015.04.004
[79] A.A. Tehrani, V. Safarifard, A. Morsali, G. Bruno, H.A. Rudbari, Ultrasound-Assisted Synthesis of Metal–Organic Framework Nanorods of Zn-HKUST-1 and Their Templating Effects for Facile Fabrication of Zinc Oxide Nanorods via Solid-State Transformation, *Inorganic Chemistry Communications* 59 (2015) 41–45. https://doi.org/10.1016/j.inoche.2015.06.028
[80] J.H. Bang, K.S. Suslick, Applications of Ultrasound to the Synthesis of Nanostructured Materials, *Advanced Materials* 22(10) (2010) 1039–1059. https://doi.org/10.1002/adma.200904093
[81] J.-L. Luche, *Synthetic Organic Sonochemistry*, Springer Science & Business Media, 2013.

[82] M.J. Bojdys, J. Jeromenok, A. Thomas, M. Antonietti, Rational Extension of the Family of Layered, Covalent, Triazine-Based Frameworks with Regular Porosity, *Advanced Materials* 22(19) (2010) 2202–2205. https://doi.org/10.1002/adma.200903436
[83] L. Xu, S.-Y. Ding, J. Liu, J. Sun, W. Wang, Q.-Y. Zheng, Highly Crystalline Covalent Organic Frameworks from Flexible Building Blocks, *Chemical Communications* 52(25) (2016) 4706–4709.
[84] N.L. Campbell, R. Clowes, L.K. Ritchie, A.I. Cooper, Rapid Microwave Synthesis and Purification of Porous Covalent Organic Frameworks, *Chemistry of Materials* 21(2) (2009) 204–206. https://doi.org/10.1021/cm802981m
[85] L.K. Ritchie, A. Trewin, A. Reguera-Galan, T. Hasell, A.I. Cooper, Synthesis of COF-5 Using Microwave Irradiation and Conventional Solvothermal Routes, *Microporous and Mesoporous Materials* 132(1) (2010) 132–136. https://doi.org/10.1016/j.micromeso.2010.02.010
[86] H. Wei, S. Chai, N. Hu, Z. Yang, L. Wei, L. Wang, The Microwave-Assisted Solvothermal Synthesis of a Crystalline Two-Dimensional Covalent Organic Framework with High CO 2 Capacity, *Chemical Communications* 51(61) (2015) 12178–12181.
[87] R.E. Morris, Ionothermal Synthesis – Ionic Liquids as Functional Solvents in the Preparation of Crystalline Materials, *Chemical Communications* (21) (2009) 2990–2998. https://doi.org/10.1039/B902611H
[88] P. Kuhn, M. Antonietti, A. Thomas, Porous, Covalent Triazine-Based Frameworks Prepared by Ionothermal Synthesis, *Angewandte Chemie International Edition* 47(18) (2008) 3450–3453. https://doi.org/10.1002/anie.200705710
[89] S. Ren, M.J. Bojdys, R. Dawson, A. Laybourn, Y.Z. Khimyak, D.J. Adams, A.I. Cooper, Porous, Fluorescent, Covalent Triazine-Based Frameworks Via Room-Temperature and Microwave-Assisted Synthesis, *Advanced Materials* 24(17) (2012) 2357–2361. https://doi.org/10.1002/adma.201200751
[90] X. Guan, Y. Ma, H. Li, Y. Yusran, M. Xue, Q. Fang, Y. Yan, V. Valtchev, S. Qiu, Fast, Ambient Temperature and Pressure Ionothermal Synthesis of Three-Dimensional Covalent Organic Frameworks, *Journal of the American Chemical Society* 140(13) (2018) 4494–4498.
[91] J. Maschita, T. Banerjee, G. Savasci, F. Haase, C. Ochsenfeld, B.V. Lotsch, Ionothermal Synthesis of Imide-Linked Covalent Organic Frameworks, *Angewandte Chemie International Edition* 59(36) (2020) 15750–15758.
[92] S. Chandra, S. Kandambeth, B.P. Biswal, B. Lukose, S.M. Kunjir, M. Chaudhary, R. Babarao, T. Heine, R. Banerjee, Chemically Stable Multilayered Covalent Organic Nanosheets from Covalent Organic Frameworks via Mechanical Delamination, *Journal of the American Chemical Society* 135(47) (2013) 17853–17861. https://doi.org/10.1021/ja408121p
[93] B.P. Biswal, S. Chandra, S. Kandambeth, B. Lukose, T. Heine, R. Banerjee, Mechanochemical Synthesis of Chemically Stable Isoreticular Covalent Organic Frameworks, *Journal of the American Chemical Society* 135(14) (2013) 5328–5331. https://doi.org/10.1021/ja4017842
[94] D.B. Shinde, H.B. Aiyappa, M. Bhadra, B.P. Biswal, P. Wadge, S. Kandambeth, B. Garai, T. Kundu, S. Kurungot, R. Banerjee, A Mechanochemically Synthesized Covalent Organic Framework as a Proton-Conducting Solid Electrolyte, *Journal of Materials Chemistry A* 4(7) (2016) 2682–2690. https://doi.org/10.1039/C5TA10521H
[95] D.A. Vazquez-Molina, G.S. Mohammad-Pour, C. Lee, M.W. Logan, X. Duan, J.K. Harper, F.J. Uribe-Romo, Mechanically Shaped Two-Dimensional Covalent Organic Frameworks Reveal Crystallographic Alignment and Fast Li-Ion Conductivity, *Journal of the American Chemical Society* 138(31) (2016) 9767–9770. https://doi.org/10.1021/jacs.6b05568

[96] D. Zhou, X. Tan, H. Wu, L. Tian, M. Li, Synthesis of C–C Bonded Two-Dimensional Conjugated Covalent Organic Framework Films by Suzuki Polymerization on a Liquid–Liquid Interface, *Angewandte Chemie International Edition* 58(5) (2019) 1376–1381. https://doi.org/10.1002/anie.201811399

[97] K. Dey, M. Pal, K.C. Rout, S. Kunjattu H, A. Das, R. Mukherjee, U.K. Kharul, R. Banerjee, Selective Molecular Separation by Interfacially Crystallized Covalent Organic Framework Thin Films, *Journal of the American Chemical Society* 139(37) (2017) 13083–13091. https://doi.org/10.1021/jacs.7b06640

[98] Q. Hao, C. Zhao, B. Sun, C. Lu, J. Liu, M. Liu, L.-J. Wan, D. Wang, Confined Synthesis of Two-Dimensional Covalent Organic Framework Thin Films within Superspreading Water Layer, *Journal of the American Chemical Society* 140(38) (2018) 12152–12158. https://doi.org/10.1021/jacs.8b07120

[99] Y. Peng, W.K. Wong, Z. Hu, Y. Cheng, D. Yuan, S.A. Khan, D. Zhao, Room Temperature Batch and Continuous Flow Synthesis of Water-Stable Covalent Organic Frameworks (COFs), *Chemistry of Materials* 28(14) (2016) 5095–5101. https://doi.org/10.1021/acs.chemmater.6b01954

[100] D.D. Medina, J.M. Rotter, Y. Hu, M. Dogru, V. Werner, F. Auras, J.T. Markiewicz, P. Knochel, T. Bein, Room Temperature Synthesis of Covalent–Organic Framework Films through Vapor-Assisted Conversion, *Journal of the American Chemical Society* 137(3) (2015) 1016–1019. https://doi.org/10.1021/ja510895m

[101] R.P. Bisbey, C.R. DeBlase, B.J. Smith, W.R. Dichtel, Two-Dimensional Covalent Organic Framework Thin Films Grown in Flow, *Journal of the American Chemical Society* 138(36) (2016) 11433–11436. https://doi.org/10.1021/jacs.6b04669

[102] G. Lin, C. Gao, Q. Zheng, Z. Lei, H. Geng, Z. Lin, H. Yang, Z. Cai, Room-Temperature Synthesis of Core–Shell Structured Magnetic Covalent Organic Frameworks for Efficient Enrichment of Peptides and Simultaneous Exclusion of Proteins, *Chemical Communications* 53(26) (2017) 3649–3652. https://doi.org/10.1039/C7CC00482F

4 Analysis and Characterization Techniques for Coordination Nanomaterials

Jasvinder Kaur, Harimohan Sharma, Monika Singh, Dipak Kumar Das, Anuj Kumar and Ghulam Yasin

4.1 INTRODUCTION

Nanoscale coordination materials often exhibit distinct features compared to their bulkier counterparts due to their higher surface-to-volume ratio, resulting in a substantial rise in molecular-level reactivity. The features of nanoparticles (NPs) can vary significantly, encompassing electrical, optical, and chemical features, as well as extensive mechanical characteristics [1]. These characteristics make them a subject of extensive research due to their academic significance and potential technical advantages in diverse areas. Nanostructures can be synthesized using several processes, including mechanical, chemical, and other approaches [2]. In recent times, there has been a significant increase in the synthesis of various nanomaterials compared to just ten years ago. Moreover, the quantities produced have also risen, mandating the development of more accurate and reliable techniques for their characterization. Nevertheless, this classification can occasionally be insufficient. This is due to the inherent challenges associated with analyzing nanoscale materials, which are more difficult to study compared to bulk materials. These challenges arise from factors such as their small size and limited quantity, particularly when produced on a laboratory scale. Furthermore, the interdisciplinary nature of nanoscience and nanotechnology poses challenges for research teams in terms of accessing a wide range of characterization capabilities. Frequently, a more extensive explanation for NPs is required, necessitating a comprehensive strategy that combines approaches in a complementary manner. It is essential to comprehend the limitations as well as advantages of several techniques to ascertain whether employing one or two of them is sufficient for obtaining reliable information when exploring a specific parameter, such as particle size. The field of nanoscience and nanotechnology is constantly developing, and scientists are well aware that there may be distinct variations in the functioning of analytical

DOI: 10.1201/9781003345886-5

characterization methods for coordination nanomaterials compared to their application in more conventional, macroscopic materials [3, 4].

In this chapter, we provide an extensive overview of various techniques used for the characterization of coordination materials. A comparative analysis of various characterization techniques, including spectroscopies, microscopies, x-ray, Raman, electrochemical, AFM, etc., considering factors such as their accessibility, expenditure, specificity, accuracy, non-invasive nature, simplicity, and compatibility with specific compositions or materials, have been highlighted.

4.2 CHARACTERIZATION OF COORDINATION NANOMATERIALS

The shape and size of coordination nanomaterials (CNs) are the two most crucial characteristics for their characterization. In addition, it is possible to measure the size distribution, level of aggregation, surface charge, surface area, and even surface chemistry to an extent [5]. All of these factors can influence the characteristics and possible applications of the CNs. Although CNs are important, their analysis poses major hurdles due to the interdisciplinary character of the area, the absence of appropriate reference materials for calibrating analytical tools, and the complexities involved in sample preparation and data interpretation. Furthermore, there are unresolved difficulties in the assessment of CNs, such as measuring their concentration in real-time and on-site, especially in large-scale production, as well as analyzing them in intricate mixtures. Therefore, as the production of coordination nanomaterials increases, there will be a need for more reliable techniques of measurement. Hence, it is imperative to thoroughly characterize nanomaterials using many approaches, with a particular emphasis on both the CNs core and the surface ligands that impact the physico-chemical properties. Furthermore, it is crucial to highlight not only conventional methods but also innovative in-situ operando techniques employed to monitor the kinetics of CNs formation and to examine recent developments in the field, such as deliberately incorporating defects that have a substantial influence on nanoparticle properties.

Microscopy-based techniques, such as TEM, HRTEM, and AFM, are light scattering methods, frequently employed to analyze the physicochemical properties of nanoparticles. Electrochemical analysis techniques, such as electrochemical impedance, potentiometry, and voltammetry, are utilized for characterization purposes. AFM is a versatile tool for analyzing the surface aspects and magnetic properties of nanomaterials, including carbon nanotubes and graphene. These techniques are essential for comprehending the physical and chemical characteristics of nanomaterials, facilitating their effective utilization in diverse applications including pharmaceutical, electronic, chemical, and agricultural applications Certain techniques have been tailored for particular categories of materials, such as magnetic techniques. Some examples of these techniques include superconducting quantum interference device (SQUID), vibrating sample magnetometer (VSM), ferromagnetic resonance (FR), and x-ray magnetic circular dichroism (XMCD). Several more techniques provide additional insights into the structure, elemental composition, optical characteristics, and other general and specific properties of the nanoparticle samples. This section provides a systematic discussion on the previously mentioned techniques for the characterization of CNs.

4.2.1 X-Ray-Based Techniques

4.2.1.1 X-Ray Diffraction (XRD)

X-ray diffraction (XRD) techniques are frequently employed to analyze and understand the properties of CNs. XRD yields insights into the structural characteristics of materials. This technique can be employed to ascertain the precise dimensions of nanoparticles in their crystalline form. XRD is utilized to acquire both morphological and structural information about nanomaterials, including nanocomposites. XRD can offer valuable insights into the structure of nanomaterials by examining the scattering of x-rays and the diffraction phenomenon known as Bragg-scattering. XRD is an important tool for thoroughly and systematically understanding the characteristics of sensor materials. From a fundamental standpoint, it is justifiable and necessary for the feasibility, sensitivity, and ability to replicate sensor materials. This method is commonly used to analyze and describe the CNs. It usually provides data about the arrangement of atoms in a material, the properties of different phases, the dimensions of the crystal lattice, and the size of individual grains. The latter parameter is calculated by employing the Scherrer equation, which utilizes the widening of the most conspicuous peak detected in an XRD measurement of a certain material. For instance, Upadhyay et al. [6] utilized x-ray line broadening to determine the average crystallite size of magnetite nanoparticles, which was observed to fall within the range of 9–53 nm. The widening of XRD peaks mostly resulted from the dimensions of particles/crystallites and lattice strains, rather than instrumental broadening. The size determined by XRD is typically larger than the magnetic size, as it includes smaller domains inside a particle where all magnetic moments are aligned in the same direction, even if the particle is the single domain. In contrast, the size determined by TEM was greater than the size computed by XRD for samples containing extremely large particles. Specifically, when the particle size exceeds 50 nm, many crystal boundaries are present on their surface. XRD is unable to differentiate between the two limits. Consequently, the true size of certain samples, as determined by transmission electron microscopy (TEM), may exceed the 50–55 nm value estimated using the Scherrer formula. Dai and his colleagues synthesized ultra-small gold nanoparticles that exhibited a higher degree of development along the (111) crystallographic direction, as evidenced by the significantly stronger peak observed in their XRD measurement compared to the (220) direction [7]. Li et al. observed that the relative intensities of the XRD peaks changed per the shape of the copper telluride nanostructures, such as cubes, plates, and rods, that they prepared [8].

XAS comprises two main techniques: extended x-ray absorption fine structure (EXAFS) and x-ray absorption near edge structure (XANES), which is also referred to as NEXAFS. XAS quantifies the x-ray absorption coefficient of a material as it varies with energy. The XAS demonstrates element selectivity by utilizing the distinct absorption edges associated with the various binding energies of electrons in each element. EXAFS is a very sensitive technology that can be used to identify the chemical state of species, even when they exist in very low concentrations. Synchrotrons are typically required for the acquisition of XAS spectra, making it a unique and not readily available technology. XANES investigates the density of states of unoccupied or partially occupied electronic states by examining the stimulation of an electron in an inner shell to those states that are allowed according to dipole selection rules.

For instance, Pugsley et al. [9] employed in-situ XAS to investigate the kinetics and mechanism of germanium NP generation through the reaction of Mg_2Ge with $GeCl_4$. The EXAFS studies and TEM results revealed the presence of both GeO_2 nanoparticles and Ge nanoparticles. The EXAFS investigation determined a Ge–Ge distance of 2.45 Å for the first neighbor, which closely matches the results obtained by XRD. In addition, Chen et al. utilized in situ EXAFS to examine the modifications in the structure around germanium atoms in GeO_2 nanoparticles. Remarkably, it was shown that under elevated temperatures, the complete conversion of germanium dioxide resulted in the formation of GeS_2, facilitated by the presence of a sulfur source. Requejo et al. examined how the interaction between sulfur and palladium interacts with the structural and electrical characteristics of alkyl thiol-capped palladium nanoparticles. The XANES and EXAFS investigations revealed that the sulfidation of Pd clusters, induced by the capping thiol molecules, occurred both on the surface and within the bulk, affecting the atomic structure and electrical characteristics of these nanoparticles [10]. Bugaev et al.[11] determined with EXAFS parameters the atomic structure of PtCu NPs in PtCu/C catalysts. EXAFS is a highly efficient method for analyzing the structure of coordination NPs that have diameters smaller than 10 nm. It has a high level of detail in determining the spatial arrangement and provides data on the nearby environment of an atom inside a compound, even when there is no long-range order. The analysis provided metrics like as partial coordination numbers, interatomic distances, and Debye-Waller factors.

Moreover, Klasovsky et al. [12] conducted a comprehensive analysis of a novel electron-conducting polymer (PANI) based PtO_2 catalyst using electron paramagnetic resonance (EPR), diffuse reflectance FTIR spectroscopy (DRIFTS), and EXAFS techniques. The significance of in-situ/operando approaches was emphasized to enhance the understanding of the functioning of oxidation catalysts. Zhang et al. [13] conducted an independent investigation in which they utilized sodium dodecylbenzene sulphonate (DBS), stearic acid, and hexadecyltrimethylammonium bromide (CTAB) surfactants to treat γ-Fe2O3 NPs by the microemulsion technique. The role of the surfactants was investigated using EXAFS analysis, which showed that all samples tended to increase the length of the Fe-O bond. These compounds display notable steric hindrance, with the CTAB molecule exhibiting the highest obstruction. Lattice distortion and disorder at the interfaces can significantly hinder the rapid development of nanoparticles. Bertagnolli et al. [14] investigated an investigation on $CuFe_2O_4$ and $CuFe_2O_4$–MO_2 (M = Sn, Ge) nanoparticles using EXAFS and XANES techniques. The authors highlight the importance of EXAFS in acquiring accurate information regarding the coordination number, the composition of the scattering atoms surrounding the absorbing atom, the distance between the absorbing and backscattering atoms, and the Debye-Waller factor, which indicates disorder caused by static displacements and thermal vibrations. Zhu et al. employed EXAFS to characterize the CeO_2 nanoparticles. The authors highlighted the appropriateness of the discussed technique for their materials, emphasizing its capacity to selectively target certain elements and its independence from the long-range arrangement of components. Based on the obtained Debye-Waller factors and the Ce-O bond lengths, it was determined that the surface or interface of the NPs coated with sodium bis(2-ethylhexyl) sulfosuccinate (AOT) surfactants exhibited a high degree of organization. However, the bond lengths were found to be extended [15]. Furthermore, Zhang et al. fabricated Co@SiO_2 core-shell nanoparticles using the sol-gel method. The oxidation process of the Co cores was monitored following heat treatment at 800°C using in-situ XRD in combination with EXAFS. The thermal treatment was conducted either in air or under an inert environment. Remarkably, it was observed that the oxidation of Co-occurred in three distinct stages, regardless of whether air or N_2 gas was used during the annealing process [16].

The EXAFS investigated Ni-P, a nanoscale material, as part of its study. Specifically, EXAFS demonstrated great resilience in examining the first crystallization behavior of amorphous NPs by investigating the atomic-level structural alteration. The utilization of XRD, HRTEM, and VSM in conjunction with each other facilitated a comprehensive examination of the structural modifications of Ni–P NPs, encompassing both short-range and long-range order, throughout the process of high-temperature heating. To be more precise, XRD provided visual representation of the crystalline phases and any alterations in their structure. HRTEM yielded data regarding the dimensions, distribution, and morphology. EXAFS analysis yielded valuable information regarding the alterations in the local atomic arrangement and the chemical valence, particularly for XRD-amorphous samples. The utilization of VSM facilitated the examination of magnetic characteristics associated with several stages of crystallization [17].

For Pt-Ru nanoclusters, the size can be determined by EXAFS analysis. This is because the number of nearest neighbors in the nanoclusters is not directly proportional to the particle diameter when the diameter is less than 3–5 nm. [18] Sokolov et al. conducted a characterization of their palladium nanoparticles, which were synthesized using two different methods. They employed x-ray reflectivity, EXAFS, and electron microscopy techniques for this purpose. If we assume a cub-octahedral fcc structural model of Pd NPs surrounded by thiol, the size derived from EXAFS analysis was less than the size determined by TEM. Lattice expansion was observed in all types of NPs as compared to bulk Pd, as confirmed by both HRTEM and EXAFS analysis [19]. Regarding cobalt nanoparticles, a study conducted by Cheng et al. demonstrated that EXAFS analysis successfully distinguished ε-Co from the fcc and hcp crystal structures [20]. Sharma et al. [21] presented a study on the utilization of XANES and EXAFS techniques to characterize CuO, Cu_2O/CuO, and CuO/TiO_2 nanoparticles. The local electrical and atomic structure of these samples was examined using these techniques, revealing the presence of several oxide phases. The XANES methodology was employed to investigate iron oxide CNs. This method yielded valuable insights into the oxidation state and local arrangement of iron atoms.

Like the XRD approach described earlier, the SAXS technique enables the measurement of elastic scattering processes inside a specific solid angle. However, the SAXS detector is designed to capture very small scattering angles, often below 1° [23]. Figure 4.1 presents a diagram of an experimental arrangement that captures real-time

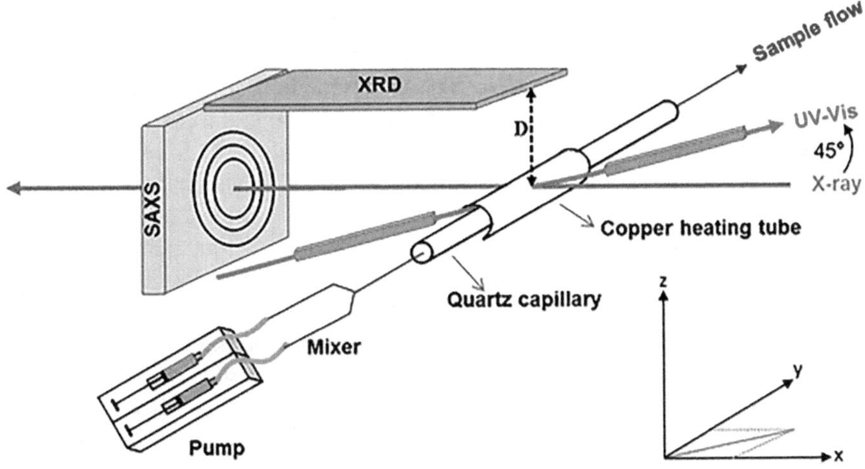

FIGURE 4.1 The schematic diagram illustrates the experimental setup used for in-situ real-time SAXS, WAXS, and UV-Vis measurements while the synthesis of Au NPs takes place. The experimental arrangement simultaneously analyses SAXS, WAXS), and the UV-Vis spectra inside the same sample volume. Taken from [22]. Copyright 2015 American Chemical Society.

SAXS/WAXS/UV-Vis observations while Au NPs are being formed. The depicted apparatus performs SAXS and WAXS while simultaneously capturing UV-Vis spectra inside a specified sample container [22]. WAXS is a technique that is similar to SAXS. However, with WAXS, the distance between the sample and the detector is reduced, resulting in the observation of diffraction maxima at greater angles. The researchers examined the process of nucleation and development of gold nanoparticles, analyzing how several factors such as concentration, temperature, ligand ratio, and solvent type influenced their kinetics.

Stuhn et al. conducted a comprehensive characterization of polystyrene-grafted SiO2 nanoparticles, employing techniques including SAXS, SANS, DLS, TGA, and TEM. SANS allowed for direct examination of the static structure of the polymer layer. Both SAXS and SANS techniques can be employed for particle size measurement. The report indicates that SAXS yielded a particle size value of 25.6 nm, while SANS generated a particle size of 23.3 nm for the nanoparticles. While SANS and SAXS have similarities in certain aspects, such as the utilization of elastic neutron scattering, SANS offers distinct advantages over SAXS. These advantages include the ability to detect light elements with high sensitivity, the potential for isotope tagging, and the intense scattering produced by magnetic moments [24]. A combination SANS and SAXS technique was used to characterize the ligand shells on tiny ZnO nanoparticles. Conventional in-situ techniques like UV-Vis and SAXS can only detect the ZnO core; however, SANS is capable of examining the organic stabilizer in the dispersion due to the strong sensitivity of neutrons to H_2. Both methodologies employed in the study enabled the simultaneous assessment of the size distribution of the cores of the nanoparticles and the distribution of the stabilizer molecules (acetate shell) in the original solution [25].

4.2.1.2 X-Ray Photoelectron Spectroscopy (XPS)

X-ray photoelectron spectroscopy (XPS) method is widely used for surface chemical analysis and to characterize CNs. The technique is based on the photoelectric effect and is a powerful quantitative tool that can determine the electronic structure, elemental composition, and oxidation states of materials. [26] It can also analyze ligand exchange interactions, surface functionalization of nanoparticles, and core/shell structures and operate under ultra-high vacuum conditions. XPS is a valuable technique for investigating the internal heterostructures of nanoparticles and differentiating between core/shell and homogeneous alloy structures. Additionally, it is capable of determining the bonding mode of ligands, such as trioctylphosphine oxide (TOPO), on the surface of metal chalcogenide nanoparticles. Unlike microscopy techniques such as TEM and TEM/EELS, which rely on lateral spatial resolution to identify elements in a path perpendicular to the probing electron beam, XPS examines the composition of the material along the same direction as the electron beam. The application of this method involves investigating the interaction between L-cysteine and Au nanoparticles, to provide experimental spectroscopic evidence to support the kinetic models of catalyst deactivation. Specifically, it focuses on examining the contribution of low-coordinated Au atoms located at the edges and corners of the nanoparticles [27]. XPS is a reliable and valuable technique for quantitative analysis of proteins and peptides adsorbed at Au surfaces. Additionally, it can describe the molecular interface of gold nanoparticles.

Analyzing the chemical information obtained from the surface of nanoparticles using XPS can be utilized to evaluate the thickness of coatings on the nanoparticles [28]. Tunc et al. introduced a straightforward technique for the XPS study of Au@SiO_2 nanoparticles, involving the application of an external voltage stress. This method enables the non-contact detection, localization, and identification of charges that form on surface structures [29]. Polzonetti et al. employed synchrotron XPS and NEXAFS techniques to investigate the molecular-metal interaction and the arrangement of ligands in the molecular shell of gold nanoparticles, which were capped with aromatic thiols. The experimental results of both techniques were corroborated by density functional theory (DFT) calculations, demonstrating the existence of a hybrid system where a metallic Au core was enveloped by a layer of aromatic thiol molecules. The thickness of this layer could be determined using XPS [30]. Castner and colleagues examined the effect of nanoparticle coatings and irregularities on XPS analysis, specifically focusing on Au@Ag core-shell nanoparticles.

4.2.2 Structural Characterization Techniques

This subsection highlights the structural characterization approaches/methods to confirm the functional groups or structural features of the CNs.

4.2.2.1 Fourier Transform Infrared Spectroscopy

Fourier transform infra-red (FTIR) spectroscopic spectrum provides the precise location of bands that are associated with the intensity and characteristics of chemical bonds, as well as certain functional groups. This information is valuable for understanding the molecular structures and interactions [31]. For instance, Feliu et al. conducted a study on the performance of Pt nanostructures in the oxidation of ethanol. They employed a hybrid technique involving in situ ATR-FTIR and differential electrochemical mass spectroscopy (DEMS). These methods facilitated the investigation of adsorbates by electrochemical means and the identification of volatile reaction by-products. Their findings corroborated prior research, indicating that the favored by-products of decomposition were associated with surface features. Specifically, CO_{ads} production was observed on (100) domains, while acetaldehyde/acetic acid generation occurred on (111) domains [32]. A separate study utilized carbon-supported platinum nanoparticles (with a size range of 3–8 nm) to catalyze the CO oxidation reaction. The progress of this catalytic process was observed through the application of DRIFTS and quadrupole mass spectrometry (QMS). FTIR measurements of adsorbed CO corroborated the fluctuations in CO_{ad} and O_{ad} in different stages of the experiment, aligning with the findings from the quadrupole mass spectrometry (QMS). Additionally, alterations in the distribution of CO across different types of Pt surface sites were observed. DRIFTS was widely recognized as a crucial instrument for assessing the surface structure of Pt NPs during in-situ settings.

4.2.2.2 Nuclear Magnetic Resonance Spectroscopy

Nuclear magnetic resonance (NMR) spectroscopy is a significant analytical technique used to quantitatively and structurally determine nanoscale materials. The phenomena on which it is based is NMR displayed by nuclei with non-zero spin in a strong magnetic field, resulting in a slight energy difference between the "spin-up" and "spin-down" states. Electromagnetic radiation in the radio wave region can be used to investigate the transitions between these states. NMR is commonly employed to investigate the interactions or coordination between the ligand and the surface of non-magnetic or anti-ferromagnetic NPs. However, it is unsuitable to describe ferri- or ferromagnetic materials due to their high saturation magnetization or spin interaction between protons and electrons. This magnetization generates fluctuations in the local magnetic field, resulting in changes in signal frequency and significant reductions in relaxation times. Consequently, there is a substantial widening of the signal peaks, rendering the data completely useless and impossible to understand [26]. The behavior of ^{1}H NMR chemical shift is very responsive to the electrical environment in its vicinity. This encompasses the electrical configurations and bonding milieu of the nucleus. Therefore, any alterations in the chirality of a molecule can be seen by adjacent spin locations and detected as variations in chemical shift. NMR is crucial for determining the chirality or lack thereof in microscopic nanoclusters that resemble molecules. NMR can also be utilized to directly monitor the migration of adsorbed gases onto the surface of metal nanoparticles. It is important to mention that the particle size that may be accurately analyzed using NMR can greatly surpass 100 nm in the case of polymer-hybrid particles. However, metallic nanoparticles need to be within the size range of 1–5 nm to obtain relevant NMR data.

Diffusion-ordered NMR (DOSY-NMR) enables the differentiation between unbound ligands and those that are bound, as well as the quantification of their distribution. NMR can aid in the identification of strongly bound ligands and the determination of their surface density on sterically stabilized colloidal nanoparticles. Scientists have utilized DOSY-NMR to examine hydrophilic heptakis(6-deoxy-6-thio)cyclomaltoheptose capped Au NPs. They discovered that this technique is fast and efficient for studying the impact of the overall gold concentration and the molar ratio of Au to capping ligand on the sizes of the nanoparticles. NMR investigations have also yielded useful insights into the drug transportation and release properties of Au NPs. The authors of the study conducted by Canzi et al. utilized DOSY-NMR to ascertain the size of Au NPs. They concluded that DOSY-NMR is a dependable and efficient alternative to TEM for calculating nanoparticle size, as it is faster and more cost-effective. NMR spectroscopy is a non-destructive and highly reproducible technique that may be employed to ascertain specific intermolecular interactions and the processes involved in immobilizing drugs and their distribution within surface-modified Au NPs coated with PEG. The integration of NMR with FTIR, UV-Vis, DLS, and TEM can provide valuable information about crucial physicochemical characteristics of drug delivery systems, which impact their therapeutic effectiveness [33].

In recent work, researchers employed deuterium (^2H) NMR to investigate the internal ligand dynamics in d_{15}-(PPh_3)-capped Au NPs. The NMR technique allowed the researchers to examine the binding characteristics of ligands on nanoparticles by analyzing chemical shifts and resonance lines in both solid and liquid phases. The straightforwardness of ^2H NMR facilitated the determination of the nature of motion in disordered and ordered regions for organic molecules labeled with deuterons. In addition, NMR was employed to examine the degree of ligand interchange among various thiolated compounds on the surface of Au NPs. Additional research has utilized multinuclear NMR to examine the surface composition and ligand binding environment of Au NPs, as well as to analyze the exchange mechanism. UV-vis and NMR techniques were employed in a study to investigate the synthesis of Au NPs using the citrate reduction method. The findings indicated that citrate may serve as a “molecular linker” during the process of particle creation. An evaluation was performed to ascertain the coordination of amine ligands on Ag NPs utilizing diverse methodologies including NMR, SERS, and DFT [34].

The investigation indicated that the amidine moiety stayed attached to the metal surface, but the hexadecylamine ligand exhibited rapid interchange between being coupled to the surface and being freely suspended in the solution. While SERS has shown effectiveness for molecules in direct proximity to the Ag surface, solution NMR spectroscopy was a potent technique for studying rapid temporal impacts. The combined utilization of surface-enhanced Raman spectroscopy (SERS) and NMR aided in the elucidation of the molecular surroundings of the synthesized nanoparticles (NPs). The study additionally discovered that amidine impeded the aggregation of NPs, whereas hexadecylamine contributed to achieving a more uniform size distribution of stabilized Ag (0) NPs. A further investigation was conducted to examine the application of 1H NMR spectroscopy in determining the size of nanoparticles (NPs) in the context of Pd NPs enclosed within dendrimers. NMR had an advantage over TEM in that it could investigate the entire population of the nanoparticles, yielding more accurate data on the average size of the nanoparticles. Furthermore, NMR allowed for in-situ operation, allowing for the observation of alterations in the dimensions and the surrounding ligand environment of the nanoparticles during catalytic events [35].

4.2.2.3 The Brunauer–Emmett–Teller Technique

The Brunauer-Emmett-Teller (BET) methodology is a method employed to analyze and describe materials at the nanoscale. The method calculates the surface area of nanostructures by quantifying the physical adsorption of gas on a solid surface and is commonly employed due to its rapidity, simplicity, and relatively high level of accuracy [26]. Sahoo et.al. created a biocompatible ferrofluid that includes Fe_3O_4 NPs functionalized with dye, which act as fluorescent indicators. The BET analysis yielded a surface area that was smaller than the expected size distribution and density of the material, potentially because smaller nanoparticles agglomerated to form larger ones, hence reducing the overall surface area. Furthermore, there is a potential for clumping together during the necessary drying procedure for these measurements [36]. Ma et al. reported to have prepared Fe_3O_4 NPs by a co-precipitation strategy, using ferrous and ferric species as precursors. This process yielded a product with a significantly high specific surface area of 286.9 m^2 g^-, surpassing the values reported in existing literature for similar particles.

4.2.2.4 Thermal Gravimetric Analysis

The thermal gravimetric analysis (TGA) offers data on the weight and makeup of stabilizers of CNs. The process is applying heat to a nanomaterial sample and documenting the alterations in mass when distinct constituents with varied degradation temperatures break down and transform into vapor. TGA can ascertain the nature and number of organic ligands of NP by considering the initial mass of the sample. This method is valuable for detecting the presence and amount of ligands attached to the surface of Au NPs and confirming the presence of polyethylene glycol (PEG) coating on silica NPs [37]. TGA offers the benefits of simplicity and straightforwardness, as it does not necessitate any specific sample preparation save from ensuring the sample is dry. Nevertheless, a limitation of traditional TGA is the requirement for a small quantity of nanomaterial samples.

4.2.2.5 UV-Vis Spectroscopy

UV-vis spectroscopy is a commonly employed and economical technique for analyzing CNs. The process entails quantifying the luminosity of light that is bounced off a specimen and contrasting it with the luminosity of light that is bounced off a substance used as a standard for comparison. This approach is important for detecting and researching nanoparticles since it allows for the determination of their optical properties, including size, shape, concentration, and refractive index [38]. The UV-vis extinction spectra of gold, silver, and copper nanostructure sols are characterized by the presence of an LSPR signal in the visible part of the spectrum, and its absorbance can be used to quantitatively characterize the colloidal stability of Au NPs. In general, UV-vis spectroscopy provides an inexpensive, non-invasive, and rapid technique for the identification and characterization of various types of nanoparticle systems.

4.2.2.6 Dynamic Light Scattering

Dynamic light scattering (DLS) is a commonly used method for determining the dimensions of NPs in colloidal solutions within the nano- and sub-micrometer size ranges. The CNs NPs distributed within a colloidal fluid exhibit uninterrupted

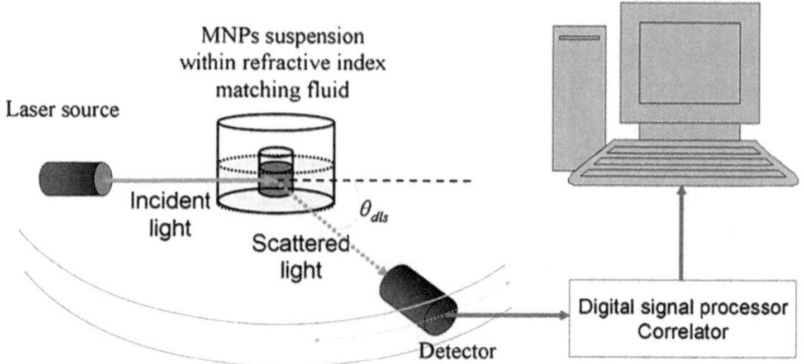

FIGURE 4.2 The optical configuration of the standard experimental setup used for performing dynamic light measurements of a suspension containing nanoparticles. The configuration can be operated from multiple angles. Taken from [39]. Copyright 2013 Springer.

Brownian motion. DLS quantifies the scattering of light over time. Lim et al. conducted a comprehensive analysis of the characterization of nanoparticles using DLS, with a specific emphasis on magnetic particles (Figure 4.2). In the case of small-sized NPs, the primary reason causing the significant discrepancy in diameter measurements obtained by TEM and DLS is the effect of radius of curvature. The Fe_3O_4 nanoparticles, which are of medium size, are coated with oleic acid and oleylamine. These nanoparticles exhibit size values that closely correspond to both DLS and TEM observations. The authors emphasize the utilization of DLS for assessing the colloidal stability of MNPs. In addition, DLS has demonstrated its utility in tracking the temporary characteristics of β-FeOOH nanorods. These nanorods assemble themselves in a parallel manner to create well-aligned 2D arrays, ultimately resulting in the development of 3D layered structures. In general, the use of DLS to screen NPs in real-time offers valuable information about their aggregation process. This technique quantitatively evaluates the size of the particle clusters that are generated. The diagnostic potential of DLS to identify aggregation is greatly dependent on its sensitivity to big particles. However, the authors emphasize the need for meticulous examination to achieve the most accurate interpretation of the DLS findings, as they can be influenced by the aforementioned parameters such as form and coating agents [39]. DLS offers several benefits, including its rapid, effortless, and accurate operation for monomodal suspensions.

Additionally, it is an ensemble measurement technique that provides a reliable statistical representation of each NP sample. This method is extremely sensitive and capable of producing consistent results for samples that have a uniform size and composition. An inherent constraint of DLS is the prerequisite conditions for the particles to remain suspended and exhibit Brownian motion. Significantly larger particles exhibit a greater scattering effect on light, and even a small quantity of these larger particles can obstruct the presence of smaller particles. Consequently, the resolution for samples that are polydisperse and heterogeneous is relatively low. DLS necessitates the use of transformational computations that include making assumptions, which need to be considered while interpreting the data, especially in the case of polydisperse samples. While DLS can occasionally measure anisotropic nanostructures, its typical assumption is that the particles are spherical. In general, DLS provides reliable measurements of the hydrodynamic radius, but it cannot identify tiny aggregates with high precision [40, 41].

4.2.2.7 Mass Spectrometry

Mass spectrometry (MS) has become a popular and dependable method for analyzing and characterizing coordination NPs. The technique offers significant insights into the elemental and molecular characteristics of coordination NPs, including their composition, shape, chemical state, and their ability to bind to certain biomolecules. MS is a highly sensitive analytical technique that may be employed to quantify bioconjugation, and it is compatible with samples of any nature. Furthermore, it can be readily used with separation techniques to acquire instantaneous data, offering diverse and innovative understandings of the characteristics of NPs and their ultimate purposes and uses. Inductively coupled plasma mass spectrometry (ICP-MS) is employed for the elemental analysis of NPs. It exhibits strong durability, heightened responsiveness, and a broad variety of capabilities, together with exceptional selectivity and near-complete autonomy from external factors. This technique enables accurate measurement and analysis of the quantity and elemental composition of metallic NPs, as well as the detection of metallic impurities in non-metallic NPs. Molecular mass spectrometry techniques, such as electrospray ionisation (ESI) and matrix-assisted laser desorption/ionization (MALDI), can offer insights into the ligands that shield the NPs and establish a connection between the whole clusters and their chemical makeup. Furthermore, specific methods of characterizing nanoparticles, such as capillary electrophoresis, hydrodynamic chromatography, ion mobility spectrometry, and field flow fractionation (FFF), provide valuable insights on the dimensions and distribution of particle sizes. MS is a highly effective method for determining the size distribution of tiny clusters and analyzing combinations of ligands with certain stoichiometry. MS also enables the discrimination between the separate ligands in the instances of NPs that are capped with several ligands [42].

4.2.2.8 Secondary Ion Mass Spectrometry

Secondary ion mass spectrometry (SIMS) is a method of analyzing surfaces that involves using primary ions to dislodge secondary ions from the topmost layer of the sample, typically measuring in nm. This approach offers precise molecular chemistry data from NPs with exceptional sensitivity and resolution. The SIMS technique is especially valuable for analyzing NPs because of its exceptional lateral and depth resolution. However, a proficient approach is necessary to decipher the analytical findings. Blanc et al. employed secondary ion mass spectrometry (SIMS) to examine the chemical makeup of dielectric nanoparticles that are situated within a silica glass matrix at the center of optical fibers. The researchers conducted high-resolution SIMS imaging and successfully illustrated the distribution of phosphorus, magnesium, and erbium into distinct zones [43]. In a similar manner, Schweikert and colleagues examined a solitary layer composed of a combination of Ag and Au NPs using cluster SIMS. They observed that the SIMS response is distinctively influenced by the physical and chemical characteristics of the NPs and their surroundings [44].

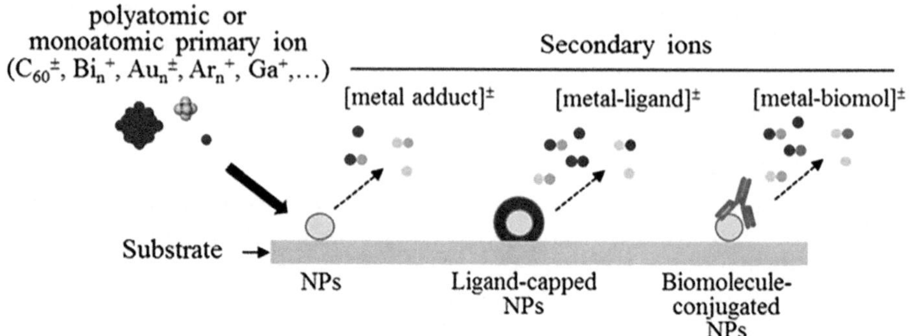

FIGURE 4.3 Strategies for investigating NPs utilizing ToF-SIMS. When polyatomic or monoatomic particles bombardment the surface, they produce various secondary ions from metal NPs, which can then be encapsulated or conjugated using ligands or biomolecules. Taken from [45]. Copyright 2015 Wiley-VCH.

4.2.2.9 Time of Flight Secondary Ion Mass Spectrometry

The time-of-flight secondary ion mass spectrometry (ToF-SIMS) is widely employed for material characterization owing to its exceptional chemical sensitivity, surface sensitivity (up to 2–3 nm), and molecular specificity. These attributes render it excellent for analyzing nanoparticle drug delivery formulations. Additionally, it can be utilized to examine the nano-regions of more substantial components such as electronic gadgets and thin films. Nevertheless, it possesses the potential to be highly damaging and induce molecular degradation problems, ultimately leading to the melting of the nanoparticles. Figure 4.3 shows a scheme which explains how ToF-SIMS is used to probe NPs [45]. The method operates by adsorbing nanoparticles onto a surface, followed by subjecting them to primary ion bombardment to release molecules. This process leads to the emission of secondary ions from the outermost layers of molecules. ToF-SIMS is valuable for drug screening without the need for labels and for monitoring the exchange of ligands. However, its ability to distinguish fine details is restricted to measurements in the range of hundreds of nano-meters, and it is not capable of detecting high mass fragments with high sensitivity.

4.2.3 Characterization Methods for Magnetic Nanostructures

Usually, magnetic CNs possess a lot of applications in pharmaceuticals, biologicals, catalysis, and separation sciences. However, their precise characterization is still challenging due to the presence of unpaired electrons over their surfaces. This section discusses the most useful approaches to characterize the magnetic CNs.

4.2.3.1 Superconducting Quantum Interference Device Magnetometry

The magnetic characteristics of nanoscale materials can be measured using superconducting quantum interference device (SQUID) magnetometry technology. When compared to bulk materials, nanoparticles' distinctive size and sensitivity cause them to exhibit various features. The material goes from having multiple domains to

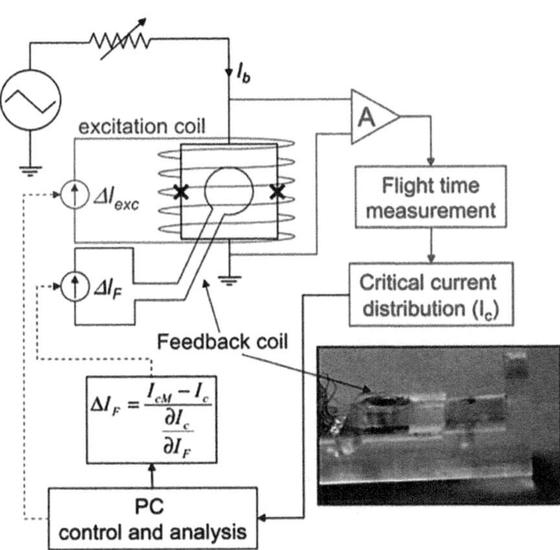

FIGURE 4.4 Schematic illustration of the experimental setup used to test the magnetization of nanoparticles [46]. Copyright 2013 Springer.

having just one, and then to being superparamagnetic, as its size decreases. Properties including blocking temperature (TB), magnetization remanence (MR), and magnetization saturation (MS) can be obtained via SQUID measurements. In addition to monitoring nanoparticles, SQUID can also measure the magnetic response of specific molecules. The use of a nanoSQUID probe in a scanning magnetic microscope has allowed for the development of imaging and spectroscopic tools applicable to the nanoscale. In Figure 4.4, we can see the experimental setup of a nanoSQUID.

4.2.3.2 Mössbauer Spectroscopy

The CNs having Mössbauer-active elements, such as Fe, can be analyzed using Mössbauer spectroscopy (MbS), which relies on the resonance fluorescence of γ-photons. Its ability to determine the oxidation state, symmetry, spin state, and magnetic ordering of Fe atoms makes identifying the magnetic phases in a sample simpler. Quantifying the thermal unblocking and determining the magnetic anisotropy energy for magnetically ordered materials are two further applications of this technology. The isomer shift, which is a significant component of MbS, is the nuclear energy shift that results from the coulombic interaction between the nucleus and the electron density close to the nucleus. The MbS is generally acknowledged as the go-to technique for calculating the oxidation number, and there are noticeable discrepancies in the isomer shift values of Fe^{2+} and Fe^{3+} The Mossbauer isomer shift in doped Fe-ZnO nanoparticles is connected to the charge of the ion in the structure, but it isn't always linked to the ion's oxidation state. When calculating the oxidation number of dopant atoms, the Mössbauer isomer shift is not the final determinant. The charge at the location of the nucleus can be studied directly using MbS. The structure and chemical features of the atom's surroundings impact its charge.

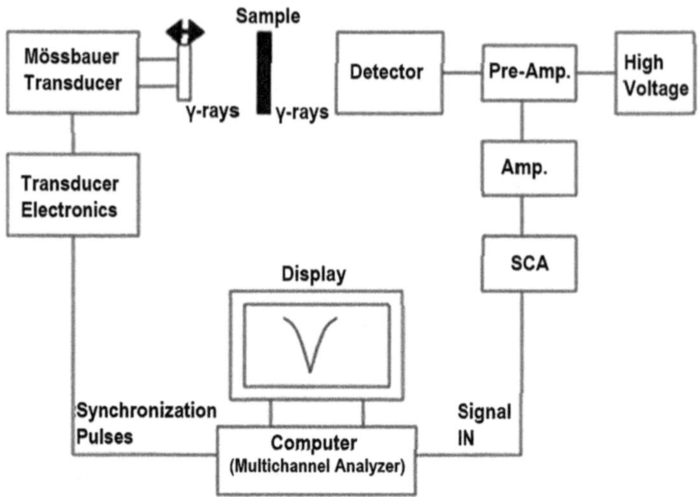

FIGURE 4.5 Schematic illustrations of a transmission Mössbauer spectrometer system. Taken from [47]. Copyright 2004 Elsevier.

One possible configuration of a gearbox Mössbauer spectrometer is shown in Figure 4.5. Research on the magnetic properties of FeCo NPs made by chemical vapor condensation was carried out by Oh et al. [47]. The Mössbauer study demonstrated that the surface of the nanoparticles contained -FeOOH, -FeOOH, and Fe_3O_4. By fitting the Mössbauer spectra precisely, the relative proportions of each phase could be determined quantitatively. According to Lange et al. [48], MbS makes use of the extremely small interactions between atomic nuclei and their local environment. Consequently, this method is quite sensitive to the setting in which a particular isotope is used as a probe. The chemical and structural environments around Fe nuclei can be better understood by the ^{57}Fe Mössbauer effect. This paves the way for the exact quantitative examination of the proportions of Fe-containing phases and their identification.

4.2.3.3 Magnetic Susceptibility

One of the methods to measure the magnetic properties of CNs is to measure their magnetic susceptibility. This property indicates whether a material is attracted to or repelled from a magnetic field, which has practical implications. The susceptibility is expressed as the ratio of magnetization to the applied magnetizing field intensity. Herrera et al. observed that magnetic NPs coated with poly(N-isopropylacrylamide) [pNIPAM] exhibited aggregation when measured using AC susceptibility, a phenomenon that was not apparent in DLS studies. The information obtained from AC susceptibility was corroborated by SANS measurements [49]. Typically, magnetic susceptibility measurements are conducted across a spectrum of temperatures instead of frequencies due to the restricted range of frequencies that most susceptometers can handle. Magnetic nanoparticles in natural materials were characterized using broadband alternating current magnetic susceptibility measurements [50]. Rinaldi et al. conducted AC susceptibility experiments on cobalt ferrite nanoparticles to ascertain the viscosity of mineral oil. Oleic acid served as a capping ligand for the suspended nanoparticles in the oil, and there was a strong correlation between the viscosities at the nanoscale and macroscale levels. [51] Lima et al. conducted susceptibility experiments to assess the dimensions and size distribution of magnetic nanoparticles within polymeric templates. In addition, they successfully assessed the magneto-crystalline anisotropy values of magnetite nanoparticles, considering the influence of the susceptibility peak's field dependence. The NP size parameters obtained from the examination of the dynamic susceptibility data were consistent with the values derived from fitting the TEM data [52].

4.2.4 Microscopy Characterization

4.2.4.1 Scanning Electron Microscopy (SEM)

Scanning electron microscopy (SEM) is a widely applied method for characterizing CNs and obtaining high-resolution images of surfaces. Imaging is achieved using electrons, identical to how a light microscope employs visible light. Mazzaglia et al. investigated supramolecular colloidal systems of Au NPs/amphiphilic cyclodextrin by combining field-emission SEM (FE-SEM) and XPS measurements. This investigation resulted in crucial insights into the morphology and characteristics of the interaction between (thiohexyl carbon chain) SC_6NH_2 and (thiohexadecyl carbon chain) $SC_{16}NH_2$ with Au NPs on the silicon surface [53]. The application of SEM in conjunction with other methodologies, including HR-SEM, AFM, T-SEM, and Nano-SIMS, is dependent upon the particular research objectives. Although these methods have unique advantages and disadvantages, they can provide a more comprehensive analysis of coordination nanomaterials when combined. On mica or silicon substrates, SEM and AFM measurements of Au NPs and SiO_2 were compared. SEM is capable of delivering adequate measurements regarding their lateral dimensions (Figure 4.6). The combination of SEM, T-SEM, EDX, and SAM was identified

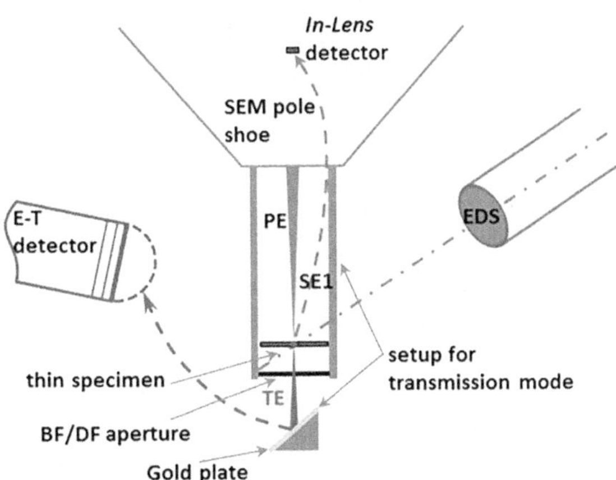

FIGURE 4.6 Illustrations of an SEM/EDS system operating in transmission mode. Taken from [54]. Copyright 2014, Royal Society of Chemistry.

as a powerful approach for conducting a comprehensive morphological and chemical analysis of individual titania and silica nanoparticles, according to a study by Rades et al. While T-SEM enabled rapid analysis of the NP morphology, its lateral resolution was confined to NPs measuring between 5 and 10 nm in length.

The ability to precisely delimit particles in the T-SEM mode was impeded due to the comparatively reduced spatial resolution attained in contrast to the conventional TEM. However, the SiO_2 NP size distributions obtained by SEM and TSEM under different conditions were highly congruent, accounting for the measurement uncertainties involved [54] According to research conducted by Hodoroaba et al., the size distribution produced by T-SEM imaging is slightly wider than that acquired through TEM. Small SiO_2 nanoparticles, on the other hand, are difficult to precisely delimit using T-SEM due to its lower spatial resolution in comparison to conventional TEM. Moreover, the detection of the particle surface layer may not always be relatively simple using T-SEM [55]. In another paper, the same authors noted that conventional SEM imaging mode is ineffective in detecting NPs on the backside of the observation-required support film. For accurate measurements, therefore, explicit knowledge of the T-SEM operator is vital. An investigation was conducted to determine whether the SiO_2 NP size distributions acquired via SEM and TSEM correlated well under different conditions while accounting for the measurement uncertainties that accompanied each technique.

4.2.4.2 Transmission Electron Microscopy

Transmission electron microscopy (TEM) is an intricate microscopy technique that interacts with a thin sample via an electron beam of uniform current density. Several factors affect the interaction between the electron beam and the sample, including the elemental composition, size, and density of the sample. This technique is extensively employed in the investigation of nanoparticles' dimensions and shape, and it offers the most precise approximation of their homogeneity. However, a few limitations should be considered, including the challenge of quantifying a significant number of particles and possibilities for misleading images caused by orientation effects. Alternative techniques, such as SAXS for larger and spherical NPs or XRD, can provide more reliable outcomes when applied to analyze greater quantities of NPs.

4.2.4.3 High-Resolution

High-resolution TEM (HRTEM) is a type of transmission electron microscopy that utilizes phase-contrast imaging to create an image by combining transmitted and scattered electrons [56]. HRTEM, in contrast to conventional TEM, necessitates the use of a wider objective aperture to utilise scattered electrons. Phase-contrast imaging is a highly sophisticated method that has achieved the greatest level of resolution to yet. It allows for the detection of atomic arrays within crystalline structures. HRTEM is an important tool for studying the structures of coordination nanoparticles, since conventional electron microscopy provides statistical assessment of nanoparticle shape and lacks the required resolution to interpret the crystal structure of individual particles. Consequently, HRTEM is extensively employed to analyze the internal composition of coordination nanoparticles.

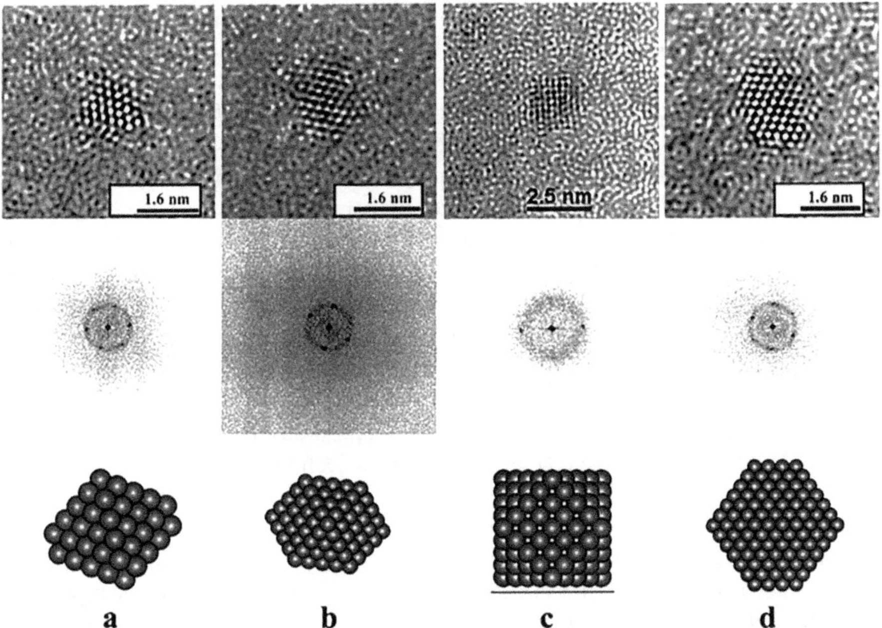

FIGURE 4.7 HR-TEM picture of Au NPs with FCC crystal geometry with (a) and (b) ⟨1 1 0⟩, and (c) ⟨1 0 0⟩ orientations, while (d) hexagonal profile with distorted ⟨1 1 0⟩ orientation. Taken from [57]. Copyright 2001 Elsevier.

4.2.4.4 High-Angle Annular Dark-Field Imaging

High-angle annular dark-field imaging-scanning transmission electron microscopy (HAADF-STEM) is a technique used in STEM to map the CNs samples. The process entails utilizing an annular dark-field detector to capture scattered electrons, which produces images that exhibit high sensitivity to fluctuations in the atomic number of atoms within the sample. The high-angle annular dark-field imaging (HAADF) approach is employed to observe local atomic structures and has proven effective in imaging different material surfaces. For example, Akita et al. employed HAADF-STEM to examine gold nanoparticles (Au NPs) on a cerium dioxide (CeO_2) support and explore the process by which the structure cyclically changes as the electron beam is activated and deactivated. HAADF-STEM images accurately depict the precise locations of Au and Ce atomic columns, without any observed artifacts, unlike HRTEM images (Figure 4.7). [58] Although multi-slice simulation may be required for quantitative structure information, the straightforward interpretability of HAADF-STEM images makes them a valuable tool in materials science research.

4.2.4.5 Atomic Force Microscopy

Atomic force microscopy (AFM) is a kind of microscopy that allows for the creation of three-dimensional surface images and offers extremely high magnification. While they were IBM employees in 1986, Gerard Binning and Heinrich Rohrer developed this technique. To work, [59] AFM must first estimate the forces that act upon the specimen as a tiny probe contact it. A cantilever constructed of silicon or silicon nitride is connected to the probe's sharp tip. The cantilever deflects during the scan due to the attractive or repulsive forces between the sample's surface and the tip. Using a laser beam reflected off the cantilever's back surface allows for the measurement of the bending. We find the forces by combining the laser fluctuation data with the known cantilever stiffness. Depending on how close the probe is to the sample, the AFM can switch between three modes: contact, non-contact, and tapping (sometimes called intermediate or oscillating mode). Despite being very sensitive to the unbounded size of the oscillating tip, the latter is commonly used to describe NPs [60, 61]. Secondary variables, such as the elasticity of the nanoparticle, surface energy, and the radius of curvature of the tip, influence the final topological properties. Plotting the particle height vs. the free amplitude of the oscillating probe, however, yields more precise results and mitigates such influences.

4.2.4.6 Magnetic Force Microscopy

Magnetic force microscopy (MFM) is a method employed in scanning probe microscopy where a magnetic probe is used to raster-scan the surface of a sample. The magnetic field emanating from the sample interacts with the magnetic tip, providing insights into the magnetic characteristics of the sample. MFM can differentiate between magnetic interactions and other forces between the tip and sample, such as van der Waals forces, which are detected in AFM. MFM is capable of operating under many different circumstances, including ambient parameters, varying temperatures, and ultra-high vacuum environments. Furthermore, it can achieve a resolution of less than 10 nm. [62] Zheng et al. published a paper on how magnetic forces of ferromagnetic nanomaterials in external magnetic fields depend on magnetic field distribution, geometry, and intrinsic magnetic characteristics of samples. Characterizing magnetic NPs (MNPs) involves understanding the magnetic properties that depend on the geometry and crystallography of the nanoparticles. This task is challenging due to its very intricate structure. Spintronic systems that rely on atomic structures can encounter unintended distortion on the surfaces of the tested samples. Therefore, it is highly valuable from the point of view of technology to offer a geometric explanation for the appearance and enhancement of magnetism in MNPs. The magnetic images obtained through MFM originate mainly from the second-order gradient of magnetic forces. This gradient is a convolution of the stray magnetic fields present on the surface of the ferromagnetic sample and the magnetic moment of the probe. However, these images are not appropriate for demonstrating microscopic magnetic structures or determining the orientations of magnetic moments within magnetic domains. As the knowledge of microscopic magnetic phenomena advances, it becomes clear that MFM is insufficient for accurately observing magnetic structures on magnetic surfaces. Rather, magnetic imaging technology will probably move from qualitative to quantitative characterizations. Analyzing magnetic force illustrations allows for the recognition of microscopic magnetic structures on the surfaces of ferromagnetic materials, such as the distribution of surface magnetic domains in terms of size and orientation.

4.3 SUMMARY AND OUTLOOK

This chapter presents a comprehensive examination of several methodologies employed in the characterization of coordination nanoparticles. The pros and cons of each technique are also briefly explained. It also shows how they can work together to make each other better, and it stresses how important it is to use a variety of techniques to fully understand the properties of a nanomaterial. Moreover, it is a valuable resource for researchers by offering a comparative examination of each approach, aiding in the selection of the most appropriate procedures for their characterization requirements. Nevertheless, there are obstacles to enhancing the precision and clarity of these methods, and we anticipate that this chapter will aid in determining which techniques need additional technical enhancements.

ACKNOWLEDGMENTS

This work was supported by the Research Fund for International Scientists (RFIS-Grant number: 52150410410) National Natural Science Foundation of China.

REFERENCES

[1] N.T. Thanh, N. Maclean, S. Mahiddine, Mechanisms of nucleation and growth of nanoparticles in solution, *Chemical Reviews*, 114 (2014) 7610–7630.
[2] K.M. Koczkur, S. Mourdikoudis, L. Polavarapu, S.E. Skrabalak, Polyvinylpyrrolidone (PVP) in nanoparticle synthesis, *Dalton Transactions*, 44 (2015) 17883–17905.
[3] Y. Wang, D. Wan, S. Xie, X. Xia, C.Z. Huang, Y. Xia, Synthesis of silver octahedra with controlled sizes and optical properties via seed-mediated growth, *ACS Nano*, 7 (2013) 4586–4594.
[4] F. Kim, S. Connor, H. Song, T. Kuykendall, P. Yang, Cover picture: platonic gold nanocrystals (Angew. Chem. Int. Ed. 28/2004), *Angewandte Chemie International Edition*, 43 (2004) 3615–3615.
[5] C. Minelli, *Talk on 'measuring nanoparticle properties: are we high and dry or all at sea?' at 'nanoparticle characterisation–challenges for the community' event–IOP*, Institute of Physics, book of abstracts, 2016.
[6] S. Upadhyay, K. Parekh, B. Pandey, Influence of crystallite size on the magnetic properties of Fe3O4 nanoparticles, *Journal of Alloys and Compounds*, 678 (2016) 478–485.
[7] W. Yan, V. Petkov, S.M. Mahurin, S.H. Overbury, S. Dai, Powder XRD analysis and catalysis characterization of ultra-small gold nanoparticles deposited on titania-modified SBA-15, *Catalysis Communications*, 6 (2005) 404–408.
[8] W. Li, R. Zamani, P. Rivera Gil, B. Pelaz, M. Ibáñez, D. Cadavid, A. Shavel, R.A. Alvarez-Puebla, W.J. Parak, J. Arbiol, CuTe nanocrystals: shape and size control, plasmonic properties, and use as SERS probes and photothermal agents, *Journal of the American Chemical Society*, 135 (2013) 7098–7101.
[9] A.J. Pugsley, C.L. Bull, A. Sella, G. Sankar, P.F. McMillan, XAS/EXAFS studies of Ge nanoparticles produced by reaction between Mg2Ge and GeCl4, *Journal of Solid State Chemistry*, 184 (2011) 2345–2352.
[10] M.A. Newton, S.G. Fiddy, G. Guilera, B. Jyoti, J. Evans, Oxidation/reduction kinetics of supported Rh/Rh2O3 nanoparticles in plug flow conditions using dispersive EXAFS, *Chemical Communications*, (2005) 118–120.
[11] V.V.E. Srabionyan, V.V. Pryadchenko, A. Kurzin, S.V.E. Belenov, L.A. Avakyan, V.E. Guterman, L. Bugaev, Atomic structure of PtCu nanoparticles in PtCu/C catalysts from EXAFS spectroscopy data, *Physics of the Solid State*, 58 (2016) 752–762.

[12] F. Klasovsky, J. Hohmeyer, A. Brückner, M. Bonifer, J.R. Arras, M. Steffan, M. Lucas, J.R. Radnik, C. Roth, P. Claus, Catalytic and mechanistic investigation of polyaniline supported PtO2 nanoparticles: a combined in situ/operando EPR, DRIFTS, and EXAFS study, *The Journal of Physical Chemistry C*, 112 (2008) 19555–19559.

[13] T. Liu, L. Guo, Y. Tao, T. Hu, Y. Xie, J. Zhang, Bondlength alternation of nanoparticles Fe2O3 coated with organic surfactants probed by EXAFS, *Nanostructured Materials*, 11 (1999) 1329–1334.

[14] V. Krishnan, R.K. Selvan, C.O. Augustin, A. Gedanken, H. Bertagnolli, EXAFS and XANES investigations of CuFe2O4 nanoparticles and CuFe2O4– MO2 (M= Sn, Ce) nanocomposites, *The Journal of Physical Chemistry C*, 111 (2007) 16724–16733.

[15] Z. Wu, L. Guo, H. Li, Q. Yang, Q. Li, H. Zhu, EXAFS study on the local atomic structures around Ce in CeO2 nanoparticles, *Materials Science and Engineering: A*, 286 (2000) 179–182.

[16] K. Zhang, Z. Zhao, Z. Wu, Y. Zhou, Synthesis and detection the oxidization of Co cores of Co@ SiO2 core-shell nanoparticles by in situ XRD and EXAFS, *Nanoscale Research Letters*, 10 (2015) 1–9.

[17] Y. Tan, D. Sun, H. Yu, B. Yang, Y. Gong, S. Yan, Z. Chen, Q. Cai, Z. Wu, Crystallization mechanism analysis of noncrystalline Ni–P nanoparticles through XRD, HRTEM and XAFS, *CrystEngComm*, 16 (2014) 9657–9668.

[18] A. Frenkel, Solving the structure of nanoparticles by multiple-scattering EXAFS analysis, *Journal of Synchrotron Radiation*, 6 (1999) 293–295.

[19] Y. Sun, A.I. Frenkel, R. Isseroff, C. Shonbrun, M. Forman, K. Shin, T. Koga, H. White, L. Zhang, Y. Zhu, Characterization of palladium nanoparticles by using X-ray reflectivity, EXAFS, and electron microscopy, *Langmuir*, 22 (2006) 807–816.

[20] G. Cheng, J.D. Carter, T. Guo, Investigation of Co nanoparticles with EXAFS and XANES, *Chemical Physics Letters*, 400 (2004) 122–127.

[21] A. Sharma, M. Varshney, J. Park, T.-K. Ha, K.-H. Chae, H.-J. Shin, XANES, EXAFS and photocatalytic investigations on copper oxide nanoparticles and nanocomposites, *RSC Advances*, 5 (2015) 21762–21771.

[22] X. Chen, J. Schröder, S. Hauschild, S. Rosenfeldt, M. Dulle, S. Förster, Simultaneous SAXS/WAXS/UV–vis study of the nucleation and growth of nanoparticles: a test of classical nucleation theory, *Langmuir*, 31 (2015) 11678–11691.

[23] C.S. Kumar, J. Hormes, C. Leuschner, *Nanofabrication towards biomedical applications: techniques, tools, applications, and impact*, John Wiley & Sons, 2006.

[24] C.J. Kim, K. Sondergeld, M. Mazurowski, M. Gallei, M. Rehahn, T. Spehr, H. Frielinghaus, B. Stühn, Synthesis and characterization of polystyrene chains on the surface of silica nanoparticles: comparison of SANS, SAXS, and DLS results: polystyrene chains on silica nanoparticle, *Colloid and Polymer Science*, 291 (2013) 2087–2099.

[25] T. Schindler, M. Schmiele, T. Schmutzler, T. Kassar, D. Segets, W. Peukert, A. Radulescu, A. Kriele, R. Gilles, T. Unruh, A combined SAXS/SANS study for the in situ characterization of ligand shells on small nanoparticles: the case of ZnO, *Langmuir*, 31 (2015) 10130–10136.

[26] L.T. Lu, *Water-dispersible magnetic nanoparticles for biomedical applications: synthesis and characterisation*, University of Liverpool, 2011.

[27] L. Caprile, A. Cossaro, E. Falletta, C. Della Pina, O. Cavalleri, R. Rolandi, S. Terreni, R. Ferrando, M. Rossi, L. Floreano, Interaction of L-cysteine with naked gold nanoparticles supported on HOPG: a high resolution XPS investigation, *Nanoscale*, 4 (2012) 7727–7734.

[28] M.Y. Smirnov, A.V. Kalinkin, A.V. Bukhtiyarov, I.P. Prosvirin, V.I. Bukhtiyarov, Using X-ray photoelectron spectroscopy to evaluate size of metal nanoparticles in the model Au/C samples, *The Journal of Physical Chemistry C*, 120 (2016) 10419–10426.

[29] I. Tunc, U.K. Demirok, S. Suzer, M.A. Correa-Duatre, L.M. Liz-Marzan, Charging/discharging of Au (core)/silica (shell) nanoparticles as revealed by XPS, *The Journal of Physical Chemistry B*, 109 (2005) 24182–24184.
[30] C. Battocchio, F. Porcaro, S. Mukherjee, E. Magnano, S. Nappini, I. Fratoddi, M. Quintiliani, M.V. Russo, G. Polzonetti, Gold nanoparticles stabilized with aromatic thiols: interaction at the molecule–metal interface and ligand arrangement in the molecular shell investigated by SR-XPS and NEXAFS, *The Journal of Physical Chemistry C*, 118 (2014) 8159–8168.
[31] C. Blanco Andujar, *Sodium carbonate mediated synthesis of iron oxide nanoparticles to improve magnetic hyperthermia efficiency and induce apoptosis*, UCL (University College London), 2014.
[32] C. Busó-Rogero, S. Brimaud, J. Solla-Gullon, F.J. Vidal-Iglesias, E. Herrero, R.J. Behm, J.M. Feliu, Ethanol oxidation on shape-controlled platinum nanoparticles at different pHs: a combined in situ IR spectroscopy and online mass spectrometry study, *Journal of Electroanalytical Chemistry*, 763 (2016) 116–124.
[33] S.C. Coelho, M. Rangel, M.C. Pereira, M.A. Coelho, G. Ivanova, Structural characterization of functionalized gold nanoparticles for drug delivery in cancer therapy: a NMR based approach, *Physical Chemistry Chemical Physics*, 17 (2015) 18971–18979.
[34] M. Doyen, K. Bartik, G. Bruylants, UV–Vis and NMR study of the formation of gold nanoparticles by citrate reduction: observation of gold–citrate aggregates, *Journal of Colloid and Interface Science*, 399 (2013) 1–5.
[35] M.V. Gomez, J. Guerra, V.S. Myers, R.M. Crooks, A.H. Velders, Nanoparticle size determination by 1H NMR spectroscopy, *Journal of the American Chemical Society*, 131 (2009) 14634–14635.
[36] Y.J. Kim, J.W. Kim, J.E. Lee, J.H. Ryu, J. Kim, I.S. Chang, K.D. Suh, Synthesis and adsorption properties of gold nanoparticles within pores of surface-functional porous polymer microspheres, *Journal of Polymer Science Part A: Polymer Chemistry*, 42 (2004) 5627–5635.
[37] E. Mansfield, K.M. Tyner, C.M. Poling, J.L. Blacklock, Determination of nanoparticle surface coatings and nanoparticle purity using microscale thermogravimetric analysis, *Analytical Chemistry*, 86 (2014) 1478–1484.
[38] M. Faraji, Y. Yamini, N. Salehi, *Characterization of magnetic nanomaterials, Magnetic nanomaterials in analytical chemistry*, Elsevier, 2021, pp. 39–60.
[39] J. Lim, S.P. Yeap, H.X. Che, S.C. Low, Characterization of magnetic nanoparticle by dynamic light scattering, *Nanoscale Research Letters*, 8 (2013) 1–14.
[40] M.P. Tn, S.P. Ng, Characterization techniques and tools for special nanomaterials, *Multidisciplinary*, 171.
[41] M.M. Sheha, A.M. Mostafa, D.G. Nasr, R.Y. Shahin, Magnetic nanoparticles: synthesis, characterization, biomedical applications and challenges, *Sphinx Journal of Pharmaceutical and Medical Sciences*, 5 (2023) 1–32.
[42] K.M. Harkness, D.E. Cliffel, J.A. McLean, Characterization of thiolate-protected gold nanoparticles by mass spectrometry, *Analyst*, 135 (2010) 868–874.
[43] W. Blanc, C. Guillermier, B. Dussardier, Composition of nanoparticles in optical fibers by secondary ion mass spectrometry, *Optical Materials Express*, 2 (2012) 1504–1510.
[44] S. Rajagopalachary, S. Verkhoturov, E. Schweikert, Examination of individual nanoparticles with cluster SIMS, *Surface and Interface Analysis*, 43 (2011) 547–550.
[45] Y.P. Kim, H.K. Shon, S.K. Shin, T.G. Lee, Probing nanoparticles and nanoparticle-conjugated biomolecules using time-of-flight secondary ion mass spectrometry, *Mass Spectrometry Reviews*, 34 (2015) 237–247.
[46] C. Granata, R. Russo, E. Esposito, A. Vettoliere, M. Russo, A. Musinu, D. Peddis, D. Fiorani, Magnetic properties of iron oxide nanoparticles investigated by nano SQUIDs, *The European Physical Journal B*, 86 (2013) 1–5.

[47] S.-J. Oh, C.-J. Choi, S.-J. Kwon, S.-H. Jin, B.-K. Kim, J.-S. Park, Mössbauer analysis on the magnetic properties of Fe–Co nanoparticles synthesized by chemical vapor condensation process, *Journal of Magnetism and Magnetic Materials*, 280 (2004) 147–157.

[48] M. Bystrzejewski, A. Grabias, J. Borysiuk, A. Huczko, H. Lange, Mössbauer spectroscopy studies of carbon-encapsulated magnetic nanoparticles obtained by different routes, *Journal of Applied Physics*, 104 (2008).

[49] A.P. Herrera, C. Barrera, Y. Zayas, C. Rinaldi, Monitoring colloidal stability of polymer-coated magnetic nanoparticles using AC susceptibility measurements, *Journal of Colloid and Interface Science*, 342 (2010) 540–549.

[50] M. Kars, K. Kodama, Rock magnetic characterization of ferrimagnetic iron sulfides in gas hydrate-bearing marine sediments at Site C0008, Nankai Trough, Pacific Ocean, off-coast Japan, *Earth, Planets and Space*, 67 (2015) 1–12.

[51] V.L. Calero-Ddelc, D.I. Santiago-Quinonez, C. Rinaldi, Quantitative nanoscale viscosity measurements using magnetic nanoparticles and SQUID AC susceptibility measurements, *Soft Matter*, 7 (2011) 4497–4503.

[52] A. Rodriguez, A. Oliveira, P. Morais, D. Rabelo, E. Lima, Study of magnetic susceptibility of magnetite nanoparticles, *Journal of Applied Physics*, 93 (2003) 6963–6965.

[53] A. Mazzaglia, L.M. Scolaro, A. Mezzi, S. Kaciulis, T.D. Caro, G.M. Ingo, G. Padeletti, Supramolecular colloidal systems of gold nanoparticles/amphiphilic cyclodextrin: a FE-SEM and XPS investigation of nanostructures assembled onto solid surface, *The Journal of Physical Chemistry C*, 113 (2009) 12772–12777.

[54] S. Rades, V.-D. Hodoroaba, T. Salge, T. Wirth, M.P. Lobera, R.H. Labrador, K. Natte, T. Behnke, T. Gross, W.E. Unger, High-resolution imaging with SEM/T-SEM, EDX and SAM as a combined methodical approach for morphological and elemental analyses of single engineered nanoparticles, *RSC Advances*, 4 (2014) 49577–49587.

[55] V.D. Hodoroaba, S. Rades, W.E. Unger, Inspection of morphology and elemental imaging of single nanoparticles by high-resolution SEM/EDX in transmission mode, *Surface and Interface Analysis*, 46 (2014) 945–948.

[56] T.R. Kim, *Magnetic imaging of magnetic recording media using transmission electron microscopy*, Stanford University, 2017.

[57] M. José-Yacamán, M. Marın-Almazo, J. Ascencio, High resolution TEM studies on palladium nanoparticles, *Journal of Molecular Catalysis A: Chemical*, 173 (2001) 61–74.

[58] T. Akita, S. Tanaka, K. Tanaka, M. Haruta, M. Kohyama, Sequential HAADF-STEM observation of structural changes in Au nanoparticles supported on CeO2, *Journal of Materials Science*, 46 (2011) 4384–4391.

[59] G. Binnig, C.F. Quate, C. Gerber, Atomic force microscope, *Physical Review Letters*, 56 (1986) 930.

[60] A. Vilalta-Clemente, K. Gloystein, N. Frangis, Principles of atomic force microscopy (AFM), *Proceedings of Physics of Advanced Materials Winter School*, (2008) 1–8.

[61] Á. Mechler, J. Kopniczky, J. Kokavecz, A. Hoel, C.-G. Granqvist, P. Heszler, Anomalies in nanostructure size measurements by AFM, *Physical Review B*, 72 (2005) 125407.

[62] G. Cordova, B.Y. Lee, Z. Leonenko, Magnetic force microscopy for nanoparticle characterization, *arXiv preprint arXiv:1704.08289*, (2017).

5 Theoretical Modeling of Coordination Materials

Sajjad Ali, Muhammad Zahoor, Shahab Khan, Chao Zeng, Qaiser Alam, Mohamed Bououdina and Ghulam Yasin

5.1 INTRODUCTION

Coordination materials represent a category of compounds characterized by the linkage of metal ions or clusters through organic ligands, resulting in extended networks with unique properties[1]. To strategically design these materials, a profound comprehension of their structure-property relationships is essential, and this understanding is attainable through theoretical modeling. This chapter explores the theoretical methodologies utilized to clarify the behavior of coordination materials, providing a glimpse into the symbiotic relationship between theoretical frameworks and experimental investigations. As computational resources continue to expand, molecular simulations and modeling have emerged as essential tools for characterizing, screening, and designing these materials. In the realm of advancing computational capabilities, molecular simulations become pivotal for characterizing and designing coordination materials. The chapter outlines recent computational studies, with a focus on metal-organic frameworks (MOFs)[2–7], as well as other coordination materials like covalent-organic framework (COFs)[8,9], coordination cages, and metal-organic polyhedra (MOPs)[10]. Molecular dynamics (MD) simulations offer dynamic insights, shedding light on critical aspects such as diffusion and stability. Integrating MD with ab initio methods provides a holistic view of material behavior, considering quantum effects and long-range interactions.

The rapid progress of Metal-Organic Frameworks (MOFs) and Covalent-Organic Frameworks (COFs) since Omar M. Yaghi's discovery of MOFs in 1995 signifies significant advancements in coordination materials[11]. Initially focused on synthesizing MOF structures, attention shifted to performance and applications, with the emergence of stable MOFs like UiO, MIL, and ZIF (Figure 5.1). Both MOFs and COFs, as subsets of coordination materials, offer unique characteristics – MOFs with open metal sites suitable for catalytic reactions and COFs with robust covalent bonds. Recent integration efforts have led to over 80,000 reported MOFs and 500 COFs with diverse topologies, promising expanded applications and novel characteristics (Figure 5.1). Challenges include slow synthesis rates due to monomer complexities, emphasizing the ongoing scientific pursuit in this field[12].

Venturing into the molecular intricacies of coordination materials, the chapter emphasizes their applications in energy, sustainability, and catalysis. It underscores

DOI: 10.1201/9781003345886-6

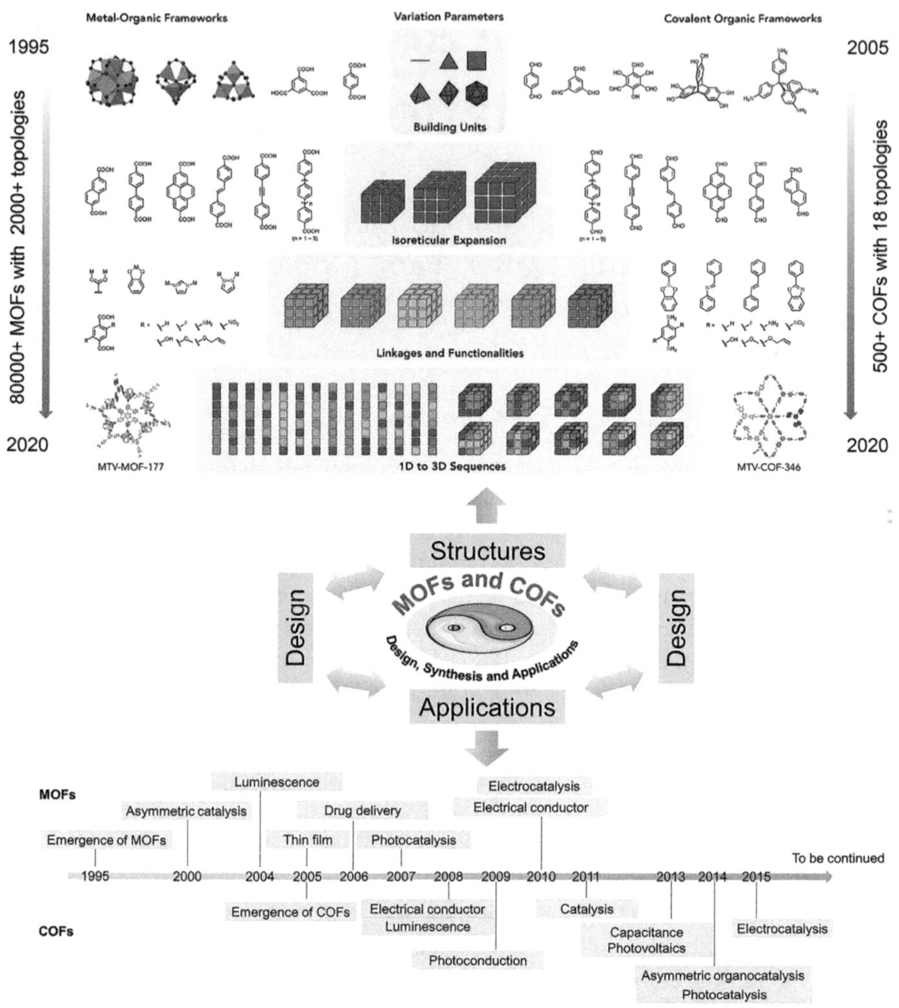

FIGURE 5.1 Exploring the diversity of MOF and COF structures: a chronological journey through the first application reports. Taken from Ref.[12] copyright 2021, Elsevier.

the inherent complexity in the building blocks of these materials. Molecular-level examinations bridge physical and engineering sciences, offering insights beyond traditional experiments and contributing to the innovative design of materials.

5.2 QUANTUM INSIGHTS INTO COORDINATION MATERIALS THROUGH DENSITY FUNCTIONAL THEORY (DFT)

Exploring coordination materials using density functional theory (DFT) offers a robust methodology for understanding their electronic structure and properties at the quantum level. DFT, extensively employed in computational research, enables researchers to analyze the intricate interactions between metal ions and ligands, providing valuable insights into the stability, reactivity, and electronic attributes of these materials. The application of DFT in studying coordination materials allows for a nuanced understanding of their behavior, thereby aiding in the deliberate design and improvement of novel compounds customized for specific functions. These applications extend from catalysis and sensing to broader contexts, showcasing the versatility and impact of this approach.

5.2.1 DFT Capabilities in Coordination Materials Analysis

Density functional theory (DFT), an essential and versatile tool in coordination chemistry, plays a pivotal role in the exploration of coordination materials. Its capabilities encompass vital operations such as geometry optimization, single-point energy calculations, and the prediction of barriers and reaction paths, aiding researchers in unraveling reaction mechanisms. DFT, as a versatile computational tool, plays a crucial role in coordination materials design by providing insights into various molecular properties. It facilitates the exploration of wave functions, detailed molecular orbitals, atomic charges, dipole moments, multipole moments, electrostatic potentials, and polarizabilities, offering a comprehensive understanding of the molecular landscape. Moreover, DFT proves indispensable in coordination chemistry by assessing spin state energetics, magnetic exchange coupling constants, vibrational frequencies, and infrared (IR) and Raman intensities. Researchers can delve into nuclear magnetic resonance (NMR) chemical shifts, ionization energies, and electron affinities. Additionally, DFT enables the conduct of time-dependent calculations for optical spectroscopy, electron paramagnetic resonance (EPR), and simulations of x-ray absorption spectra. In the design and exploration of coordination materials, DFT's capabilities further extend to considering electrostatic effects on solvation and investigating noncovalent interactions in extended molecular systems. Its multifaceted applications make DFT an invaluable asset, providing essential tools for a comprehensive understanding of coordination materials and their properties.

5.2.2 Strategic DFT Selection for Coordination Materials

Boarding on the journey of "Strategic DFT Selection for Coordination Materials" reveals a critical challenge for the meticulous choice of density functionals (DFs). In the realm of coordination materials, the absence of a systematic method for precise DF selection necessitates a nuanced approach, guided by comparisons with experimental data. Navigating the landscape of spin state computations for transition metal complexes, inherent to coordination materials, demands caution in DF choices. Among options, local or semi-local DFs, particularly those on Jacob's ladder's initial three rungs, show efficacy by offsetting nonlocal effects. A caveat arises against using the local density approximation (LDA) in coordination materials. Conversely, BP86's proficiency in delivering accurate geometries and vibrational frequencies makes it strategic for spectroscopic investigations. In long-range corrected DFs, B97XD stands out, strategically outpacing conventional DFT methods across coordination materials metrics[13]. CAM-B3LYP excels in predicting charge transfer energies, aligning with coordination material considerations. DFT-D3 by Grimme enriches predictions for reaction energies and activation barriers, offering strategic insights[14]. In double hybrid functionals, DSD-BLYP showcases proficiency, particularly in systems with significant nondynamical correlation – an essential factor in understanding coordination materials. Introducing the long-range-corrected hybrid LC-vPBE captivates with remarkable accuracy across diverse molecular properties pertinent to coordination materials. Expanding the horizon of functionals, DFT-dDXDM strategically augments interactions within coordination materials, while DFT-dDsC, incorporating a density-dependent energy correction, finds its niche in modeling redox reactions and charged species. As we navigate the world of density functional theory in coordination materials, the quest for precision unfolds as both a challenge and an opportunity, guiding us towards enhanced understanding in this scientific domain.

5.3 MODELLING COORDINATION MATERIALS FOR GAS MOLECULES ADSORPTION AND SEPARATION

Coordination materials are widely recognized for their effective gas adsorption, making them valuable for gas separation and storage applications. Theoretical modeling plays a crucial role in helping us understand how gas molecules interact with these porous materials. By predicting key factors like adsorption isotherms, selectivity, and diffusion rates, researchers gain valuable insights. This information acts as a guide, allowing for the precise customization of materials to excel in specific gas separation processes. To enhance the ability of materials to adsorb gases, scientists use molecular modeling techniques to closely study the complex interactions within coordination materials. This approach provides a detailed understanding of how these materials selectively capture and interact with gas molecules. By doing so, it opens the door for the strategic design of advanced adsorption systems tailored to specific properties, catering to a range of applications. Since the discovery of MOFs in 1995, more than 45,000 articles have been published up to now, emphasizing the rapid progress. MOFs are significant in gas adsorption and storage, with over 10,000 MOF structures designed for various gas storage and separation applications. Molecular modeling, particularly using density functional theory (DFT), plays a vital role in understanding gas-MOF interactions[15].

The integration of experimental and theoretical efforts is crucial for more efficient MOF and gas selection, reducing reliance on trial-and-error approaches. The urgent need for sustainable energy conversion methods has driven research into clean and viable alternatives to fossil fuels. Fossil fuels dominate global energy sources, contributing to greenhouse gas emissions. Shifting to non-fossil fuel energy is vital to mitigate environmental impacts. Various energy conversion technologies, including fuel cells, water splitting, and gas-based energy carriers, offer promising solutions. However, the low energy density and flammability of gases present challenges. Designing lightweight materials for efficient gas storage and separation at ambient conditions is essential. Zeolites, COFs, and MOFs are explored for gas storage, with MOFs showing promise due to their porous nature, tunability, and crystalline structure. Molecular

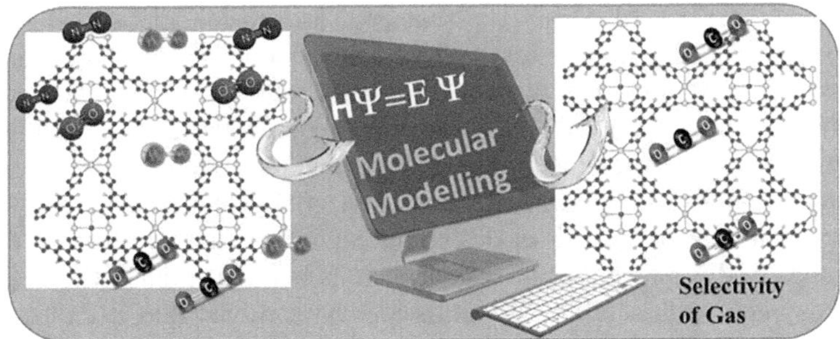

FIGURE 5.2 Utilizing density functional theory-based modeling to enhance gas adsorption in existing MOFs and design new MOFs with increased capacity when coupled with experimental synthesis. Taken from Ref.[15] copyright 2023 with permission of Springer.

modeling helps enhance gas uptake by understanding host-guest interactions, but selectivity remains a challenge. Advancements in MOF materials have achieved remarkable gas storage records, but a systematic protocol for selectivity remains elusive. Computational tools, particularly density functional theory (DFT), provide insights into MOF modifications and gas adsorption capabilities, considering factors like metal ions and ligand linkers. Combining DFT with classical simulations aids in designing next-generation MOFs with improved gas adsorption properties. Theoretical and computational methods, including DFT calculations, are employed to study gas adsorption in MOFs. These approaches involve periodic DFT simulations that mimic real systems, cluster models focusing on repeating units of MOFs, and other methods like quantum Monte-Carlo and quantum mechanics/molecular mechanics (QM/MM). The discussion focuses on DFT studies of gas binding in MOFs containing first-row transition elements, emphasizing the role of metal ions and organic-ligands in achieving selectivity and gas uptake capacity. MOFs are considered a promising material for adsorptive gas storage, providing a balance between gravimetric and volumetric capacity. Tuning MOF-based gas adsorptive systems holds potential for clean energy production, climate change mitigation, and efficient gas separation and storage.[15]

5.3.1 Modelling of Carbon-Dioxide Capture in Coordination Materials

Theoretical modeling serves as an insightful tool for unraveling the complexities inherent in catalytic mechanisms and the reactivity exhibited by coordination materials[16–27]. By simulating intricate reaction pathways and discerning transition states, researchers gain a comprehensive understanding of the pivotal role played by active sites. This process involves meticulous assessments of the energetics governing reactions, enabling insightful predictions about catalytic activity. This in-depth comprehension proves crucial in strategically designing and developing highly efficient catalytic materials capable of facilitating a wide array of diverse and complex chemical transformations.

The knowledge amassed through theoretical modeling not only enhances our understanding of the fundamental principles dictating catalysis but also propels advancements in the field of material science. Through the application of sophisticated computational methods, researchers can explore various potential reaction scenarios, leading to the identification of optimal catalysts and reaction conditions. This holistic approach not only augments our predictive capabilities but also expedites the discovery of groundbreaking catalytic materials, thereby contributing to the sustainable evolution of chemical processes. Essentially, theoretical modeling stands as a cornerstone in the pursuit of designing catalytic materials that not only meet the demands of diverse chemical transformations but also align with the principles of efficiency and sustainability. CO_2 emissions have surged due to industrialization and fossil fuel use, exacerbating global warming and climate change[28–30]. Research on CO_2 capture and conversion has gained significance. Material-based storage, particularly in MOFs, offers an efficient and low-energy technique for CO_2 capture and sequestration. Modeling CO_2 adsorption in MOFs requires rigorous benchmarking of exchange-correlation functionals. Siegel and colleagues compared binding energies (BE) predicted by various density functionals with experimental data for CO_2 adsorption on different MOFs. The revPBE-vdW functional emerged as promising, although it overestimated metal-oxygen (CO_2) distances. Dispersion effects, crucial for thermodynamics, showed potential for density functional methods in rapid screening of CO_2 adsorbents. CO_2 capture and its subsequent conversion into useful gases are essential due to the rising greenhouse gas levels. Material-based storage, particularly in MOFs, offers an energy-efficient approach. Siegel and colleagues[31] compared various density functionals to predict CO_2 adsorption on MOFs, finding the revPBE-vdW functional promising despite some limitations. Long-range van der Waals interactions were critical for thermodynamics. Further research involving improved dispersion corrections, hybrid functionals, and ab initio methods holds potential. Ligand functionalization in MOFs, such as the BTT family, influences CO_2 adsorption. Ali et al. examined Cu-MOFs and showed the role of ligand functionalization in tuning CO_2 adsorption and selectivity. Overall, these studies emphasize the importance of accurate modeling in understanding and enhancing CO_2 adsorption in MOFs, shedding light on factors affecting adsorption efficiency and selectivity[32]. Pengyan Wu et al[33]. designed Zn-DPA·$2H_2O$, a two-fold interpenetrated framework with propeller-like ligands (Figure 5.3). The ligands undergo rotational rearrangement in response to guest molecules, leading to subtle channel changes. This PCP displays high CO_2 affinity, confirmed by its crystal structure in CO_2-adsorbed phases. Additionally, it exhibits efficient catalysis and size selectivity in CO_2 cycloaddition to epoxides.

In another work Omid T. Qazvini et al[34]. presented MUF-16 (Massey University Framework), a MOF with inverse selectivity favoring carbon dioxide over hydrocarbons (Figure 5.4). Efficient carbon dioxide capture through hydrogen bonding and favorable interactions enables high selectivities for gas mixtures applicable to natural gas and industrial feedstocks. MUF-16, economically produced at scale and stable, stands out as an attractive and recyclable adsorbent.

5.3.2 Modelling of Hydrogen Capture in Coordination Materials

Hydrogen holds considerable potential as a clean energy source, given its high gravimetric and volumetric capacity. This quality presents a promising solution for

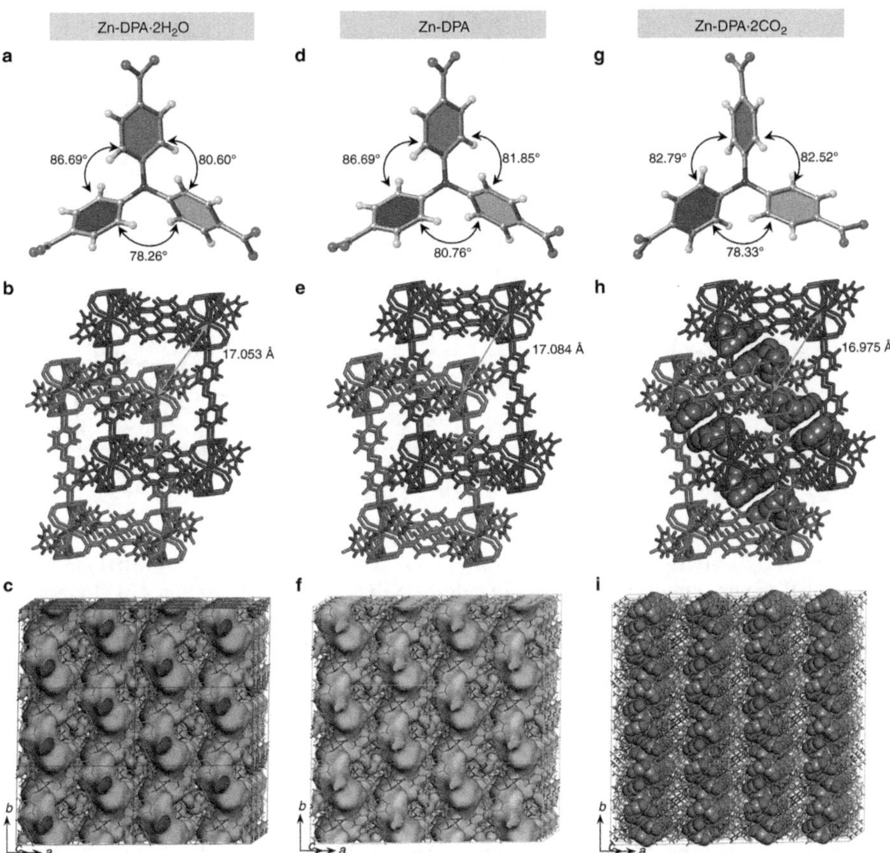

FIGURE 5.3 Crystal structures of Zn-DPA·2H2O (a–c), guest-free Zn-DPA (d–f), and CO_2-containing Zn-DPA·2CO_2 (g–i). Dihedral angle between tca^{3-} ligand phenyl rings in PCPs (a, d, g). Interpenetrated frameworks with trinuclear Zn clusters, tca^{3-}, and dpa ligands, showing interlayer distances (b, e, h). 3D connected channels in PCPs with Connolly surfaces (Connolly radius: 1.6 Å). Inner channel surfaces in blue (c), cyan (f), and green (i), outer surfaces in grey. Water molecules lost for clarity in Zn-DPA·2H_2O. Adopted from Ref.[33] with 2019 copyright permission from Springer Nature.

replacing fossil fuels and advancing fuel cell technology. Nevertheless, the challenge lies in developing cost-effective hydrogen storage systems, especially as the predominant method involves cryogenic conditions, leading to significant energy consumption and infrastructure demands. MOFs stand out as an appealing option for hydrogen storage due to their remarkable versatility. They allow precise adjustments in chemical composition, topological structure, surface chemistry and offer substantial surface area. Consequently, there is a keen interest in investigating hydrogen adsorption within MOFs through modeling studies. Existing literature boasts a variety of studies, employing approaches such as DFT, coupled-cluster methods, and symmetry-adapted perturbation theory[35,36], to address numerous aspects of

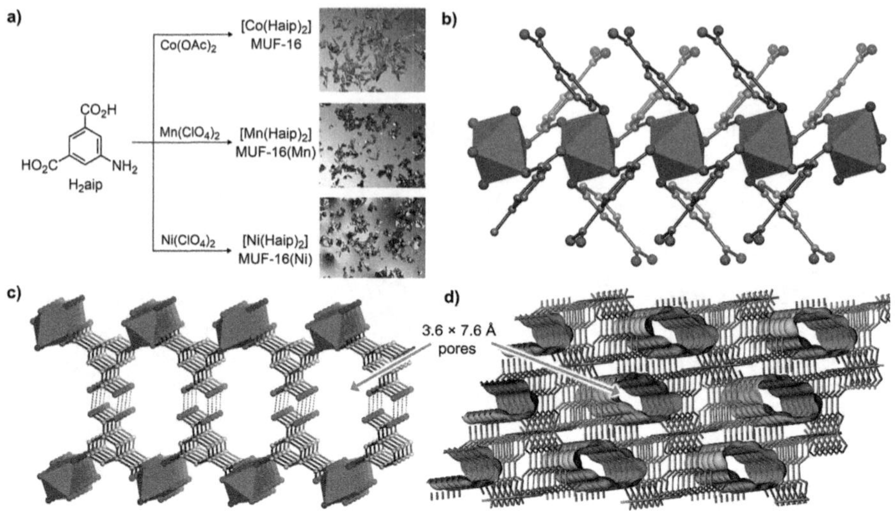

FIGURE 5.4 a) MUF-16 synthesis pathways and resulting micrographs.b) iSBUs in MUF-16: cobalt(II) chains connected by Haip ligands. c) iSBUs form 2D sheets and a 3D framework through hydrogen bonding. d) MUF-16 features 3.6×7.6 Å channels with a Connolly surface. Color code: Co = magenta; O = red; C = grey, N = blue. Adopted from Ref.[34] with 2021 copyright permission from Springer Nature.

hydrogen binding, uptake, and the exploration of potential hydrogen-binding sites within MOFs.[36,37]

Ihm and colleagues[38] conducted pioneering electronic structure calculations focused on modifying MOF-5 by introducing calcium atoms and substituting boron for carbon within the benzene ring of the BDC linker. These calculations employed spin-polarized density functional theory (pDFT) utilizing the PW91 Generalized Gradient Approximation (GGA) functional, with the computational framework executed using the VASP software. Figure 5.5 illustrates the structural alterations, wherein calcium atoms were introduced onto the BDC (benzene dicarboxylate) linker, forming associations with hydrogen. The investigation encompassed an assessment of binding energies both before and after the boron substitution process[38]. Notably, the results indicated that boron substitution led to enhancements in the binding strength between calcium and the BDC linker, as well as between hydrogen molecules and calcium. Specifically, this modification facilitated the binding of eight H_2 molecules in proximity to the calcium atom, yielding an average binding energy of 20 kJ/mol. This outcome represents a favorable improvement conducive to enhanced hydrogen adsorption.

1. The key findings and insights from various studies related to hydrogen adsorption and its implications in metal-organic frameworks (MOFs) are summarized here. These studies provide valuable insights into enhancing hydrogen adsorption in MOFs through various modifications and highlight the importance of factors such as metal decoration, anion substitution, and

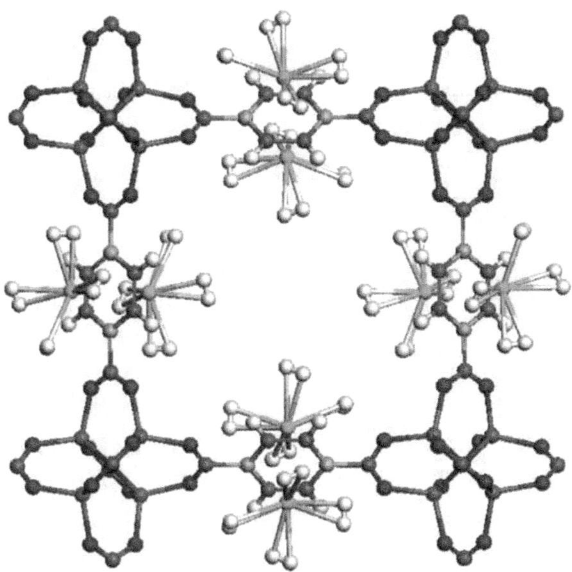

FIGURE 5.5 The optimized configuration of MOF-5 with B-substitution and Ca decoration, accommodating eight H_2 molecules per linker. In the illustration, carbon, boron, hydrogen, oxygen, calcium, and zinc atoms are denoted by gray, pink, white, red, green, and dark violet spheres, respectively. Taken from Ref.[38] copyright 2009 Elsevier.

structural configuration in optimizing MOF materials for hydrogen storage applications.

2. Calcium and boron substitution in MOF-5: Ihm and colleagues conducted electronic structure calculations involving the introduction of calcium atoms and substitution of boron for carbon in the MOF-5 framework. Their findings revealed enhanced binding between calcium and the BDC linker, as well as improved hydrogen binding due to boron substitution. This modification facilitated the binding of eight H_2 molecules with an average energy of 20 kJ/mol, which is favorable for hydrogen uptake.[38]
3. Light metal decoration in MOF-5: Pal and co-workers explored the impact of decorating MOF-5 with light metals (Li, Be, Mg, Al) using ab initio and pDFT-based calculations (Figure 5.6). Their goal was to comprehend the changes in charge transfer and interaction energy linked to the adsorption of hydrogen on the metal-decorated surface. Among the studied metals, Li and Al decorated MOF-5 exhibited a favorable HOMO (highest occupied molecular orbital) interaction, with electron density transferred from metal to the organic linker (Figure 5.7). This decoration enhanced hydrogen adsorption, especially in the case of Li and Al, making them suitable for hydrogen storage at room temperature.[39]
4. Anion substitution in M-BTT Series MOFs: Long and colleagues investigated the effect of anion substitution on hydrogen adsorption in M-BTT series MOFs (Cu, Mn, Fe, Zn) using DFT calculations. They found that the

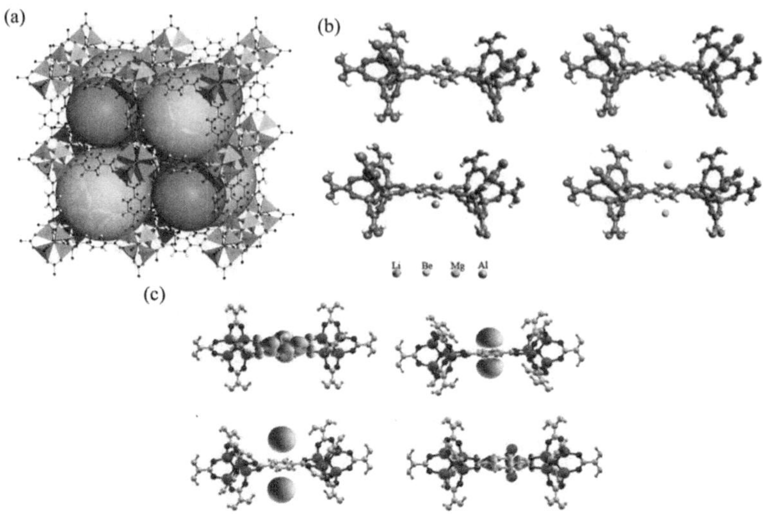

FIGURE 5.6 (a) MOF-5 repeating unit, (b) optimized structures of Li-, Be-, Mg-, and A1-decorated MOF-5 cells, and (c) frontier molecular orbitals of metals and MOF-5. Taken from Ref.[39] copyright 2011 Elsevier.

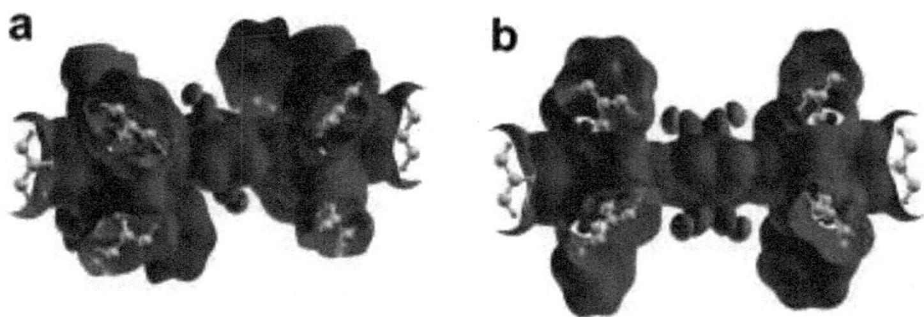

FIGURE 5.7 Charge density difference plots for (a) MOF-5:Li_2:$2H_2$ and (b) MOF-5:Li_2:$4H_2$. Taken from Ref.[39] copyright 2011 Elsevier.

size of the anion affected the position of the metal center in the framework, influencing H_2 binding strength. Moving metal centers away from the plane of nitrogen atoms created more accessible binding pockets for H_2, leading to stronger electrostatic interactions with metal centers. This study identified Zn-BTT as a promising candidate for hydrogen storage.[40]

Hu et al.[41] introduced a pioneering theoretical framework to investigate the binding interactions between molecular hydrogen (H_2) and a MOF, building upon the seminal work by the Yaghi group[42] in 2003, which highlighted the remarkable H_2 absorption capabilities of MOF-5. Their study systematically explored the interaction of H_2 with various aromatic systems, including C_6H_5X (where X = H, F, OH, NH_2, CH_3, and CN), $C_{10}H_8$ (comprising naphthalene and azulene), C14H10 (anthracene),

$C_{24}H_{12}$ (coronene), p-$C_6H_4(COOH)_2$ (terephthalic acid), and p-$C_6H_4(COOLi)_2$ (dilithium terephthalic acid). According to their rigorous calculations, the binding energies of H_2 to benzene and naphthalene were quantified as 3.91 and 4.28 kJ mol-1, respectively. These findings elucidate a direct correlation, indicating an augmentation in interaction energy with the increasing complexity of the aromatic system. Similar investigations were conducted in the same year (2004) by Sagara et al.[43] and Hamel et al.[44]. Sagara et al. employed the MP_2 methodology to calculate H_2 binding energies to the $Zn_4O(HCO_2)_6$ cluster and to H_2-1,4-benzenedicarboxylate-H_2, which are components of MOF-5 consisting of a metal oxide and an organic linker. Their results indicated a higher H2 binding energy for the zinc oxide cluster compared to the organic linker component. Hamel and Cote[44] utilized ab initio techniques, including MP_2 and coupled cluster with noniterative triple excitation (CCSD(T)), as well as density functional theory (DFT) methods like local-density approximation (LDA) and generalized gradient approximation (GGA). They calculated the H_2 binding energy to benzene while considering various H_2 configurations on the molecule. Additionally, they estimated hypothetical rotational spectra of adsorbed H_2, revealing results comparable to experimental findings obtained using inelastic neutron scattering (INS) for H_2 adsorption in MOFs[45].

Sang Soo Han et al. investigated H_2 binding in $Zn_4O(HCO_2)_6$, $Mg_4O(HCO_2)_6$, and $Be_4O(HCO_2)_6$ clusters within IRMOF types. Metal substitution (M=Zn to Mg and Be) doesn't alter the $M_4O(HCO_2)_6$ cluster's fundamental configuration, and Mg shows the strongest H_2 binding energy[46]. MP_2 calculations, currently, cannot account for periodic crystals. Consequently, DFT calculations are essential for determining H_2 binding energies in MOF crystals. Mulder et al.[47] and Mueller and Ceder[48] attempted DFT applications to the MOF-5 crystal. Both studies confirmed strong hydrogen interactions near Zn_4O clusters, although they reported different H_2 binding energies (70 meV per H_2 by Mulder et al. and 20 meV per H_2 by Mueller and Ceder), supported by various experimental and theoretical evidence[49,50]. Ab initio and DFT approaches are essential for determining H_2 binding energies in MOFs, offering valuable insights into H_2 adsorption sites in these materials. It is critical to compare ab initio and DFT methods when calculating van der Waals interactions, especially for H_2 interactions with MOFs and covalent organic frameworks (COFs). Van der Waals interactions, characterized by induced dipole-dipole forces, require the use of ab initio methods, at least at the MP_2 (second-order Møller–Plesset perturbation theory) and CCSD (coupled cluster singles and doubles) levels, due to the need to encompass double excitations from a Slater determinant reference corresponding to single excitations of each sub-system. However, DFT methods lack consideration for these excitations, making them less accurate in modeling long-range dispersion interactions. Additionally, triple excitations significantly contribute to the dispersion energy, rendering CCSD(T) (coupled cluster singles, doubles, and perturbative triples) more precise than MP_2, albeit with increased computational time[51].

5.3.3 Modelling Oxygen Adsorption in Coordination Materials

Oxygen holds immense significance in the medical field, and the COVID-19 pandemic highlighted shortages of oxygen in various countries. This shortage arose because the industrial-scale separation of oxygen from nitrogen (O_2/N_2) predominantly relies on cryogenic distillation, a time-consuming and expensive process. Additionally, the growing levels of carbon dioxide in the atmosphere have spurred global research efforts into this area, as high-purity oxygen is essential for several post-carbon dioxide capture applications. In 2015, Nenoff and colleagues[52] conducted an extensive computational screening of MOFs for selective oxygen gas binding. This study focused on $M_3(BTC)_2$ and $M_2(DOBDC)$ type frameworks as depicted in Figures 5.8a and 5.8b, where *M* represents a range of metals including Sc, Ti, V, Cr, Mn, Fe, Co, Ni, Cu, Zn, Mo, Ru, Be, and Mg. The objective was to investigate the influence of different metals on O_2/N_2 separations. The accuracy of their DFT calculations was confirmed by matching the experimental trends in gas adsorption binding energies. Structural data for $Fe_3(BTC)_2$,[53] $Cu_3(BTC)_2$,[54] $Fe_2(DOBDC)$,[55] $Co_2(DOBDC)$,[56] $Ni_2(DOBDC)$,[57] $Zn_2(DOBDC)$,[58] and $Mg_2(DOBDC)$[59] were adopted

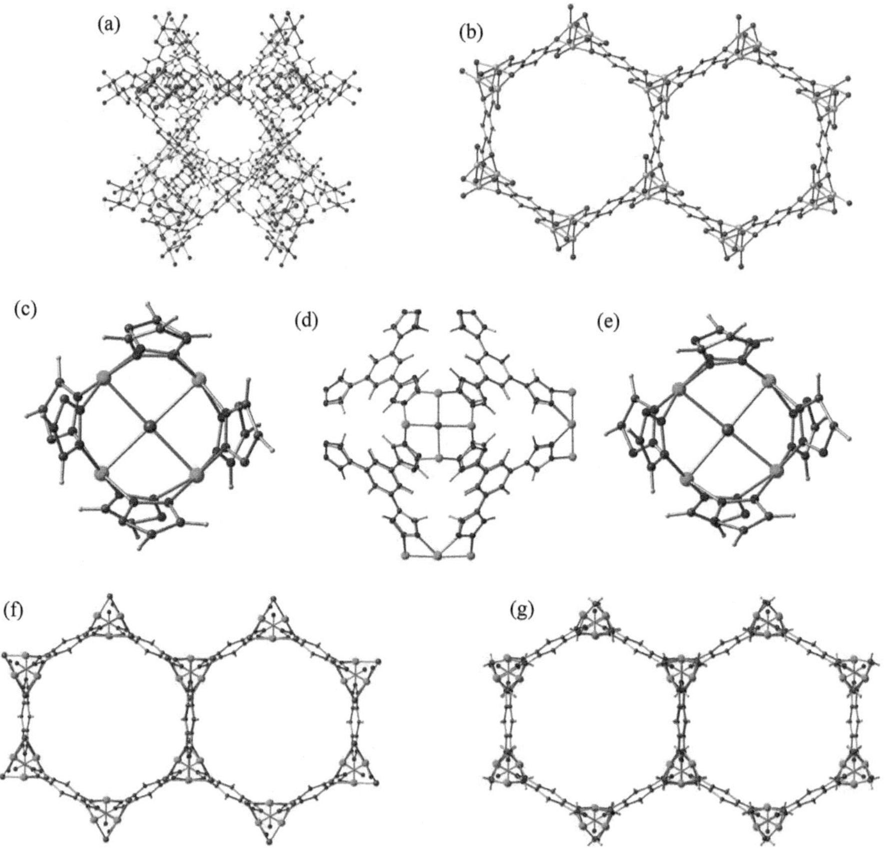

FIGURE 5.8 O_2 adsorption modeling in MOFs with representative structures: (a) $Cr_3(BTC)_2$, (b) $Fe_2(DOBDC)$, (c) Co-BTTri tetramer, (d) Co-BTTri unit, (e) Co-BDTrip tetramer, (f) $Co_2Cl_2(BBTA)_2$ unit, and (g) $Co_2(OH)_2(BBTA)_2$ unit. Element colors: Cr (dark green), Fe (orange), Co (light green), N (blue), Cl (purple), O (red), C (gray), H (white).

from existing literature. These structures endured modifications, which involved eliminating solvents and water molecules while introducing hydrogen atoms for saturation. In the case of other MOFs, new structures were generated by substituting metal atoms in their corresponding model structures. The calculations made use of the PBE functional with DFT-D2 dispersion corrections, maintaining a convergence threshold of 10^{-4} eV and ensuring that the force on each atom remained below 0.03 eV/Å, employing the VASP software. Structural optimizations were achieved by utilizing a plane wave cutoff energy of 500 eV and a projector-augmented wave formalism, resulting in cell lengths accurate within 5% of experimental values. The estimated binding energy for O_2 varied within the range of -354 to -21 kJ/mol, while for nitrogen, it ranged from -188 to -7 kJ/mol. Interestingly, no apparent correlation was observed between binding energies and MOF structural types. The calculated binding energies were validated by assessing experimental gas adsorption at low pressure and low temperature in $Cu_3(BTC)_2$, $Ni_2(DOBDC)$, and $Co_2(DOBDC)$.[15] The study revealed that 4d transition metals generally exhibited higher binding energies than 3d metals, except for $Mo_3(BTC)_2$. This variation was attributed to the diffuse nature of 4d orbitals, allowing for stronger interactions with oxygen compared to 3d metals. Furthermore, the study emphasized the significance of d-orbital occupancy, highlighting that late transition metals were not well-suited for O_2/N_2 separation due to their similar binding affinities for O_2 and N_2.[15] Early transition metal-based MOFs, such as $Sc_3(BTC)_2$, Ti3$(BTC)_2$, $V_3(BTC)_2$, $Sc_2(DOBDC)$, $Ti_2(DOBDC)$, $V_3(BTC)_2$, $Cr_2(DOBDC)$, and Mo2(DOBDC), emerged as promising candidates for O_2/N_2 separation. Notably, it was the type of metal, rather than the framework structure, that primarily dictated a MOF's capability to separate oxygen from nitrogen. The study demonstrated that ligands played a relatively minor role in O_2/N_2 separation, as illustrated in Figure 5.9.[52]

5.4 SIMULATIONS TO MACHINE LEARNING

Recently, machine learning (ML) techniques have proven indispensable in accelerating the exploration and comprehension of coordination materials. Utilizing models trained on databases with known materials, these techniques can predict properties like adsorption energies, band gaps, and crystal structures for novel compounds. This approach streamlines high-throughput material screening, guiding experimental efforts to identify promising candidates efficiently.

Porous materials, existing in either an amorphous or crystalline state, manifest a rich variety of structures encompassing MOFs[1,3,4], COFs[8,9], zeolites, activated carbons, and coordination materials. This diverse spectrum of structures offers a wealth of physical and chemical properties, facilitating selective molecule separation based on pore size and chemistry, thus finding applications in versatile fields such as catalysis, medicine, and beyond. The ability to precisely tailor the chemistry, porosity, and surface area of these materials holds immense promise for the development of energy-efficient, environmentally friendly, and cost-effective technologies. In particular, MOFs represent a relatively recent generation of crystalline porous materials distinguished by highly advantageous characteristics, including exceptionally high surface areas (up to 10,000 m^2/g)[60], a broad range of porosity and pore sizes, low densities,

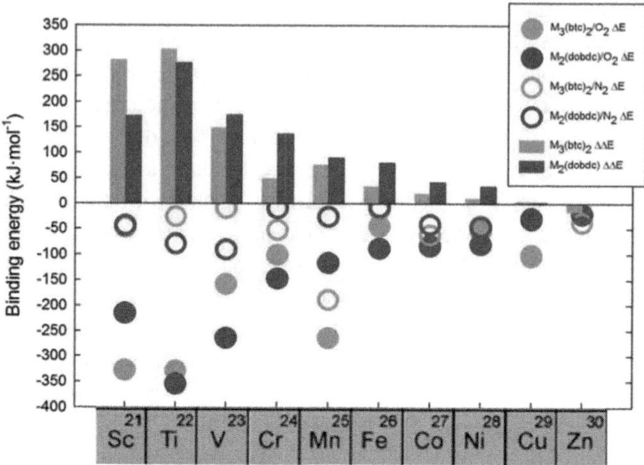

FIGURE 5.9 Illustrating oxygen and nitrogen binding energy differences in $M_3(BTC)_2$ (orange) and M_2-(DOBDC) (purple) structures. Taken from Ref.[52]. Copyright 2015 American Chemical Society.

and remarkable chemical tunability. This has propelled MOFs into the forefront of research and garnered extensive attention across various domains, including gas storage and separation, catalysis, biomedicine (e.g., drug storage and delivery), chemical sensing, electrical conductivity, and light harvesting. Additionally, coordination materials, characterized by well-defined and controlled coordination environments, contribute significantly to the rich landscape of porous materials. These materials, often involving metal-ligand interactions, exhibit intriguing properties and functionalities that further expand the potential applications in fields like catalysis, sensing, and electronic devices. Thus, the synergistic exploration of diverse porous materials, including MOFs and coordination materials, continues to drive innovation and advance our understanding of their versatile applications in materials science and technology.

Recent advancements in experimental synthesis and modification techniques have significantly eased the manipulation of both structural and chemical properties within MOFs. These modifications encompass a wide range, including the introduction of functional groups, the exchange of metals and/or ligands, the encapsulation of nano-sized guests, and the interpenetration of frameworks. As a result, the number of synthesized MOFs has surpassed 100,000, and the corresponding crystallographic information files (CIFs) have been deposited into the Cambridge Structural Database (CSD). Simultaneously, researchers have explored computationally generated hypothetical MOFs (hMOFs) by investigating various combinations of building units and topologies. Despite the abundant existence of hMOFs in comparison to their synthesized counterparts, only a fraction of them have been successfully realized experimentally. The extensive MOF material space, depicted in Figure 5.10(a), encompasses both experimentally synthesized and computationally generated MOFs. Many of these structures hold the potential to address critical societal challenges such as climate

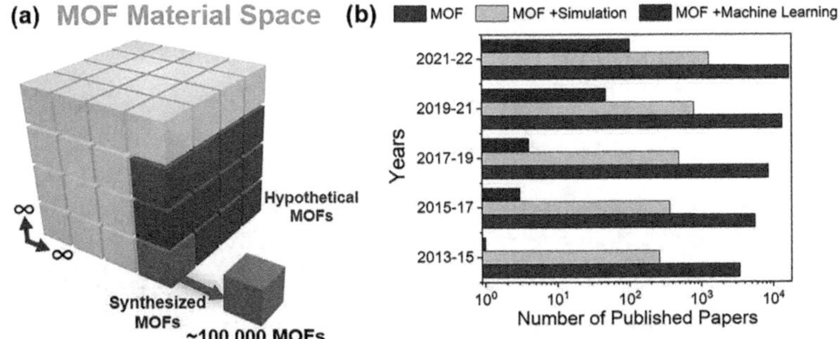

FIGURE 5.10 (a) The large cube signifies the extensive domain of MOF space, encompassing a diverse array of materials, both synthesized and theoretical MOFs, approaching infinity. (b) The data retrieved on October 17, 2022, from the Web of Science presents the count of published papers featuring keywords (i) exclusively "MOF," (ii) "MOF" and "Simulation," (iii) "MOF" and "Machine Learning" in their titles and abstracts[62].

change, environmental pollution, energy efficiency, and drug therapy. Nevertheless, the vast number of conceivable MOFs exceeds practical limits for traditional trial-and-error experiments and brute-force molecular simulations within a reasonable timeframe. To navigate this complexity, traditional molecular simulation methods such as Grand Canonical Monte Carlo (GCMC), equilibrium molecular dynamics (EMD), and nonequilibrium molecular dynamics (NEMD) simulations have been utilized for calculating gas adsorption and separation in MOFs. Initially applied to a limited set of MOFs, recent studies have expanded these approaches to encompass a much larger number, thanks to high-throughput computational screening (HTCS) methods[61]. HTCS efficiently evaluates various structures for specific applications, identifying promising materials to guide experimental endeavors. Nevertheless, the rapid expansion of metal-organic frameworks (MOFs) renders simulating each one impractical. To enhance cost-effective research, imperative data-driven approaches reduce reliance on extensive simulations. The incorporation of data science marks a paradigm shift, employing machine learning (ML) to predict properties and cluster datasets. The growth of ML in MOF research is notable, evident in publications rising from one in 2013–2015 to nearly 100 in 2021–2022 (as illustrated in Figure 5.10(b)). This review offers a comprehensive perspective, initiating with the chronological framework for computational modeling of MOFs, progressing from early molecular simulations to HTCS approaches and ML-guided modeling. Recent innovative studies are explored, contributing to material design through the definition of new representations, the creation of diverse databases, the application of novel learning methods, and the construction of ML workflows. Challenges and opportunities in AI-aided MOF research are discussed, underscoring the potential for hastening the discovery of novel, high-performance materials that address critical societal issues[62].

In recent decades, significant advancements have occurred in both experimental and computational approaches to MOFs. Figure 5.11 illustrates key developments, starting from early experiments and molecular simulations to HTCS and the

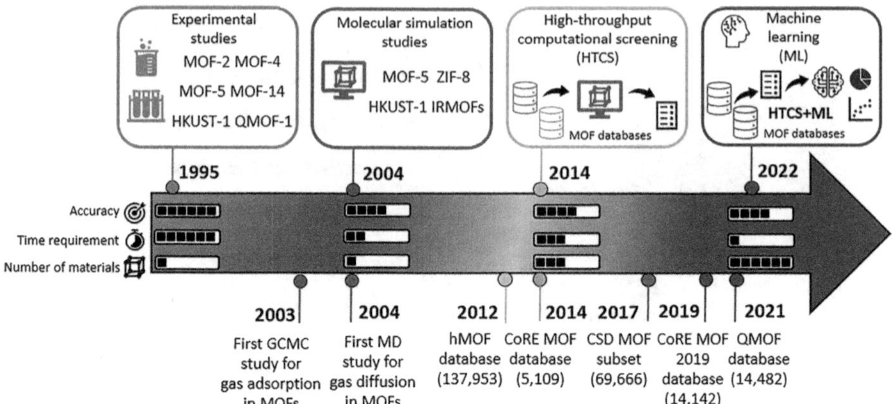

FIGURE 5.11 Timeline of Computational MOF Research: Black boxes indicate method accuracy, time needed for target properties, and materials studied (experiments, simulations, HTCS, HTCS + ML)[62].

integration of machine learning (ML)[62]. The late 1990s saw a surge in MOF research, driven by breakthroughs in synthesis and confirmation of their permanent porosity. Highly porous MOFs like HKUST-1 and MOF-177 demonstrated exceptional gas uptake capacities, surpassing traditional materials. Computational methods, including DFT, have become crucial for assessing MOFs' catalytic activities. The integration of data science, such as the Computation-Ready and Experimental MOF (CoRE MOF) database and high-throughput computational screening has identified promising MOFs with outstanding adsorption and selectivity properties. This multidisciplinary approach has led to the discovery of novel, high-performing MOFs with applications in gas storage, separation, catalysis, drug delivery etc[62].

5.5 CONCLUSION

Theoretical modeling, from electronic structure calculations to machine learning, enables the prediction and design of coordination materials with tailored properties. The synergy between theory and experimentation accelerates the development of innovative materials, advancing science and technology. Our journey through theoretical modeling, simulations, and computational methods has significantly advanced our understanding of coordination materials. From unraveling the principles of density functional theory (DFT) to exploring the intricate world of metal-organic frameworks (MOFs) and beyond, we've addressed critical gaps in knowledge. This book chapter not only shed light on the nuanced relationship between water and coordination-materials based gas separation but also offered valuable insights into optimizing materials for hydrogen storage and clean energy technologies. The integration of accurate force fields derived from ab initio calculations in our simulations showcased the precision of our predictions, particularly in the context of hydrogen storage. Our exploration of various strategies for enhancing hydrogen storage capacity highlighted promising pathways, aligning with DOE targets and advancing the field. Moreover, this book chapter underscored the pivotal role of computational techniques, including simulations and machine learning, in propelling coordination chemistry, gas adsorption, and materials science into the future. As these methodologies evolve, their applications hold the promise of catalyzing groundbreaking advancements to address pressing global challenges. In essence, our collective exploration across diverse topics has not only enriched our understanding of the molecular world but also opened doors to addressing critical environmental and energy challenges. The evolving landscape of simulations and machine learning, coupled with the promising applications of coordination materials, positions computational techniques as indispensable tools for the future of materials science and its contributions to global advancements.

ACKNOWLEDGMENTS

This work was supported by Prince Sultan University, and the author would like to express gratitude for their support.

REFERENCES

[1] Shah, R.; Ali, S.; Raziq, F.; Ali, S.; Ismail, P. M.; Shah, S.; Iqbal, R.; Wu, X.; He, W.; Zu, X.; Zada, A.; Adnan; Mabood, F.; Vinu, A.; Jhung, S. H.; Yi, J.; Qiao, L., Exploration of metal organic frameworks and covalent organic frameworks for energy-related applications. *Coordination Chemistry Reviews* **2023,** *477*, 214968.

[2] Iqbal, R.; Ali, S.; Saleem, A.; Majeed, M. K.; Hussain, A.; Rauf, S.; Rehman Akbar, A.; Xu, H.; Qiao, L.; Zhao, W., Electrically conductive Pt-MOFs for acidic oxygen reduction: Optimized performance via altering conjugated ligands. *Chemical Engineering Journal* **2023**, *455*, 140799.

[3] Shah, R.; Ali, S.; Ali, S.; Xia, P.; Raziq, F.; Adnan; Mabood, F.; Shah, S.; Zada, A.; Ismail, P. M.; Hayat, A.; Rehman, A. U.; Wu, X.; Xiao, H.; Zu, X.; Li, S.; Qiao, L., Amino functionalized metal-organic framework/rGO composite electrode for flexible Li-ion batteries. *Journal of Alloys and Compounds* **2023,** *936*, 168183.

[4] Ali, S.; Ismail, P. M.; Wahid, F.; Kumar, A.; Haneef, M.; Raziq, F.; Ali, S.; Javed, M.; Khan, R. U.; Wu, X.; Xiao, H.; Yasin, G.; Qiao, L.; Xu, H., Benchmarking the two-dimensional conductive $Y_3(C_6X_6)_2$ (Y = Co, Cu, Pd, Pt; X = NH, NHS, S) metal-organic framework nanosheets for CO2 reduction reaction with tunable performance. *Fuel Processing Technology* **2022,** *236*, 107427.

[5] Iqbal, R.; Akbar, M. B.; Ahmad, A.; Hussain, A.; Altaf, N.; Ibraheem, S.; Yasin, G.; Khan, M. A.; Tabish, M.; Kumar, A.; Majeed, M. K.; Saleem, A.; Ali, S., Exploring the synergistic effect of novel Ni-Fe in 2D bimetallic metal-organic frameworks for enhanced electrochemical reduction of CO2. *Advanced Materials Interfaces* **2022,** *9* (1), 2101505.

[6] Qadir, S.; Gu, Y.; Ali, S.; Li, D.; Zhao, S.; Wang, S.; Xu, H.; Wang, S., A thermally stable isoquinoline based ultra-microporous metal-organic framework for CH4 separation from coal mine methane. *Chemical Engineering Journal* **2022,** *428*, 131136.

[7] Zuhra, Z.; Ali, S.; Ali, S.; Xu, H.; Wu, R.; Tang, Y., Exceptionally amino-quantitated 3D MOF@CNT-sponge hybrid for efficient and selective recovery of Au(III) and Pd(II). *Chemical Engineering Journal* **2021**, 133367.

[8] Ali, S.; Yasin, G.; Iqbal, R.; Huang, X.; Su, J.; Ibraheem, S.; Zhang, Z.; Wu, X.; Wahid, F.; Ismail, P. M.; Qiao, L.; Xu, H., Porous Aza-doped graphene-analogous 2D material a unique catalyst for CO_2 conversion to formic-acid by hydrogenation and electroreduction approaches. *Molecular Catalysis* **2022,** *524*, 112285.

[9] Ali, S.; Iqbal, R.; Wahid, F.; Ismail, P. M.; Saleem, A.; Ali, S.; Raziq, F.; Ullah, S.; Ullah, I.; Tahir; Zahoor, M.; Wu, X.; Xiao, H.; Zu, X.; Qiao, L., Cobalt coordinated two-dimensional covalent organic framework a sustainable and robust electrocatalyst for selective CO2 electrochemical conversion to formic acid. *Fuel Processing Technology* **2022,** *237*, 107451.

[10] Yasin, G.; Ali, S.; Ibraheem, S.; Kumar, A.; Tabish, M.; Mushtaq, M. A.; Ajmal, S.; Arif, M.; Khan, M. A.; Saad, A.; Qiao, L.; Zhao, W., Simultaneously engineering the synergistic-effects and coordination-environment of dual-single-atomic iron/cobalt-sites as a bifunctional oxygen electrocatalyst for rechargeable zinc-air batteries. *ACS Catalysis* **2023**, 2313–2325.

[11] Yaghi, O. M.; Li, G.; Li, H., Selective binding and removal of guests in a microporous metal–organic framework. *Nature* **1995,** *378* (6558), 703–706.

[12] Li, Y.; Karimi, M.; Gong, Y.-N.; Dai, N.; Safarifard, V.; Jiang, H.-L., Integration of metal-organic frameworks and covalent organic frameworks: Design, synthesis, and applications. *Matter* **2021,** *4* (7), 2230–2265.

[13] Chai, J.-D.; Head-Gordon, M., Long-range corrected double-hybrid density functionals. *The Journal of Chemical Physics* **2009,** *131* (17).

[14] Grimme, S.; Antony, J.; Ehrlich, S.; Krieg, H., A consistent and accurate ab initio parametrization of density functional dispersion correction (DFT-D) for the 94 elements H-Pu. *The Journal of Chemical Physics* **2010,** *132* (15), 154104.

[15] Jose, R.; Bangar, G.; Pal, S.; Rajaraman, G., Role of molecular modelling in the development of metal-organic framework for gas adsorption applications. *Journal of Chemical Sciences* **2023,** *135* (2), 19.

[16] Ali, S.; Zuhra, Z.; Ali, S.; Han, Q.; Ahmad, M.; Wang, Z., Ultra-deep removal of Pb by functionality tuned UiO-66 framework: A combined experimental, theoretical and HSAB approach. *Chemosphere* **2021,** *284*, 131305.

[17] Bao, L.; Ren, X.; Liu, C.; Liu, X.; Dai, C.; Yang, Y.; Bououdina, M.; Ali, S.; Zeng, C., Modulating the doping state of transition metal ions in ZnS for enhanced photocatalytic activity. *Chemical Communications* **2023,** *59*, 11280–11283.

[18] Ali, S.; Liu, T.; Lian, Z.; Li, B.; Su, D. S., The tunable effect of nitrogen and boron dopants on a single walled carbon nanotube support on the catalytic properties of a single gold atom catalyst: A first principles study of CO oxidation. *Journal of Materials Chemistry A* **2017,** *5* (32), 16653–16662.

[19] Ali, S.; Fu Liu, T.; Lian, Z.; Li, B.; Sheng Su, D., The effect of defects on the catalytic activity of single Au atom supported carbon nanotubes and reaction mechanism for CO oxidation. *Physical Chemistry Chemical Physics* **2017,** *19* (33), 22344–22354.

[20] Liu, T.; Ali, S.; Lian, Z.; Li, B.; Su, D. S., CO_2 electoreduction reaction on heteroatom-doped carbon cathode materials. *Journal of Materials Chemistry A* **2017,** *5* (41), 21596–21603.

[21] Ali, S.; Olanrele, S.; Liu, T.; Lian, Z.; Si, C.; Yang, M.; Li, B., Single Au anion can catalyze acetylene hydrochlorination: Tunable catalytic performance from rational doping. *The Journal of Physical Chemistry C* **2019,** *123* (48), 29203–29208.

[22] Ali, S.; Qiu, Y.; Lian, Z.; Olanrele, S.; Lan, G.; Li, Y.; Su, D. S.; Li, B., Screening of active center and reactivity descriptor in acetylene hydrochlorination on metal-free doped carbon catalysts from first principle calculations. *Applied Surface Science* **2019,** *478*, 574–580.

[23] Qiu, Y.; Ali, S.; Lan, G.; Tong, H.; Fan, J.; Liu, H.; Li, B.; Han, W.; Tang, H.; Liu, H.; Li, Y., Defect-rich activated carbons as active and stable metal-free catalyst for acetylene hydrochlorination. *Carbon* **2019,** *146*, 406–412.

[24] Ali, S.; Iqbal, R.; Khan, A.; Rehman, S. U.; Haneef, M.; Yin, L., Stability and catalytic performance of single-atom catalysts supported on doped and defective graphene for CO_2 hydrogenation to formic acid: A first-principles study. *ACS Applied Nano Materials* **2021,** *4* (7), 6893–6902.

[25] Ali, S.; Haneef, M.; Akbar, J.; Ullah, I.; Ullah, S.; Samad, A., Single Au atom supported defect mediated boron nitride monolayer as an efficient catalyst for acetylene hydrochlorination: A first principles study. *Molecular Catalysis* **2021,** *511*, 111753.
[26] Ali, S.; Xie, Z.; Xu, H., Stability and catalytic performance of single-atom supported on Ti2CO2 for low-temperature CO oxidation: A first-principles study. *ChemPhysChem* **2021,** *22*, 2352–2361.
[27] Muhammad Ismail, P.; Ali, S.; Raziq, F.; Bououdina, M.; Abu-Farsakh, H.; Xia, P.; Wu, X.; Xiao, H.; Ali, S.; Qiao, L., Stable and robust single transition-metal atom catalyst for CO2 reduction supported on defective WS2. *Applied Surface Science* **2023**, 157073.
[28] Ismail, P. M.; Ali, S.; Ali, S.; Li, J.; Liu, M.; Yan, D.; Raziq, F.; Wahid, F.; Li, G.; Yuan, S.; Wu, X.; Yi, J.; Chen, J. S.; Wang, Q.; Zhong, L.; Yang, Y.; Xia, P.; Qiao, L., Photoelectron "bridge" in van der Waals heterojunction for enhanced photocatalytic CO2 conversion under visible light. *Advanced Materials* **2023,** *35* (38), 2303047.
[29] Wu, L.; Li, Y.; Zhou, B.; Liu, J.; Cheng, D.; Guo, S.; Xu, K.; Yuan, C.; Wang, M.; Hong Melvin, G. J.; Ortiz-Medina, J.; Ali, S.; Yang, T.; Kim, Y. A.; Wang, Z., Vertical graphene on rice-husk-derived SiC/C composite for highly selective photocatalytic CO2 reduction into CO. *Carbon* **2023,** *207*, 36–48.
[30] Sun, M.; Ali, S.; Liu, C.; Dai, C.; Liu, X.; Zeng, C., Synergistic effect of Fe doping and oxygen vacancy in AgIO3 for effectively degrading organic pollutants under natural sunlight. *Environmental Pollution* **2024,** *344*, 123325.
[31] Rana, M. K.; Koh, H. S.; Hwang, J.; Siegel, D. J., Comparing van der Waals density functionals for CO2 adsorption in metal organic frameworks. *The Journal of Physical Chemistry C* **2012,** *116* (32), 16957–16968.
[32] Ali, S.; Ismail, P. M.; Humayun. M.; Bououdina, M.; Qiao, L., Tailoring 2D metal-organic frameworks for enhanced CO_2 reduction efficiency through modulating conjugated ligands. *Fuel Processing Technology* **2024,** *225*, 108049.
[33] Wu, P.; Li, Y.; Zheng, J.-J.; Hosono, N.; Otake, K.-I.; Wang, J.; Liu, Y.; Xia, L.; Jiang, M.; Sakaki, S.; Kitagawa, S., Carbon dioxide capture and efficient fixation in a dynamic porous coordination polymer. *Nature Communications* **2019,** *10* (1), 4362.
[34] Qazvini, O. T.; Babarao, R.; Telfer, S. G., Selective capture of carbon dioxide from hydrocarbons using a metal-organic framework. *Nature Communications* **2021,** *12* (1), 197.
[35] Goings, J. J.; Ohlsen, S. M.; Blaisdell, K. M.; Schofield, D. P., Sorption of H2 to open metal sites in a metal–organic framework: A symmetry-adapted perturbation theory analysis. *The Journal of Physical Chemistry A* **2014,** *118* (35), 7411–7417.
[36] Kumar, R. M.; Subramanian, V., Interaction of H2 with fragments of MOF-5 and its implications for the design and development of new MOFs: A computational study. *International Journal of Hydrogen Energy* **2011,** *36* (17), 10737–10747.
[37] Kancharlapalli, S.; Gopalan, A.; Haranczyk, M.; Snurr, R. Q., Fast and accurate machine learning strategy for calculating partial atomic charges in metal–organic frameworks. *Journal of Chemical Theory and Computation* **2021,** *17* (5), 3052–3064.
[38] Zou, X.; Cha, M.-H.; Kim, S.; Nguyen, M. C.; Zhou, G.; Duan, W.; Ihm, J., Hydrogen storage in Ca-decorated, B-substituted metal organic framework. *International Journal of Hydrogen Energy* **2010,** *35* (1), 198–203.
[39] Dixit, M.; Adit Maark, T.; Ghatak, K.; Ahuja, R.; Pal, S., Scandium-decorated MOF-5 as potential candidates for room-temperature hydrogen storage: A solution for the clustering problem in MOFs. *The Journal of Physical Chemistry C* **2012,** *116* (33), 17336–17342.
[40] Asgari, M.; Semino, R.; Schouwink, P.; Kochetygov, I.; Trukhina, O.; Tarver, J. D.; Bulut, S.; Yang, S.; Brown, C. M.; Ceriotti, M., An in-situ neutron diffraction and DFT study of hydrogen adsorption in a sodalite-type metal–organic framework, Cu-BTTri. *European Journal of Inorganic Chemistry* **2019,** *2019* (8), 1147–1154.

[41] Hübner, O.; Glöss, A.; Fichtner, M.; Klopper, W., On the interaction of dihydrogen with aromatic systems. *The Journal of Physical Chemistry A* **2004,** *108* (15), 3019–3023.
[42] Yaghi, O. M.; O'Keeffe, M.; Ockwig, N. W.; Chae, H. K.; Eddaoudi, M.; Kim, J., Reticular synthesis and the design of new materials. *Nature* **2003,** *423* (6941), 705–714.
[43] Sagara, T.; Klassen, J.; Ganz, E., Computational study of hydrogen binding by metal-organic framework-5. *The Journal of Chemical Physics* **2004,** *121* (24), 12543–12547.
[44] Hamel, S.; Côté, M., First-principles study of the rotational transitions of H2 physisorbed over benzene. *The Journal of Chemical Physics* **2004,** *121* (24), 12618–12625.
[45] Eddaoudi, M.; Kim, J.; Rosi, N.; Vodak, D.; Wachter, J.; O'Keeffe, M.; Yaghi, O. M., Systematic design of pore size and functionality in isoreticular MOFs and their application in methane storage. *Science* **2002,** *295* (5554), 469–472.
[46] Han, S. S.; Mendoza-Cortés, J. L.; Goddard Iii, W. A., Recent advances on simulation and theory of hydrogen storage in metal–organic frameworks and covalent organic frameworks. *Chemical Society Reviews* **2009,** *38* (5), 1460–1476.
[47] Mulder, F.; Dingemans, T.; Schimmel, H.; Ramirez-Cuesta, A.; Kearley, G., Hydrogen adsorption strength and sites in the metal organic framework MOF5: Comparing experiment and model calculations. *Chemical Physics* **2008,** *351* (1–3), 72–76.
[48] Mueller, T.; Ceder, G., A density functional theory study of hydrogen adsorption in MOF-5. *The Journal of Physical Chemistry B* **2005,** *109* (38), 17974–17983.
[49] Rowsell, J. L.; Spencer, E. C.; Eckert, J.; Howard, J. A.; Yaghi, O. M., Gas adsorption sites in a large-pore metal-organic framework. *Science* **2005,** *309* (5739), 1350–1354.
[50] Rowsell, J. L.; Eckert, J.; Yaghi, O. M., Characterization of H2 binding sites in prototypical metal–organic frameworks by inelastic neutron scattering. *Journal of the American Chemical Society* **2005,** *127* (42), 14904–14910.
[51] Van Mourik, T.; Gdanitz, R. J., A critical note on density functional theory studies on rare-gas dimers. *The Journal of Chemical Physics* **2002,** *116* (22), 9620–9623.
[52] Parkes, M. V.; Sava Gallis, D. F.; Greathouse, J. A.; Nenoff, T. M., Effect of metal in M3 $(btc)_2$ and M2 (dobdc) MOFs for O2/N2 separations: a combined density functional theory and experimental study. *The Journal of Physical Chemistry C* **2015,** *119* (12), 6556–6567.
[53] Xie, L.; Liu, S.; Gao, C.; Cao, R.; Cao, J.; Sun, C.; Su, Z., Mixed-valence Iron (II, III) trimesates with open frameworks modulated by solvents. *Inorganic Chemistry* **2007,** *46* (19), 7782–7788.
[54] Chui, S. S.-Y.; Lo, S. M.-F.; Charmant, J. P.; Orpen, A. G.; Williams, I. D., A chemically functionalizable nanoporous material $[Cu_3(TMA)_2\ (H2O)_3]_n$. *Science* **1999,** *283* (5405), 1148–1150.
[55] Bloch, E. D.; Queen, W. L.; Krishna, R.; Zadrozny, J. M.; Brown, C. M.; Long, J. R., Hydrocarbon separations in a metal-organic framework with open iron (II) coordination sites. *Science* **2012,** *335* (6076), 1606–1610.
[56] Dietzel, P. D.; Morita, Y.; Blom, R.; Fjellvåg, H., An in situ high-temperature single-crystal investigation of a dehydrated metal–organic framework compound and field-induced magnetization of one-dimensional metal–oxygen chains. *Angewandte Chemie* **2005,** *117* (39), 6512–6516.
[57] Roger, I.; Shipman, M. A.; Symes, M. D., Earth-abundant catalysts for electrochemical and photoelectrochemical water splitting. *Nature Reviews Chemistry* **2017,** *1* (1), 0003.
[58] Wang, W.; Xu, X.; Zhou, W.; Shao, Z., Recent progress in metal-organic frameworks for applications in electrocatalytic and photocatalytic water splitting. *Advanced Science* **2017,** *4* (4), 1600371.
[59] Li, X.; Yu, J.; Low, J.; Fang, Y.; Xiao, J.; Chen, X., Engineering heterogeneous semiconductors for solar water splitting. *Journal of Materials Chemistry A* **2015,** *3* (6), 2485–2534.

[60] Furukawa, H.; Ko, N.; Go, Y. B.; Aratani, N.; Choi, S. B.; Choi, E.; Yazaydin, A. Ö.; Snurr, R. Q.; O'Keeffe, M.; Kim, J., Ultrahigh porosity in metal-organic frameworks. *Science* **2010,** *329* (5990), 424–428.
[61] Himanen, L.; Geurts, A.; Foster, A. S.; Rinke, P., Data-driven materials science: status, challenges, and perspectives. *Advanced Science* **2019,** *6* (21), 1900808.
[62] Demir, H.; Daglar, H.; Gulbalkan, H. C.; Aksu, G. O.; Keskin, S., Recent advances in computational modeling of MOFs: From molecular simulations to machine learning. *Coordination Chemistry Reviews* **2023,** *484*, 215112.

Part II

Multifunctional Coordination Materials for Energy Storage Technologies

6 Coordination Nanomaterials for Energy Storage

An Introduction

Safana Haqani, Mohammad Tabish, Saira Ajmal and Ghulam Yasin

6.1 INTRODUCTION

Concerns over electrical energy generation and consumption have been raised because of the growing global population and the global energy crisis. Furthermore, air pollution, water contamination, and an aging population issue are continuously increasing, causing a boost in the popularity of wind and solar energy. Therefore, an electrochemical energy storage system (EES) is highly demanded [1]. EES technologies, such as batteries and supercapacitors with high energy densities and power densities, respectively, have already made their mark in the commercial market. In these technologies, lithium-ion batteries (LIBs) and sodium-ion batteries (SIBs) are suitable examples in terms of portable electronic devices, electric vehicles, and large-scale energy storage systems. A supercapacitor also called a polymer electrochemical capacitor or an ultra-capacitor has a thin dielectric layer and high surface area electrodes [2] due to which supercapacitors have a tendency to store more electrical energy than traditional capacitors. On the other hand, batteries with higher energy storage capacity can store more energy per unit volume or weight by delivering a sustainable power output. The chemistry and specific surface area of an electrode material are critical factors that play a role in the storage of energy. Coordination nanomaterials, however, have seen progress in this scenario. Kinetics and rates are not the only factors that determine surface impacts; they include the energy and chemistry of coordination nanomaterials. These factors exert important or notable effects on the heterogeneous reaction thermodynamics at interfaces, as well as on the processes of nucleation and growth during phase transitions [3]. Coordination nanomaterials have found their application in batteries and supercapacitors due to their tremendous electrochemical performance.

In order to improve energy storage devices, the development of coordination nanomaterials such as porous coordination polymers (PCPs), covalent organic frameworks (COFs), and metal-organic frameworks (MOFs) play a vital role by offering many advantages in energy storage applications. These materials equipped with pores

DOI: 10.1201/9781003345886-8

have numerous properties depending on their arrangement, structure, porosity, size, and composition, and these properties vary with their use in different electrochemical applications [4]. The electrochemical properties of MOFs and COF derivatives, including transition metal oxides, sulfides, and phosphides, make them superior in energy storage. These materials are composed of metal ions or clusters that are coordinated with organic molecules to form porous crystalline MOFs [5]. PCPs or MOFs are promising materials for creating nanopores with customizable dimensions and interior surfaces. PCPs offer superior molecular-level control over their internal pore structure, unlike zeolites and mesoporous silica [6]. The PCP is designed with a highly regular channel structure, a channel size that matches molecular dimensions, and a surface that can be customized. Furthermore, they have responsive frameworks that adapt to the molecules of guests. Consequently, these properties enable PCPs to be used for energy storage [7]. Aside from their unique properties, nanomaterials are also a cost-effective alternative to electrochemical devices because of their unique characteristics.

This chapter is intended to provide information about the most important aspects and latest developments in coordination nanomaterials for energy storage applications. In addition, due to the nano-scaled size of coordination nanomaterials, they are also capable of forming architectures with significantly larger internal surfaces. There are several applications for coordination nanomaterials, including sodium-ion batteries, lithium-ion batteries, lithium-oxygen batteries, lithium-sulfur batteries, and supercapacitors. This chapter will discuss PCP/MOFs/COFs-based materials for anodes, cathodes, electrolytes separators, interlayers, and their current challenges and future prospects. Its aim is to illuminate the multiple roles of porous-based materials in EES and to pave the way for future developments.

6.2 RISE OF COORDINATION NANOMATERIALS

Technological developments and materials that are capable of storing clean and sustainable energy are in high demand, therefore manufacturing large-scale energy storage devices with nanotechnology has become a revolutionary trend [8]. Nanomaterials involve measuring, regulating substances, and manipulating on a nanoscale level, employing scientific skills from a range of industrial applications. As the fundamental component of nanotechnology, nanoparticles are nanoscale objects with three external dimensions, so nanomaterials are materials that contain nanoscale structures on their surface or internally [9]. Coordination nanomaterials possess unique physical and chemical properties making them promising candidates for revolutionizing energy storage. Coordination of nanomaterials and their potential to transform energy storage systems stand out among these emerging technologies [10].

These types of nanomaterials have emerged in several stages during the rise of storage applications. (a) As coordination nanomaterials began to emerge, researchers identified their advanced properties, such as high porosity to synthesize them, including porous coordination polymers (PCPs), metal-organic frameworks (MOFs) and covalent organic frameworks (COFs). (b) Storage of energy is essential for the transition to renewable energy sources and for maintaining grid stability. These nanomaterials played an important role in improving the energy density, rate of charge/discharge, and lifespan of conventional batteries and supercapacitors [11]. (c) Metallic ions or clusters with organic ligands in coordination nanomaterials present new possibilities for energy storage. Metal clusters or ions coordinated with organic ligands make coordination nanomaterials an exciting new energy storage technology. (d) Coordination nanomaterials are characterized by their remarkable tunability. These materials can be fine-tuned using specific metal ions and ligands [12]. Having control over properties such as capacity, charge/discharge rates, and cycle life is beneficial for energy storage. A sustainable and greener future may be unlocked by coordination nanomaterials with their tunable properties and transformative potential.

6.3 COORDINATION NANOMATERIALS FOR ENERGY STORAGE

6.3.1 Coordination Nanomaterials for Rechargeable Batteries

Coordination nanomaterials have emerged as a transformative technology in the field of rechargeable batteries potentially offering a sustainable and energy-efficient answer to the ever-increasing demand for energy storage [13]. Rechargeable batteries can be enhanced by using these innovative materials, characterized by their precise atomic structures and nanoscale dimensions [14]. As a result of their unique properties, coordination nanomaterials can significantly improve battery capacity, cycle life, and energy storage efficiency, whether they are MOFs, COFs, PCPs, or molecular clusters [15]. In PCPs, MOFs, and COFs organic linkers are combined with metal ions to form organic nanoporous materials. Developing coordination nanomaterials using nanotechnology is an explosion to exciting new possibilities in the engineering and design of rechargeable batteries, making them a focus of research and innovation for greener and more reliable energy storage [16]. Recently, due to their potential for improving energy storage, nanomaterial-based rechargeable batteries have gained attention. A few types of rechargeable batteries that incorporate coordination nanomaterials for energy storage purposes are lithium-ion batteries (LIBs), lithium-sulfur batteries (LSBs), solid-state batteries (SSBs) and sodium-ion batteries (SIBs).

6.3.1.1 MOFs-Based Coordination Nanomaterials for Batteries

Considering their high energy density, low self-discharge rate, portability, and compact size, lithium-ion batteries (LIBs) are a vital energy storage device for electric vehicles, portable electronics, and renewable power stations [17]. To improve LIB performance, it is necessary to find electrode materials with high surface areas and corresponding energy levels [18]. As electrode materials for lithium-ion batteries, MOFs have attracted considerable interest due to their ability to be tailored based on organic ligands [19]. Pyridine-based MOFs have been known to provide a capacity of 200 mA h g^{-1}, but there are few active ligand studies on MOFs in this case [20]. At 100 cycles, Ni-Me_4bpz-based MOFs with flexible bipyrazole ligands demonstrated a stable 120 mA h g^{-1} capacity, establishing the precedent for N-donor ligands in MOFs. MOFs with N-donor ligands have been developed into LIBs anodes as a result of these pioneering studies. As a lithium storage electrode material, MOFs hold a great promise, since Febri Baskoro and colleagues [21] have demonstrated a stable capacity of 200 mA h g^{-1} over 1,000 cycles. A diagram showing the proposed lithiation/delithiation mechanism is shown in Figure 6.1a These findings suggest that dual-ligand MOFs could provide long-term lithium storage that is reversible and efficient. Using the Co-MOF-D@Si@C structure, Zhilin Yan et al. [22] have found a capacity of 1,493 mAh.g^{-1} and a specific capacity of 957 mAh.g^{-1}. Furthermore, after 1,200 cycles, the structure maintained its capacity at 648 mAh.g^{-1}, showing the applicability of Si anodes for electrochemical energy storage applications [23]. Despite poor initial efficiency, MOF-177 can act as a LIBs anode material providing lithium intercalation and deintercalation sites and comparable redox activity [24]. Using MOFs, homogeneous TMOs formed within a carbon content matrix maintain a constant volume and improve the electrical conductivity of LIBs [25].

Anode materials for sodium-ion batteries (SIBs) can be built with MOFs with longer linear ligands. Furthermore, SIBs are more durable than LIBs because of the porous carbon microstructure provided by MOF. By using reduced graphene oxide and C_{60}, soft-template methods have been developed to synthesize hierarchical porous carbon. Recently, a graphitization composite combining polyacrylonitrile fibers with zinc and cobalt acetate has been developed by Goodenough et al. [26] as shown in Figure 6.1b, this material has a good rate capability and long-lasting cycling performance. Mai et al. [29] developed novel nano architectures based on CoS nanoparticles embedded in porous N-doped carbon to demonstrate their potential for SIBs. With sulfurization, coating, cation exchange, and annealing, Yin et al. [30] produced ZnS-Sb_2S_3@C core-double shell structures. In 120 cycles, the composites showed an excellent sodium storage property with 630mAh g^{-1} reversible capacity.

Due to their superior performance, LSBs are becoming an increasingly popular energy storage technology [31]. MOFs are highly porous materials that are gaining attention for their controllable structure and potential for use in LSBs in the future. Based on a unique GNS network structure, Zhao et al. [32] developed graphene sheets for (GNS)-MIL-101(Cr)/S cathodes, enhancing composite stability over conventional MIL-101(Cr)/S cathodes. Using ZIF-8 nanosheets, Jiang et al. [27] created ZIF-8-NS-C, a sulfur immobilizer for LSBs. It was observed that the composites had been impregnated with sulfur and that sulfur, carbon, and nitrogen were distributed uniformly within the pores. As shown in Figure 6.1c, after 100 cycles, 65.2% of the material retained its initial capacity. The Co-N-C hybrid nanostructure, which Zhao et al. [33] used to generate high-performance batteries with sulfur cathodes, demonstrated its potential over traditional LSBs.

A solid-state battery (SSB) with a metallic anode is commonly used because of its high energy density and safety. Solid-state electrolyte (SSE) with excellent mechanical strength and non-flammability reduces short circuit risks and suppresses dendrite growth as in Figure 6.1d. A solid-like electrolyte based on Li-IL@MOF has been developed by Ziqi Wang et al. [28] that is compatible with $LiFePO_4$ cathodes and anodes, making it a potent application to a variety of battery systems [34]. Through modifying the composition of ionic liquid, the research presents a promising approach for high-energy-density SSBs. From modified MOF-808, Pan and co-workers reported [35] a new solid-state electrolyte containing single-ion Zn^{2+} (WZM SSE). This material offers high conductivity, low activation energy, and mechanical strength, suitable for Zn batteries and SSE based on MOF modifications.

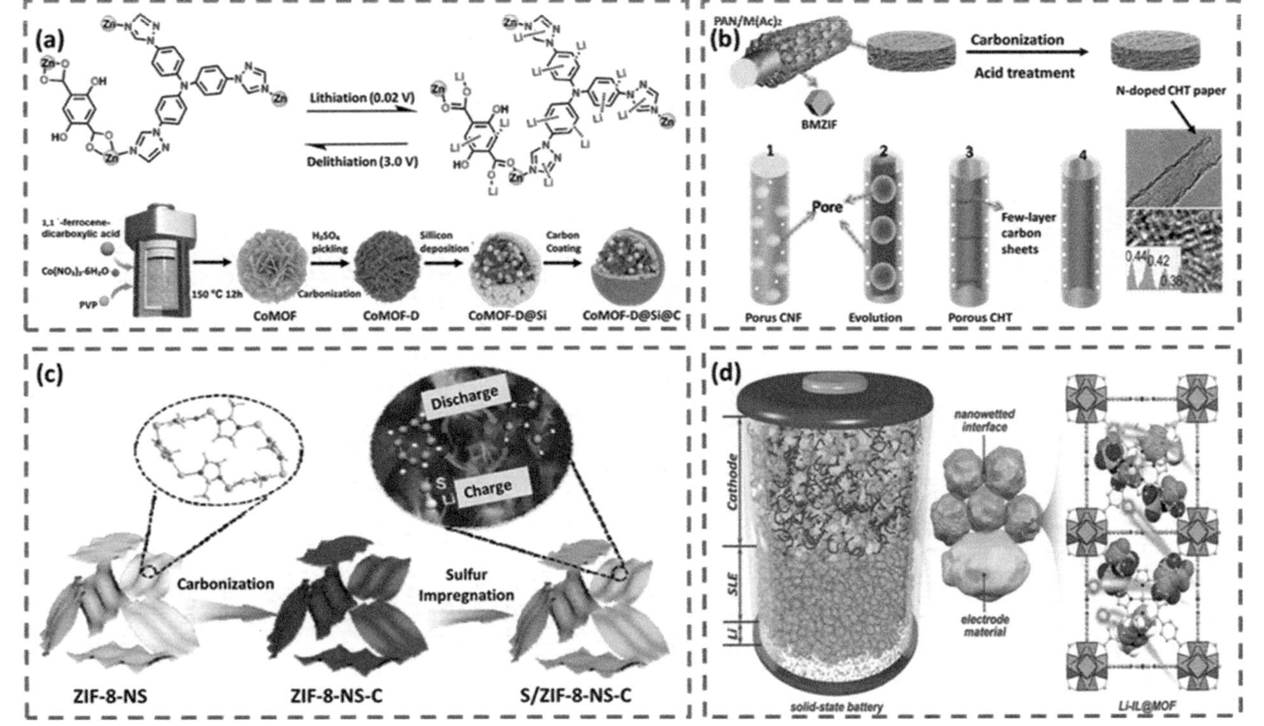

FIGURE 6.1 (a) At the charge (lithiation) and discharge (delithiation) stages in a Zn-MOF@180 electrode, schematic of Li storage. Reproduced with permission from reference [22] copyright claim from Elsevier, Composite CoMOF-D@Si@C synthesis schematic. Reproduced with permission from reference [24] copyright claim from Elsevier. (b) A diagram depicting the manufacture of porous carbon hollow tubules. Reproduced with permission from reference [26] copyright claim from cell press. (c) Preparation of S/ZIF-8-NS-C composites and S/ZIF-8-P-C cell cycling efficiencies at 0.2ºC. Reproduced with permission from reference [27] copyright claim from ACS Publications. (d) This diagram shows how a solid-state battery is put together, namely its interfacial mechanism, the crystal structures of a MOF are revealed, revealing a random distribution of [TFSI] and $[EMIM]^+$ ions within the pores. To better see the movement of the Li^+ ions, the hydrogen atoms have been left out of the picture and replaced by luminous pink spheres. Reproduced with permission from reference [28] copyright claim from Willey online library.

6.3.1.2 COFs-Based Coordination Nanomaterials for Batteries

The anode and cathode of a rechargeable battery transfer metal ions during charging and discharging. Due to their good conductivity and hierarchy of intercalation structures, carbon materials are commonly used as anodes in LIBs [36]. It is generally acknowledged that graphite anodes have a low specific capacity and safety concerns due to which alternative anodes (such as nickel, silver, aluminum, gallium, and magnesium) and transition metals are being studied.

The porous and layered structures enable COFs as potential anode materials in LIBs. By designing COFs, mass transport, and electrocatalysis are utilized through materials with tunable pore sizes, defined chemical structures, and geometries [37]. By incorporating metals into carbon nanotubes (COFs), nitrogen-containing ligands serve as single-atom catalysts. In addition to hybrid materials based on COF mass transport, electrocatalysis, and energy storage technologies are rapidly evolving. Two conjugated COFs have been studied by Zhao et al. [38] as LIBs anodes, and the N_2-COF and N_3-COF showed high charge capacity and retention after 500 cycles. In Figure 6.2a, carbon-based conductive materials such as CNTs and graphene demonstrated high reversible capacity and good retention [39].

Because of their low price and low cycle stability, sodium-ion batteries (SIBs) are gaining attention nowadays. COF materials are primarily used for two-dimensional materials in practical applications. Recent progress has been made on SIB anodes by 3D COFs. The three-dimensional COF material that has a C_3-C_3 symmetric topology and can store Na+ ions reversibly was synthesized by Patri et al. [41] as shown in Figure 6.2c.

Potassium-ion batteries (PIBs) can replace LIBs as they offer abundant resources, higher voltage plateaus, high energy density, and better conductivity. According to Wolfson et al. [40] alkynyl functional groups promote potassium binding enthalpically and geometrically, resulting in high reversible capacities through alkynyl functional groups. Using two-dimensional phthalocyanines, Yang et al. [43] produced an aromatic-conjugated COF (FAC-Pc-COF) with excellent stability, high conductivity, and high K^+ storage capacity. In order to develop SSBs, researchers are looking for materials that can act as electrolytes specifically COFs. Yan Yang et al. [42] used a Celgard separator to address Li_2S_6 dissolution and migration issues in LSBs. A high-power application such as an electric vehicle requires a high conductivity, electrochemical stability, and transference number as shown in Figure 6.2d.

6.3.1.3 PCPs-Based Coordination Nanomaterials for Batteries

Ascribing to the low solubility in organic solvents, 3D amorphous structure and good reversibility, PCPs are ideal electrode materials for LIBs. Using the 2,3,5,6-tetraphthalimido-1,4-benzoquinone TPB organic electrode, Zhiqiang Luo et al. [44] discovered that it can discharge 223.2 mAh per gram over a several-hour period Figure 6.3a. As reported by Cai et al. [45] in MOFs, Lewis acid metal sites are combined with Lewis base ligands to protect lithium polysulfides. The MOF-74 structure swapped out sixteen different types of metal sites so that Siegel et al. [46] can study the electrostatic interaction between metal ions and MOFs. It has been shown in Figure 6.3b that ligands with aromatic rings absorb sulfur and facilitate the transfer of charge.

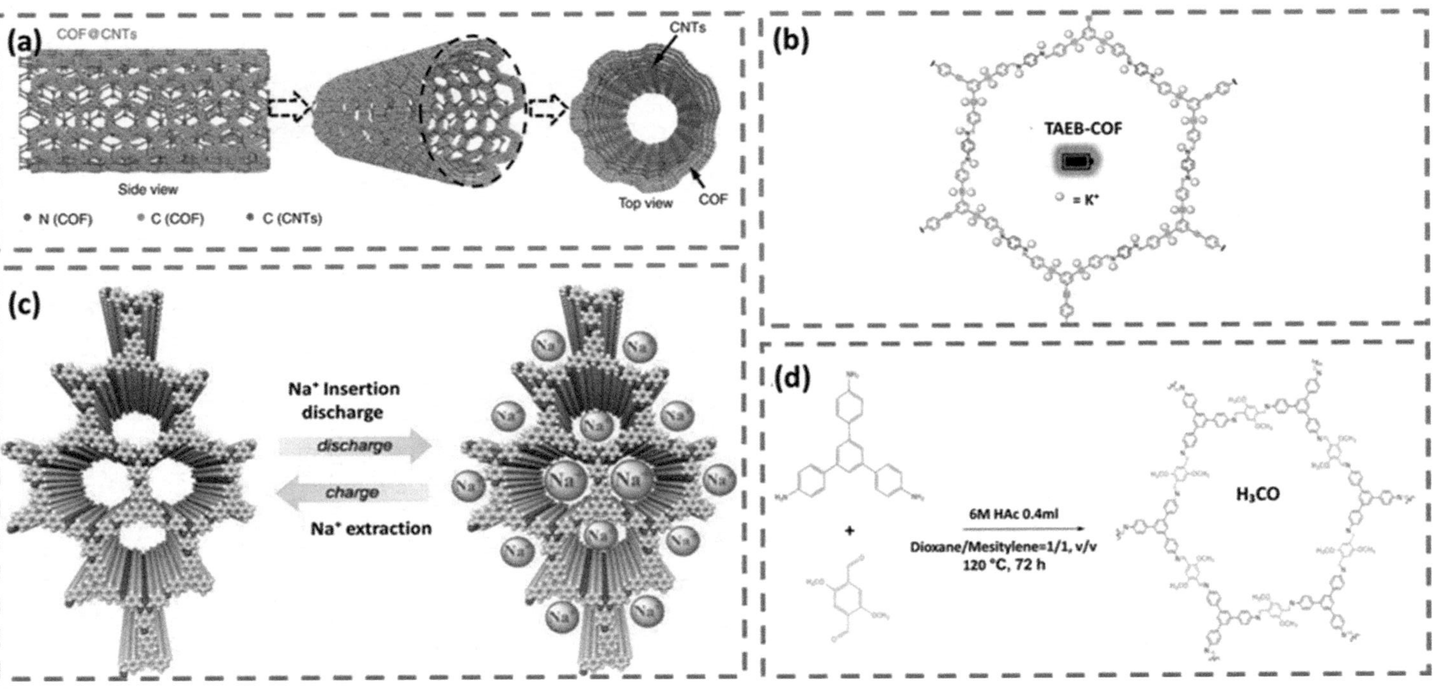

FIGURE 6.2 (a) Constructing COFs and COF@CNTs, building COFs. Reproduced with permission from reference [39] copyright claim from Nature Communications. (b) Scheme of synthesis for TAEB-COF. Reproduced with permission from reference [40] copyright claim from ACS publications. (c) Cycle Model for Reversible Storage of Na^+ Ions in COF. Reproduced with permission from reference [41] copyright from The Royal Society of Chemistry. (d) The diagram demonstrates how the bifunctional TPB-DMTP-COF separator coating blocks polysulfide transport to build a high-performance lithium-selenium sulfide battery. Reproduced with permission from reference [42] copyright claim from ACS publications.

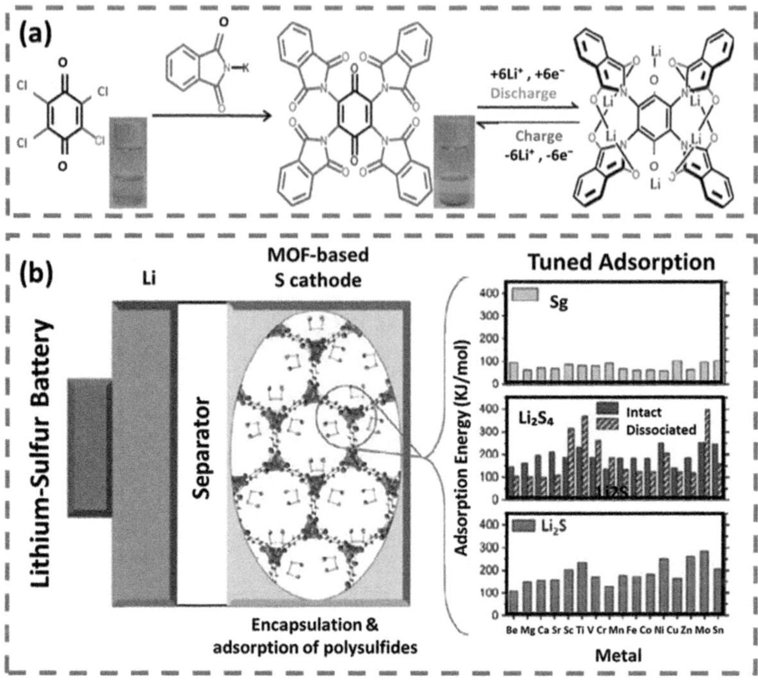

FIGURE 6.3 (a) The TPB/Li_6TPB has a reversible electrochemical redox mechanism with a predicted specific capacity of 233 mAh g^{-1} and a MW of 689 g mol^{-1}. A computer image demonstrates that TPB is insoluble and TCB is extremely soluble. Reproduced with permission from reference [44] copyright claim from Wiley online library. (b) Adsorption energies of sulfur on various metal sites. Reproduced with permission from reference [46] copyright claim from ACS publications.

6.3.2 Coordination Nanomaterials for Supercapacitors

In sustainable nanotechnology, energy is increasingly stored and converted by supercapacitors; however, graphite electrodes have insufficient specific capacitances for that purpose. Energy storage is made more convenient by using supercapacitors, due to their short storage period, large heat range, ease of packaging, low weight, and low maintenance cost. Supercapacitors differ from batteries in that they store charge in double electric layers created at the electrode-electrolyte interface. A supercapacitor is used in technologies such as industrial power management, consumer electronics, and memory backup systems. A high-performance supercapacitor's electrode fabrication requires materials such as MOFs, COFs, and PCPs with large surface areas, temperature stability, corrosion resistance, and high conductivity.

6.3.2.1 MOFs-Based Coordination Nanomaterials for Supercapacitors

The use of MOFs and hybridized nanostructures in supercapacitors has recently been explored. Figure 6.4a shows a high energy density and specific capacitance from Salunkhe et al.'s study [47] [32] converted ZIF-8 to NPC due to its high

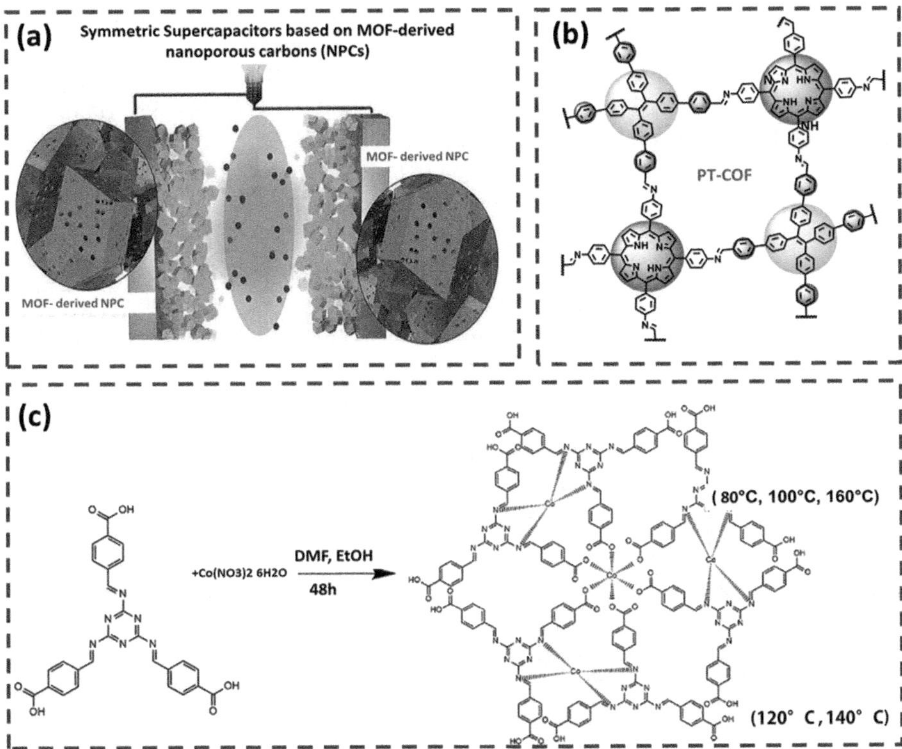

FIGURE 6.4 (a) An example of a supercapacitor using nanoporous carbons (NPCs) generated from MOF. Without the need for separators, symmetric supercapacitors are constructed with two electrodes of NPCs made from MOFs with a comparable charge capacity. Reproduced with permission from reference [47] copyright claim from royal society of chemistry. (b) Alternative path to PT-COF. Reproduced with permission from reference [52] copyright claim from ACS publication. (c) Synthesis of schematic representation for the Co-PCP. Reproduced with permission from reference [53] copyright claim from Elesvier.

specific capacitance and energy density. From ZIF-7 MOF, Cao et al [48] produced nanoporous carbon (NPC); the specific capacitances of the battery are 228 F/g and 178 F/g, respectively, and they have a 94% cycling capacity. MOFs combined with polyoxometalates (POM) create supercapacitor electrodes, with specific capacitances of 248 F/g and 85.2% cycle life [49]. Xiao-Feng Guo and colleagues used in-situ chemical oxidative polymerization to synthesize carbonized MOF/PANI composites [50]. The carbonized Zn-MOF/PANI composite demonstrated a high specific capacitance of 477 F g^{-1} at 1 A g^{-1} as an electrode for SCs. In order to make supercapacitors more efficient, Ehsani et al. [51] developed a nanocomposite based on p-type conductive polymers Cu-MOF to create a novel electrode.

6.3.2.2 COF-Based Coordination Nanomaterials for Supercapacitors

Covalent organic frameworks (COFs), crystalline porous materials with covalent bonds have been extensively studied in terms of synthetic strategies and molecular design. Due to their large surface area, porosity, ability to be modified, conjugation, sufficient building blocks, and low density, these materials are promising [54]. In addition to building extended framework structures, linking atoms in two and three dimensions showed the possibility of application in the future [55]. Haldar and colleagues [56] developed solid-state supercapacitors using pyridyl hydroxyl functionalized carbon nanotubes (COFs). In a study highlighting COFs for rapid power delivery, Schiff-bonded COFs achieved excellent specific capacitance.

The tetraphenyl-ethylene moiety of the porphyrin-based COF was synthesized by Patra and Santanu Bhattacharya [52]. Ions are transported and stored in the COF by porphyrin rings. It exhibits excellent energy storage properties, high capacitance, and high capacity retention. Its structural periodicity retention post-GCD analysis as shown in Figure 6.4b illustrates the material's robustness. Furthermore, PTCOF may be an attractive material for future applications in energy storage.

6.3.2.3 PCPs-Based Coordination Nanomaterials for Supercapacitors

As promising electrodes in supercapacitors, porous coordination polymers (PCPs) display large surface areas, tunable pore sizes, crystalline ordered structures, and different functional groups. An electrode material made of a PCP with a large specific surface area and high nitrogen content enhances the electrical performance of supercapacitors.

A Co-PCP electrode material with high nitrogen content was developed by Li Wang et al. [53] to achieve 512 F g^{-1} specific capacitance in a 3M KOH electrolyte. As a result of their large specific surface area and their interconnected pores, electrodes can be used as energy storage devices Figure 6.4c. As electrode materials for supercapacitors, Meng-Ke Wu, et al [40, 57] used Cu-Asp nanowires and porous CuO nanotubes. Due to their large surface area, nanowires showed high capacitance and retention, but their performance was limited by a lack of electrical conductivity. Future studies of transition metal ions, organic ligands, and electrolytes will use PCP nanowires as self-sacrificial templates [58].

6.4 SUMMARY AND PROSPECTS

The purpose of this chapter is to summarize recent developments in the field of coordination materials (MOFs, COFs, PCPs) as electrode materials for batteries and supercapacitors. With respect to energy storage, especially for batteries and supercapacitors, coordination nanomaterials have gained significant attention. Energy storage technologies can be improved by using these materials because they have unique structural properties and tunable chemical compositions. Coordination nanomaterials such as metal organic frameworks are highly porous materials containing metal ions or clusters linked by organic compounds. Their large specific surface area and adjustable pore size enable them to accommodate a variety of guest molecules. Because of their good electrical conductivity, high surface area, and potential to accommodate different ions and molecules, a new generation of MOF materials has shown promise as electrode materials for batteries and supercapacitors. However, their low stability in aqueous environments and limited electrical conductivity should be considered. Another type of coordination nanomaterial is COFs, which consist of organic building blocks linked by covalent bonds. They are highly porous and maintain a tunable chemical structure, which makes them suitable for the storage of a wide variety of energy. Considering their superior stability in aqueous environments and their ability to be functionalized, they are promising for long-term, high-performance energy storage. Coordination nanomaterials with porous structures include porous coordination polymers. Last, In PCPs, metal ions or clusters are combined with organic ligands to create materials with high surface areas and variable pore sizes. Their applications include gas storage, catalysis, and electrochemical energy storage.

Batteries and supercapacitors offer several promising prospects using coordination nanomaterials: (a) Due to their large specific surface areas, these materials can enhance the capacity of batteries and supercapacitors by providing high energy density. (b) A MOF, COF, or PCP can be tailored to optimize its performance for a specific energy storage application by modifying its chemical composition and structure. (c) A key consideration for energy storage devices is that COFs are highly stable in aqueous environments. (d) Developing more eco-friendly energy storage technologies often involves the synthesis of coordination nanomaterials. (e) Energy storage and catalysis can both be performed simultaneously with these materials, enhancing their efficiency.

Planning strategies and preparation methods for coordination nanomaterial composites could be a promising approach in order to develop advanced materials. The enhancement in the electrical conductivity and stability of MOF in water, furthermore, demonstrates that scaling up the production of these materials for commercial applications is still a necessary approach. To gain insights into the utilization of these coordination nanomaterials in energy storage applications is more important now than ever.

ACKNOWLEDGMENTS

This work was supported by the Research Fund for International Scientists (RFIS-Grant number: 52150410410) National Natural Science Foundation of China.

REFERENCES

[1] Yasin, G., et al., Chapter 1 – Introduction to electrochemical energy storage technologies, in *Lithium-Sulfur Batteries*, R.K. Gupta, et al., Editors. 2022, Elsevier. p. 3–10.

[2] Tanwar, S. and A.L. Sharma, Insight into use of biopolymer in hybrid electrode materials for supercapacitor applications – A critical review. *Journal of Applied Physics*, 2023. **133**(18): p. 180701.

[3] Haridas, R., et al., Indoor light-harvesting dye-sensitized solar cells surpassing 30% efficiency without co-sensitizers. *Materials Advances*, 2021. **2**(23): p. 7773–7787.

[4] Zhang, P., J. Zhang, and S. Dai, Mesoporous carbon materials with functional compositions. *Chemistry – A European Journal*, 2017. **23**(9): p. 1986–1998.

[5] Cao, X., et al., Hybrid micro-/nano-structures derived from metal–organic frameworks: preparation and applications in energy storage and conversion. *Chemical Society Reviews*, 2017. **46**(10): p. 2660–2677.

[6] Yasin, G., et al., Simultaneously engineering the synergistic-effects and coordination-environment of dual-single-atomic iron/cobalt-sites as a bifunctional oxygen electrocatalyst for rechargeable zinc-air batteries. *ACS Catalysis*, 2023. **13**(4): p. 2313–2325.

[7] Kim, C.R., T. Uemura, and S. Kitagawa, Inorganic nanoparticles in porous coordination polymers. *Chemical Society Reviews*, 2016. **45**(14): p. 3828–3845.
[8] Yasin, G., et al., Self-templating synthesis of heteroatom-doped large-scalable carbon anodes for high-performance lithium-ion batteries. *Inorganic Chemistry Frontiers*, 2022. **9**(6): p. 1058–1069.
[9] Jeevanandam, J., et al., Review on nanoparticles and nanostructured materials: History, sources, toxicity, and regulations. *Beilstein Journal of Nanotechnology*, 2018. **9**: p. 1050–1074.
[10] Yasin, G., et al., Chapter 8 – Advanced carbon nanomaterial–based anodes for sodium-ion batteries, in *Advanced Nanomaterials and Their Applications in Renewable Energy* (Second Edition), J.L. Liu, T.-H. Yan, and S. Bashir, Editors. 2022, Elsevier. p. 251–272.
[11] Asif, H.M., et al., Chapter 12 – Polyoxometalate-based metal organic frameworks (POMOFs) for lithium-ion batteries, in *Metal-Organic Framework-Based Nanomaterials for Energy Conversion and Storage*, R.K. Gupta, T.A. Nguyen, and G. Yasin, Editors. 2022, Elsevier. p. 245–268.
[12] Yasin, G., et al., Chapter 1 – MOF-based nanostructures and nanomaterials for next-generation energy storage: An introduction, in *Metal-Organic Framework-Based Nanomaterials for Energy Conversion and Storage*, R.K. Gupta, T.A. Nguyen, and G. Yasin, Editors. 2022, Elsevier. p. 3–10.
[13] Yasin, G., et al., Defective/graphitic synergy in a heteroatom-interlinked-triggered metal-free electrocatalyst for high-performance rechargeable zinc–air batteries. *Journal of Materials Chemistry A*, 2021. **9**(34): p. 18222–18230.
[14] Matsuyama, K., Supercritical fluid processing for metal–organic frameworks, porous coordination polymers, and covalent organic frameworks. *The Journal of Supercritical Fluids*, 2018. **134**: p. 197–203.
[15] Yasin, G., et al., Chapter 9 – Nanostructured anode materials in rechargeable batteries, in *Nanobatteries and Nanogenerators*, H. Song, et al., Editors. 2021, Elsevier. p. 187–219.
[16] Yasin, G., et al., Chapter 11 – Nanostructured cathode materials in rechargeable batteries, in *Nanobatteries and Nanogenerators*, H. Song, et al., Editors. 2021, Elsevier. p. 293–319.
[17] Wang, M., et al., A high-performance tin phosphide/carbon composite anode for lithium-ion batteries. *Dalton Transactions*, 2020. **49**(46): p. 17026–17032.
[18] Baskoro, F., et al., Dual-ligand Zn-based metal–organic framework as reversible and stable anode material for next generation lithium-ion batteries. *Energy Technology*, 2021. **9**(11): p. 2100212.
[19] Fernández de Luis, R., et al., Electrochemical behavior of [{Mn(Bpy)}(VO3)2]≈(H2O)1.24 and [{Mn(Bpy)0.5}(VO3)2]≈(H2O)0.62 inorganic–organic Brannerites in lithium and sodium cells. *Journal of Solid State Chemistry*, 2014. **212**: p. 92–98.
[20] An, T., et al., A flexible ligand-based wavy layered metal–organic framework for lithium-ion storage. *Journal of Colloid and Interface Science*, 2015. **445**: p. 320–325.
[21] Sharma, N., et al., Dual-ligand Fe-metal organic framework based robust high capacity Li ion battery anode and its use in a flexible battery format for electro-thermal heating. *ACS Applied Energy Materials*, 2019. **2**(6): p. 4450–4457.
[22] Yan, Z., et al., Metal-organic frameworks-derived CoMOF-D@Si@C core-shell structure for high-performance lithium-ion battery anode. *Electrochimica Acta*, 2021. **390**: p. 138814.
[23] Muhammad, N., et al., Volumetric buffering of manganese dioxide nanotubes by employing 'as is' graphene oxide: An approach towards stable metal oxide anode material in lithium-ion batteries. *Journal of Alloys and Compounds*, 2020. **842**: p. 155803.

[24] Li, T., et al., Advances in transition-metal (Zn, Mn, Cu)-based MOFs and their derivatives for anode of lithium-ion batteries. *Coordination Chemistry Reviews*, 2020. **410**: p. 213221.
[25] Ullah, S., et al., Construction of well-designed 1D selenium–tellurium nanorods anchored on graphene sheets as a high storage capacity anode material for lithium-ion batteries. *Inorganic Chemistry Frontiers*, 2020. **7**(8): p. 1750–1761.
[26] Chen, Y., et al., Nitrogen-doped carbon for sodium-ion battery anode by self-etching and graphitization of bimetallic MOF-based composite. *Chem*, 2017. **3**(1): p. 152–163.
[27] Jiang, Y., et al., Monoclinic ZIF-8 nanosheet-derived 2D carbon nanosheets as sulfur immobilizer for high-performance lithium sulfur batteries. *ACS Applied Materials & Interfaces*, 2017. **9**(30): p. 25239–25249.
[28] Wang, Z., et al., A metal–organic-framework-based electrolyte with nanowetted interfaces for high-energy-density solid-state lithium battery. *Advanced Materials*, 2018. **30**(2): p. 1704436.
[29] Zhou, L., et al., Structural and chemical synergistic effect of CoS nanoparticles and porous carbon nanorods for high-performance sodium storage. *Nano Energy*, 2017. **35**: p. 281–289.
[30] Dong, S., et al., ZnS-Sb2S3@C core-double shell polyhedron structure derived from metal–organic framework as anodes for high performance sodium ion batteries. *ACS Nano*, 2017. **11**(6): p. 6474–6482.
[31] Yasin, G., et al., Understanding and suppression strategies toward stable Li metal anode for safe lithium batteries. *Energy Storage Materials*, 2020. **25**: p. 644–678.
[32] Zhao, Z., et al., Graphene-wrapped chromium-MOF(MIL-101)/sulfur composite for performance improvement of high-rate rechargeable Li–S batteries. *Journal of Materials Chemistry A*, 2014. **2**(33): p. 13509–13512.
[33] Zhao, J., et al., In-situ catalytic growth carbon nanotubes from metal organic frameworks for high performance lithium-sulfur batteries. *Materials Today Energy*, 2018. **8**: p. 134–142.
[34] Wang, H., et al., Reviewing the current status and development of polymer electrolytes for solid-state lithium batteries. *Energy Storage Materials*, 2020. **33**: p. 188–215.
[35] Wang, Z., et al., A MOF-based single-ion Zn2+ solid electrolyte leading to dendrite-free rechargeable Zn batteries. *Nano Energy*, 2019. **56**: p. 92–99.
[36] Yasin, G., et al., A novel strategy for the synthesis of hard carbon spheres encapsulated with graphene networks as a low-cost and large-scalable anode material for fast sodium storage with an ultralong cycle life. *Inorganic Chemistry Frontiers*, 2020. **7**(2): p. 402–410.
[37] Yasin, G., et al., Facile and large-scalable synthesis of low cost hard carbon anode for sodium-ion batteries. *Results in Physics*, 2019. **14**: p. 102404.
[38] Bai, L., Q. Gao, and Y. Zhao, Two fully conjugated covalent organic frameworks as anode materials for lithium ion batteries. *Journal of Materials Chemistry A*, 2016. **4**(37): p. 14106–14110.
[39] Lei, Z., et al., Boosting lithium storage in covalent organic framework via activation of 14-electron redox chemistry. *Nature Communications*, 2018. **9**.
[40] Wolfson, E.R., et al., Alkynyl-based covalent organic frameworks as high-performance anode materials for potassium-ion batteries. *ACS Applied Materials & Interfaces*, 2021. **13**(35): p. 41628–41636.
[41] Patra, B.C., et al., Covalent organic framework based microspheres as an anode material for rechargeable sodium batteries. *Journal of Materials Chemistry A*, 2018. **6**(34): p. 16655–16663.
[42] Wang, Z., W. Zheng, W. Sun, L. Zhao, and W. Yuan, Covalent organic frameworks-enhanced ionic conductivity of polymeric ionic liquid-based ionic gel electrolyte for lithium metal battery. *ACS Applied Energy Materials*, 2021. **4**: p. 2808–2819.

[43] Yang, X., et al., Ionothermal synthesis of fully conjugated covalent organic frameworks for high-capacity and ultrastable potassium-ion batteries. *Advanced Materials*, 2022. **34**(50): p. 2207245.
[44] Luo, Z., et al., An insoluble benzoquinone-based organic cathode for use in rechargeable lithium-ion batteries. *Angewandte Chemie International Edition*, 2017. **56**(41): p. 12561–12565.
[45] Hong, X.-J., et al., Confinement of polysulfides within bi-functional metal–organic frameworks for high performance lithium–sulfur batteries. *Nanoscale*, 2018. **10**(6): p. 2774–2780.
[46] Park, H. and D.J. Siegel, Tuning the adsorption of polysulfides in lithium–sulfur batteries with metal–organic frameworks. *Chemistry of Materials*, 2017. **29**(11): p. 4932–4939.
[47] Salunkhe, R.R., et al., Fabrication of symmetric supercapacitors based on MOF-derived nanoporous carbons. *Journal of Materials Chemistry A*, 2014. **2**(46): p. 19848–19854.
[48] Zhang, P., et al., ZIF-derived porous carbon: A promising supercapacitor electrode material. *Journal of Materials Chemistry A*, 2014. **2**(32): p. 12873–12880.
[49] Gao, Y., et al., Synthesis of nickel oxalate/zeolitic imidazolate framework-67 (NiC2O4/ZIF-67) as a supercapacitor electrode. *New Journal of Chemistry*, 2015. **39**(1): p. 94–97.
[50] Guo, S., et al., (Metal-organic framework)-polyaniline sandwich structure composites as novel hybrid electrode materials for high-performance supercapacitor. *Journal of Power Sources*, 2016. **316**: p. 176–182.
[51] Ehsani, A., et al., Nanocomposite of p-type conductive polymer/Cu (II)-based metal-organic frameworks as a novel and hybrid electrode material for highly capacitive pseudocapacitors. *Ionics*, 2017. **23**(1): p. 131–138.
[52] Patra, B.C. and S. Bhattacharya, New covalent organic square lattice based on porphyrin and tetraphenyl ethylene building blocks toward high-performance supercapacitive energy storage. *Chemistry of Materials*, 2021. **33**(21): p. 8512–8523.
[53] Wang, L., et al., Superstable porous co-coordination polymer as the electrode material for supercapacitor. *Journal of Solid State Chemistry*, 2019. **277**: p. 630–635.
[54] Kumar, A., et al., Chapter 21 – Battery-supercapacitor hybrid systems: An introduction, in *Nanotechnology in the Automotive Industry*, H. Song, et al., Editors. 2022, Elsevier. p. 453–458.
[55] Mehtab, T., et al., Metal-organic frameworks for energy storage devices: Batteries and supercapacitors. *Journal of Energy Storage*, 2019. **21**: p. 632–646.
[56] Haldar, S., et al., Pyridine-rich covalent organic frameworks as high-performance solid-state supercapacitors. *ACS Materials Letters*, 2019. **1**(4): p. 490–497.
[57] Wu, M.-K., et al., High-performance supercapacitors of Cu-based porous coordination polymer nanowires and the derived porous CuO nanotubes. *Dalton Transactions*, 2017. **46**(48): p. 16821–16827.
[58] Ali, H.H., et al., Rationally designed Mo-based advanced nanostructured materials for energy storage technologies: Advances and prospects. *Sustainable Materials and Technologies*, 2023. **38**: p. e00738.

7 Coordination Nanostructures for Metal-Ion Batteries

Ghulam Yasin, Mahnoor Ahmed, Mohammad Tabish, Muhammad Arif, Anuj Kumar, Tuan Anh Nguyen and Saira Ajmal

7.1 INTRODUCTION

In the last few years, the increasing demand for energy around the world has led scientists to develop technologies for the generation of green energy and sustainable storage from renewable energy resources. Renewable and nonrenewable are the two main energy resources that are classified based on their inherent characteristics and intrinsic natures. For instance, fossil fuels like coal and natural gas, regarded as nonrenewable energy resources, exist around the globe to a limited extent that cannot be replenished in a short time [1]. In this regard geothermal, wind, biothermal, solar, and tidal are some of renewable energy resources. Electrochemical energy storage (EES) technologies are important tools to store the obtained energy from renewable energy sources [2]. The lithium-ion battery (LIB) in terms of EES has paid a large interest and commercial success because of its long service life, high power density, and energy density [3]. However, extensive research has been stimulated on new batteries due to the urgent scarcity of lithium resources, increasing lithium compounds cost, and the emerging safety issues that have restricted its application on large-scale storage systems [4]. As an alternate to LIBs, an ideal energy storage and conversion system is metal-ion batteries (MIBs) because of their lower cost and plentiful resources of zinc, magnesium, aluminum, potassium, and sodium for these battery systems [5]. The electrode material is the pivotal component of MIBs. The wider applications of MIBs in smart renewable energy grids and electric vehicles are restricted due to the slow charge/discharge rate, low power density, and short cycle life [6]. To achieve high-performance in MIBs, the key challenge is the design and development of electrode materials with tunable nanostructures [7]. In recent years, MIBs have shown a good rate of performance, high reversible capacity, and ultralong cycle life by using metal sulfides and advanced tunable porous carbon frameworks bridged with metal oxides as anode and cathode materials [8]. The commercial utilization of these electrode materials is restricted due to low cycling performance and complicated synthesis methods [9]. Therefore electrode materials with easy synthesis methods and high-performance are needed [10]. In the typical energy storage and

DOI: 10.1201/9781003345886-9

conversion technologies, porous coordination polymers (PCPs), covalent-organic frameworks (COFs), and metal-organic frameworks (MOFs) are essential classes of coordination nanomaterials. First, as a significant class of nanoporous materials, PCPs have emerged owing to their tunable pore surface, structure with high preparation and designability, and large surface area with potential applications in energy conversion and storage [11]. Second, MOFs are an important family of coordination nanomaterials formed via coordination bonds by self-assembling metal-containing organic linkers and nodes. Because of their porous structure, large surface area, and wide pore volume, they have been extensively utilized in the field of energy storage and conversion, catalysis, gas adsorption, and drug delivery applications [12]. Metal-containing nodes combined with organic linkers enable MOFs to have a wide range of structural possibilities. As a result, MOFs have even more intriguing properties due to the controlled connectivity of the vertices within a given structure [13]. Third, another important and dynamic member of porous organic materials is the COF which is constructed using reticular chemistry and is connected with covalent bonds. The fascinating characteristics of COFs, like their chemical stability, structural versatility, large specific surface area, and tunable surface, have led to their widespread application in gas storage, sensing, adsorption, and catalysis as well as in energy storage and conversion systems [14].

On the basis of their compositional and structural characteristics PCPs, COFs, and MOFs have undergone notable progress in the realm of energy storage and conversion, specifically in their capacity as electrode materials. This comprehensive examination of coordination nanomaterials in the context of MIBs has unveiled a pathway leading to more sustainable and efficient solutions for energy storage. Encompassing a spectrum of battery families, from lithium-ion batteries (LIBs), potassium-ion batteries (PIBs), magnesium-ion batteries (MIBs), sodium-ion batteries (SIBs), aluminum-ion batteries (AIBs), and zinc-ion batteries (ZIBs), coordination nanomaterials have exhibited considerable potential in shaping the future landscape of energy storage. To conclude this chapter, we eagerly anticipate innovative and exploratory endeavors poised to establish an era characterized by cleaner, more potent, and environmentally friendly energy technologies.

7.2 METAL-ION BATTERIES (LITHIUM-ION, SODIUM-ION, POTASSIUM-ION, ZINC-ION, MAGNESIUM-ION, AND ALUMINUM-ION BATTERIES)

The prominent energy storage systems, because of their potentially long cycle life and high energy densities for utilization in various gadgets and today's automobiles, are named rocking chair batteries (RCBs). During operation, between the (positive and negative electrodes) the charge carriers shuttle back and forth which are separated via electrolyte and the separator is the rocking mechanism in RCBs. Currently, RCBs based on alkali-ions have been extensively studied by researchers worldwide. These alkali metal ions are called metal-ion batteries (MIBs) [15]. MIBs feature (positive and negative electrodes) separated by an electrolyte that is ionically conductive and usually consists of a liquid solution of alkali metal salts. Negative electrodes (termed anodes) contain extractable metal cations that can be added into positive electrodes (termed cathodes) upon discharge [16]. An electric current is induced due to the difference in electrochemical potentials between these electrodes. The metal cations from the cathode migrate back to the anode when the electric current is applied in the opposite direction and resultantly the energy is stored. There are 3 types of metal-ion batteries including monovalent cation batteries (Li^+, Na^+, and K^+), divalent cation batteries (Mg^{+2}, Zn^{+2}), and trivalent cation batteries (Al^{+3}). Safety might be increased, expenses might be reduced, and materials, as well as metals, might be used without resource limitations [17].

LIBs have played an essential role in the transformation of microelectronics and have appeared as the preferred power source for electronic gadgets. Their dominance in the portable electronics sector stems from the high gravimetric and large volumetric energy densities that they provide in comparison to other rechargeable systems. Beyond the realm of portable electronics, LIBs have begun to enter the electric vehicle market and are actively being pursued for applications in grid energy storage [18]. By employing in-situ polymerization on graphene, Chen et al. [19] developed a 2D COF material by using poly(imide-benzoquinone) at active sites and denoted as PIBN-G. The as-obtained PIBN-G material was utilized as a cathode for LIBs and showed a good rate of performance due to the high availability of carbonyl groups for electrons and Li^+ ions. Aside from the aforementioned uses, LIBs also have some drawbacks, like their high price, inadequate safety, shortage of resources, and energy density. The unresolved issues have driven scientists to evolve non-LIBs with high-performance and sustainability [20].

It is noteworthy that sodium (Na) is characterized by plentiful resources and widespread geographical distribution, and SIBs are like LIBs in terms of storing high capacities of electricity. Due to less Lewis acidity in sodium ions, the solvated ions are smaller than Li^+ ions [21]. Desolvation energy lower than that of the electrolyte/electrode interface will facilitate ion diffusion [22]. The volumetric and gravimetric capacities of Na (1,129 mAh cm^{-3} and 1,165 mAh g^{-1}) are lower than that of Li. As an effective anode material for SIBs [23] Zhenan Bao and coworkers reported a high intrinsic conductive cobalt-based MOF. In both non-aqueous and aqueous media, the obtained cobalt-based hexaaminobenzene (Co-HAB) reported good chemical stability, high thermal stability, abundant porosity, and bulk conductivity. Some obvious differences were observed between these systems. Transport characteristics, the formation of interphase, and the stability of phase are affected due to larger Na^+ ions (1.02 Å). So, there is a need for scientists to study alternative storage devices.

Because of the relatively low reduction potential of potassium (K), cost-effective, and plentiful resources, PIBs are an alternative to SIBs and LIBs [24]. K^+/K redox couples have the lowest potential in organic electrolytes, and K^+ ions have a lower Lewis acidity than Na and Li ions, which results in smaller solvated ions. Also, K^+ ions have a low desolvation energy, resulting in faster diffusion through the electrolyte/electrode interface [25]. As anode material for PIB Qijiu Deng et al. [26] synthesized a macroporous Fe-based MOF (MOF-235). A good rate capability and reversible capacity were shown by the obtained material in PIB. There are still a number of challenges that slow down the spread of PIB systems. For PIBs, the large K^+ ion causes the volumetric changes throughout the charging and discharging processes to be more substantial than for different alkali MIBs, resulting in electrode structural instability. Therefore, the development of alternatives to PIBs is also urgently needed [27].

ZIBs containing a zinc-comprising electrolyte, zinc metal anode, and a cathode for storing Zn^{+2} ions are gaining interest due to the wide variety of electrolytes available, including aqueous as well as non-aqueous electrolytes. ZIBs have a high redox potential, which is hard to achieve with other portable ion batteries, enhanced safety and decreased toxicity, and higher volumetric densities than LIBs [28]. A high specific capacity as cathode for aqueous zinc ion batteries (AZIBs) was achieved by MOF-73, which was synthesized by using a manganese-based MOF reported by Gou et al. [29] and it showed a high specific capacity. Because ZIBs have a high freezing point, lower ionic conductivity, and less interfacial freezing of the water-based electrodes and electrolyte, they undergo slow migration of ions decreased charge transfer kinetics, and waning of capacity at low temperatures. Therefore these problems need to be addressed [30].

MIBs are promising candidates due to their inexpensive incorporation into electrode materials, the plentiful occurrence of magnesium (Mg) reserves, its high melting point, stability in the atmosphere, and ease of handling during synthesis [31]. In addition, MIBs will not develop adverse dendrites during the charging/discharging processes, which will improve security – together with a density of 1.74 g/cm^{-3} – and will decrease overall weight [32]. PSHATN poly(hexaazatrinaphthalene sulfide), which is a sulfur-linked porous polymer, was reported by Wang et al.[33] to exhibit a reversible capacity of 196 mA h g^{-1} in Mg electrolytes and among all polymer electrodes achieves material utilization of 98%. In the development of MIBs there are two major roadblocks. The cathode materials pose a major problem. The 2^+ charge of Mg^{+2} ions causes a strong polarization effect, which slows down the diffusion in the solid state through the cathode materials. Consequently, the power output is reduced and the reversible capacity is low. The advancement of MIBs is further hampered by the irreversible creation of passivating layers on Mg anode surface in most standard electrolytic solutions (e.g., $Mg(ClO_4)_2$, $Mg(PF_6)_2$, or $Mg(TFSI)_2$ salts in ether or carbonate solvents), which prevents the plating/stripping of Mg^{+2} from being reversible. Therefore, uncovering an alternative cation battery may help in prevailing against this obstacle [34].

Aluminum (Al), owing to being the third most plentiful element in the universe, easy handling, longer cyclic life, faster-charging efficiency, low reactivity and flammability, and more stable electrochemical performance, AIBs are the most studied trivalent cation MIBs, with more stable electrochemical performance [35]. Al atoms can transfer three electrons at a time; consequently, high energy density and excellent specific capacity can be obtained compared to Li [36]. In rechargeable aluminum ion batteries (RAIBs), 2D copper-based MOF is used as a cathode material reported by Guo et al. [37] with bipolar ligand. The obtained electrodes exhibit a maximum reversible capacity of 184 mA h g^{-1} at 50mA g^{-1}. AIBs also face some drawbacks of corrosion during the discharging and charging rates and development of a passivation layer (Al_2O_3), small ionic radius, and with three positive charges and an increased charge density, which considerably enlarge the desolvation energy barrier of Al^{3+} ion in the solution of electrolytes. These issues obstruct the widespread utilization of the AIBs and provide a new gap for researchers [5].

7.3 COORDINATION NANOMATERIALS FOR METAL-ION BATTERIES

The problem of energy storage can be resolved by searching on metal-ion batteries (MIBs) such as lithium-ion batteries (LIBs), sodium-ion batteries (SIBs), potassium-ion batteries (PIBs), zinc-ion batteries (ZIBs), magnesium-ion batteries (MIBs), and aluminum-ion batteries (AIBs). Although the MIBs have accomplished much progress, the performance of existing materials can be improved by implying coordination nanomaterials such as porous coordination polymers (PCPs), metal-organic frameworks (MOFs), and covalent-organic frameworks (COFs). We will discuss some of the coordination nanomaterials applications in the metal-ion batteries.

7.3.1 Coordination Nanomaterials for Lithium-Ion Batteries

The promising energy storage system with a longer storage cycle, lighter weight, and high specific capacity for widespread utilization is the LIBs. LIBs have been applied in digital cameras, mobile phones, and laptops. The electrode material is the key point of LIBs [38]. Different materials for LIBs have been explored. The physicochemical properties of LIBs are represented in Table 7.1. For instance, in the application field of LIBs, PCPs have attracted attention for their variety of architectures and interesting molecular topologies resulting from the self-assembly of organic and inorganic units using well-established principles of coordination chemistry [39]. A coordination polymer reported by Luo et al. [40] called Co-btca (H_4btca = 1,2,4,5-benzenetetracarboxylic acid) was applied as anode material in LIBs with high purity synthesized hydrothermal method. During charge/discharge processes, the availability of Li ions as well as a suitable contact area between electrolyte and active sites was advantageous due to the specific area of the electrode and the special porous layer structure of the electrode. An 801.3 mA h g^{-1} reversible capacity of Co-btca was observed after 50 cycles at 200 mA g^{-1} current density Figure 7.1a (i). A reversible capacity of 535.4 mA h g^{-1} was achieved by the Co-btca electrode at 1A g^{-1} rate after 400 cycles along with coulombic efficiency (CE) of about 100% exhibited perfect cyclability Figure 7.1a (ii). A high-capacity carbonyl compound of chloranilic acid

TABLE 7.1
LIBs Coordination Nanomaterials Physiochemistry

Coordination Nanomaterials	Morphology	Current Density (mA g^{-1})	Cycle Performance	Reversible Capacity (mA h g^{-1})	Cycles	Reference
Co-btca	Lamellar structures	100 mA g^{-1}	935.6 mA h g^{-1}	801.3 mA h g^{-1}	50	[40]
CuCA	Nanosheets	50 mA g^{-1}	165.5 mA hg^{-1}	297.0 mA hg^{-1}	50	[41]
$V(PO_3)_3$	Nanosphere structure	100 mAg^{-1}	280 mAhg^{-1}	475 mAhg^{-1}	100	[46]
N_2-COF	Ordered stacking channels	0.2–5 A g^{-1}	500 mA h g^{-1}	400 mAh g^{-1}	500	[44]

(CA) was selected as an organic ligand for the coordination of Cu^{2+} as a transition metal node by Luo et al. [41] synthesizing copper(II) chloranilate (CuCA) with π–d conjugation, monocrystalline nature, and layered structure. With quasi-solid-state electrolyte, CuCA exhibited a high discharge capacity of 297.0 mA h g^{-1} at 50 mA g^{-1}, good rate capability of 160.6 mA h g^{-1} at 1000 mA g^{-1}, cycling stability of 165.5 mA h g^{-1} at 500 mA g^{-1} after 50 cycles presenting fast Li^+ diffusion kinetics, organic and inorganic redox centers, and high electronic conductivity.

MOFs are attracting attention in LIBs and are formed by coordination bonds between metal as the center atom and ligand as organic matter [45]. In a study by Li et al. [46] vanadium (III) metaphosphate $V(PO_3)_3$ was successfully synthesized by adding red phosphorus and calcining the synthesized V-MOF. With a current density of 100 mA g^{-1}, $V(PO_3)_3$ demonstrated a high reversible capacity of 465 mA h g^{-1} and a high initial discharge specific capacity of 1,058 mA h g^{-1} Figure 7.1b (i). After 100 cycles the diameter of the semicircle in the high-frequency region of the $V(PO_3)_3$ electrode decreases representing faster charge transfer and lower impedance value after cycling was achieved for the $V(PO_3)_3$ electrode Figure 7.1b (ii). Consequently, $V(PO_3)_3$ has been found to be an electrochemically promising anode material. In a solvothermal process with octylamine as a trigger, Mingdeng Wei et al. [47] synthesized hierarchical Co-based MOF nanoflakes (H-Co-MOF). After 700 cycles, H-Co-MOF exhibited good cycling stability (retained 828 mA h g^{-1} at 2 A g^{-1}) and a high reversible capacity of 1,345 mA h g^{-1} at 0.1 A g^{-1} as an anode for LIBs. Due to uniform porosity, easy preparation, and smart structural assembly, two-dimensional layered materials such as COFs are an intriguing family and have attracted attention in LIBs. By using aldehyde and amine in a condensation reaction in a solvothermal environment, Bai et al.[44] synthesized the COF structures Figure 7.1c (i). As anodes material in LIBs, N_2-COFs after 500 cycles showed a high charge capacity (600 to 700 mAh g^{-1}) and an 82% capacity retention rate Figure 7.1c (ii). In energy storage, these COF materials showed good performance due to their conjugated structure and good stability. By using naphthalenetetracarboxylic diimide with 2D-COF represented as NTCDI-COF comprising high stability, rich redox active sites, and crystallinity, Luo et al. [48] reported a cathode material for LIBs. After 1,500 cycles at 2 A g^{-1}, the NTCDI-COF demonstrated a high capacity retention of 125 mA h g^{-1}, along with a good discharge capacity of 210 mA h g^{-1} at 0.1 A g-1.

Coordination nanomaterials with porous architectures are well-suited to LIBs. Even though coordination nanomaterials can be used in LIBs, they have several limitations. The majority of them rely too heavily on unstable and inefficient COFs, MOFs, and PCPs. Additionally, more research needs to be conducted on promising coordination nanomaterial catalysts that meet the aforementioned requirements.

7.3.2 Coordination Nanomaterials for Sodium-Ion Batteries

Owing to the low prices of sodium element and abundant resources, SIBs are considered as the advanced power sources for widespread energy storage and conversion [49]. In SIB, capacity and rate performance decrease due to sluggish reaction kinetics, large volume changes in electrodes, and slow ion transport caused by the large radius of Na^+ ions. To solve this problem, PCP emerged as a new frontier in the energy

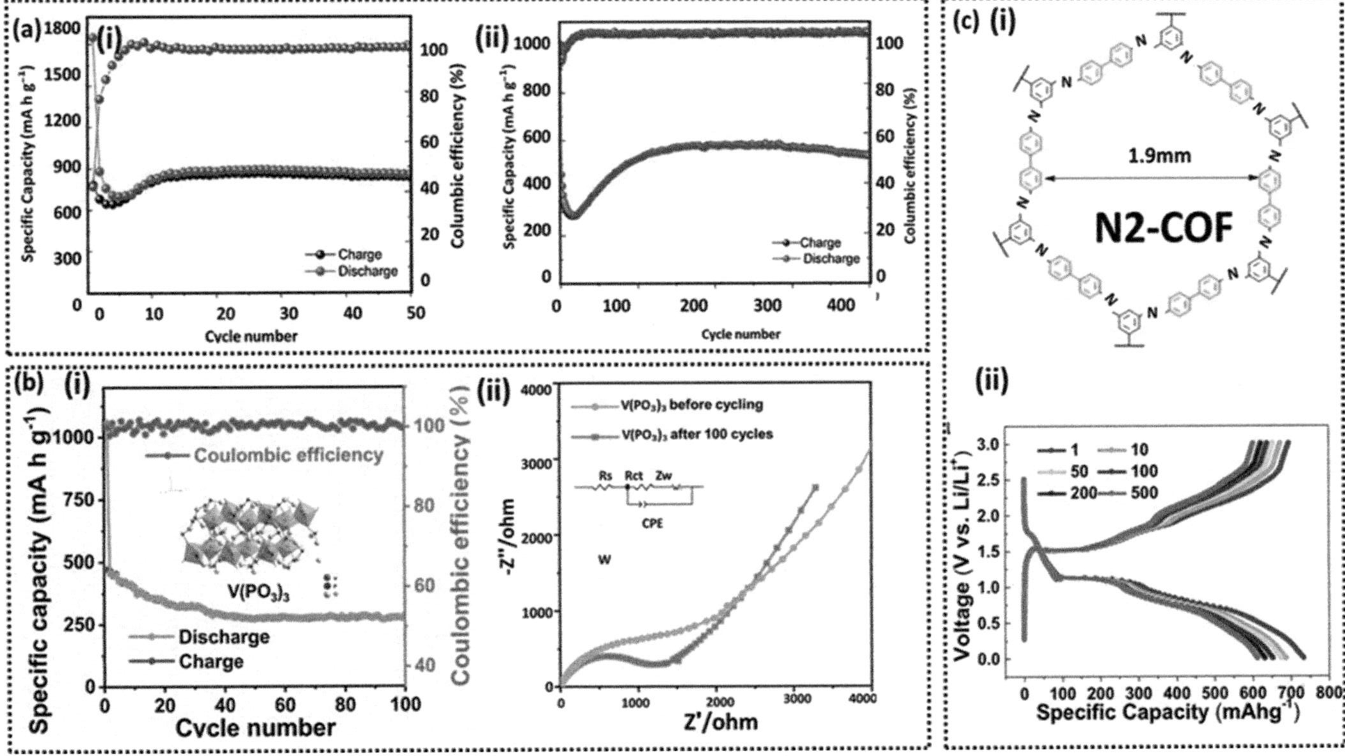

FIGURE 7.1 (a) (i) Cyclic performance of Co-btca electrode, (ii) Co-btca electrode cyclic performance at 1A g^{-1}. Reproduced with permission from reference [42] copyright claim from Elsevier. (b) (i) Cycle performance and coulombic efficiency of $V(PO_3)_3$ (ii) Electrochemical impedance spectra of $V(PO_3)_3$ electrode. Reproduced with permission from reference [43] copyright claim from Chemistry Europe. (c) (i) Schematic illustration of N_2-COF (ii) Charge/discharge capacity of N_2-COF at a current density of $1Ag^{-1}$. Reproduced with permission from reference [44] copyright claim from Royal Society of Chemistry.

TABLE 7.2
SIBs Coordination Nanomaterials Physiochemistry

Coordination Nanomaterials	Morphology	Current Density (mA g^{-1})	Cycle Performance	Reversible Capacity (mA h g^{-1})	Cycles	Reference
Co-MA CPs	Flower-like porous microsphere	100 mAg^{-1}	372 mAhg^{-1}	554 mAh g^{-1}	300	[52]
CPL-4	–	50 mA g^{-1}	200 mA h g^{-1}	91 mA h g^{-1}	100	[51]
7-CoS/C	Porous nanorods	5 A g^{-1}	542 mAh g^{-1}	352 mAh g^{-1}	2000	[53]
TFPB-TAPT COF	Spherical structure	0.04 A g^{-1}	–	246 mA h g^{-1}	500	[56]

storage application [50]. The physiochemical properties of SIBs are represented in Table 7.2. A layered pillar framework synthesized of a two-dimensional sheet of [Cu(pzdc)] and an azpy pillar ligand developed by Shimizu et al. [51] presenting as CPL-4 ([$Cu_2(pzdc)_2$(azpy)], pzdc = pyrazine-2,3-dicarboxylate, azpy = 4,4-azopyridine, pore size: 10 × 6 Å2) exhibited reversible guest insertion/deinsertion stability in the large space. A discharge capacity of 169 mA h g^{-1} equivalent to 82% of its theoretical capacity was exhibited by CPL-4 in Figure 7.2a (i). CPL-4 showed good performance in SIBs as a cathode-active material. The Na^+ ions were deeply added into the CPL-4 particles, exhibiting high capacity and good rate capability due to the reduction of Cu^{2+} to Cu^+ and N double bond N/N–N in the first cycle. SIB cathode-active materials can be designed based on the results of this study. A group of Co-based coordination polymers (CPs) reported by Yin et al. [52] was prepared by using the maleic acid (MA) organic ligand representing as Co-MA CPs exhibiting a flower-like porous microsphere investigated for Na+ storage performances. Excellent cycle performance of 460 mA h g^{-1} was achieved after 300 cycles and a high initial reversible Na^+ extraction capacity of about 554mA h g^{-1} was obtained at 100 mA g^{-1} exhibited by Co-MA CPs as a result of its porous structure and large surface area. A facile heating method of Co-MOF in a sulfide sublimation atmosphere was reported by Zhou et al. [53] to synthesize ultrafine CoS nanoparticles (7 nm) embedded in porous carbon nanorods (7-CoS/C). Embedded porous architecture and partial pseudocapacitive behaviors of the redox reactions, as well as the usage of ether-based electrolytes, all contribute to the outstanding performance Figure 7.2a (i). A discharge capacity of 504 mA h g^{-1} after 1,000 cycles was exhibited by -CoS/C at 5 A g^{-1} Figure 7.2a (ii). Thus there is significant possibility for advanced energy-storage materials based on this MOF-driven approach.

Using an expanded linear ligand, Zhang et al. [55] designed an anode material for SIB based on MOF. It exhibited a capacity retention of 79.0% after 1,000 cycles and a specific capacity of about 269 mA h g^{-1} with an anticipated voltage plateau. By using 1, 3, 5 tris-(4-formyl phenyl) benzene (TFPB) and 1, 3, 5 tris-(4- amino phenyl)-triazine (TAPT), Patra et al. [56] reported a new COF material designated as TFPB-TAPT

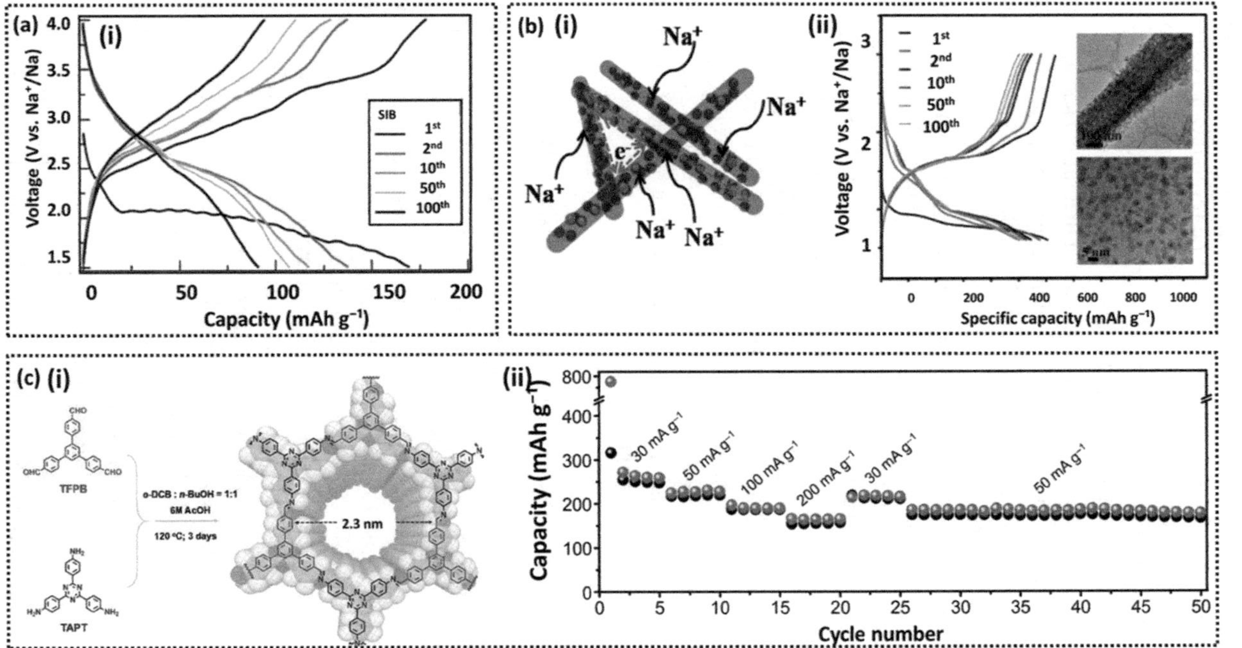

FIGURE 7.2 (a) (i) Charge-discharge capacities of CPL-4 in the SIB. Reproduced with permission from reference [51] copyright claim from ACS. (b) (i) Synthesis process of the 7-CoS/C, (ii) Galvanostatic discharge/charge profiles of the 7-CoS/C electrode. Reproduced with permission from reference [53] copyright claim from Elsevier. (c) (i) A scheme for TFPB-TAPT COF, (ii) the rate of performance of TFPB-TAPT COF anode at different current rates. Reproduced with permission from reference [54] copyright claim from Royal Society of Chemistry.

with C_3-C_3 symmetrical topology and used as an anode material in SIBs Figure 7.2c (i). In Figure 7.2c (ii), the storage performance of Na^+ ions by TFPB-TAPT COF is shown. With the same current rate, TFPB-TAPT COF can retain about 80% of its initial capacity, indicating an excellent Na^+ ion storage capability. As a result of its open-ordered nanoporous framework, Na^+ ions can be reversibly accommodated. Using COFs as electrode materials in SIBs offers a promising new approach. By using hexaketocyclohexane and 1,2,4,5-benzenetetramine, Shehab et al. [57] synthesized the aza-COF, a microporous-based 2D COF by using a condensation reaction, which comprised phenazine channels and a comprised a theoretical specific capacity of 603 mA h g^{-1}. A power density of 1,182 W kg^{-1} at 40 C, an average specific capacity of about 550 mA h $g^{-1,}$ and an energy density of 492 W h kg^{-1} at 0.1 C was exhibited by an Aza-COF-based electrode.

Overall, PCP/MOF/COF materials have proven to have potential for use in SIBs, comparable to carbon materials and metal compounds. Inherently tunable porous structures and accessible catalytic sites give PCP/MOFs/COFs an ultrahigh discharge-specific capacity. Additionally, PCP, MOFs, and COFs perform differently counting close to the size and morphology of their crystals. Anyhow, the use of PCPs, MOFs, and COFs as electrode catalysts in SIBs is still in its inception. It may be attributed to the inherent disadvantages of PCP, MOF, and COF electrodes, including the inadequate ability to decompose emissions and unfavorable electrical conductivity, leading to inadequate rate capabilities, high overpotentials, and insufficient cycling stability as a consequence. Thus, it is imperative that materials are designed and built to fulfill future requirements.

7.3.3 Coordination Nanomaterials for Potassium-Ion Batteries

In view of the abundance and low cost of potassium, the small Stokes' radius of K^+ resulting in rapid ionic diffusion in the electrolyte, and the higher discharge potential of PIBs, they are regarded as a promising alternative to LIBs that have been widely used for decades [58]. A wide range of materials was proposed as electrodes for PIBs such as PCPs, MOF, and COF. Table 7.3 shows the physiochemical properties of KIB.

A 1D Ni-DAB (derived from 3,3'-diaminobenzidine)-based coordination polymer was reported by Kraevaya et al. [59] as an anode material for PIBs. A specific capacity of 158 mA h g^{-1} at 1 A g^{-1} was achieved with Ni-DAB Figure 7.3a (i, ii), resulting in an 8.6% capacity fade after 600 cycles (0.014% per cycle), which represents an outstanding performance for PIB anode materials. As a fast and stable PIB anode, Ni-based coordination polymers derived from 1,2,4,5-tetraaminobenzene (P1) have been reported by Kapaev et al. [60]. A reversible capacity of 220 mA h g^{-1} was obtained at 0.1 A g^{-1} and after 200 cycles the capacity fade was only 4.4%. Based on the obtained results, metal-organic coordination polymers derived from amines have the potential to be used as energy storage materials. Li et al. [61] reported a high pyridine N-doped porous carbon (NPC-600) derived from MOFs that could be used as anode materials with PIBs. The cycling performance is depicted in Figure 7.3b (i). In comparison to NPC-500 (438.8 mAh g^{-1}) and NPC-700 (372.8 mAh g^{-1}), NPC-600 delivered a higher initial reversible specific capacity of 578.2 mAh g^{-1}. The cycling performance of NPC-600 was demonstrated over a long period of time. A reversible

TABLE 7.3
PIBs Coordination Nanomaterials Physiochemistry

Coordination Nanomaterials	Morphology	Current Density (mA g^{-1})	Cycle Performance	Reversible Capacity (mA h g^{-1})	Cycles	Reference
Ni-DAB	Uniform surfaces	50 mA g^{-1}	158 mA h g^{-1}	183 mA h g^{-1}	600	[59]
NPC-600	Smooth polyhedral structure	500 mA g^{-1}	231.6 mAh g^{-1}	587.6 mAh g^{-1}	2000	[61]
$_D$-COF	Multi-layered bulk	0.2 A g^{-1}	106.1 mA h g^{-1}	229.5 mA h g^{-1}	2000	[63]
COF-JLU_2	Porous structure	500 mA g^{-1}	–	85 mA h g^{-1}	2500	[64]

specific capacity of 231.6 mAh g^{-1} was achieved by NPC-600 at 500 mA g^{-1} after 2,000 cycles as in Figure. 7.3b (ii). In potassiation/depotassiation of NPC-600, pyridine N increases, and interlayer space changes, which can provide additional adsorption sites to "capture" more K ions.

A low-price and macroporous Fe-based MOF (MOF-235) was developed using a solvothermal method by Qijiu Deng et al. [26]. By using carbon nanotubes (CNTs), a reversible capacity of 132mAh g^{-1} at 200mA g^{-1} over 200 cycles was achieved.

The robust and porous crystalline structure of COFs makes them ideal anode materials for PIBs. Through a simple solvothermal process, Su et al. [63] synthesized a multi-layer structural COF connected by double functional groups, including imine and amidogent. $_D$-COF delivered a charge-specific capacity of 116.2, 229.4, 135.5, 176.1, and 152.7 mAh g^{-1} at current densities from 0.2 to 5.0 A g^{-1}. Because of the imine groups, $_D$-COF has excellent structure stability. It is still possible for D-COF to deliver 195.8 mA h g^{-1} when the current density falls to 0.2 A g^{-1} Figure 7.3c (i). The cycling performances of the (double functional group) $_D$-COF and the single functional group (S-COF) are shown in Figure 7.3c (ii). $_D$-COF and S-COF capacities are increasing after about 200 cycles, indicating gradual activation of electrode materials. $_{D-}$COF has a high specific capacity of 127.8 mAh g^{-1} after 800 cycles, higher than S-COF (87.2 mAh g^{-1} after 600 cycles), indicating that double functional groups enhance electrode construction stability. By using the carbonization method, a COF-derived nitrogen-doped porous carbon (NPC) as anode material was developed by Zhang et al. [64] by using hydrazine-1,3,5-triformylphloroglucinol copolymer (COF-JLU_2) as a precursor. Even after 2,500 cycles, it exhibited excellent stability and high coulombic efficiency.

PIBs' electrodes based on PCP/MOF/COF have made significant advances, but there will always be some challenges. These materials do not reversibly break down discharge products; thus, they must be combined with simple metallic substances or materials derived from them. In PIBs, this restriction may hinder the development of PCP, MOF, and COF materials.

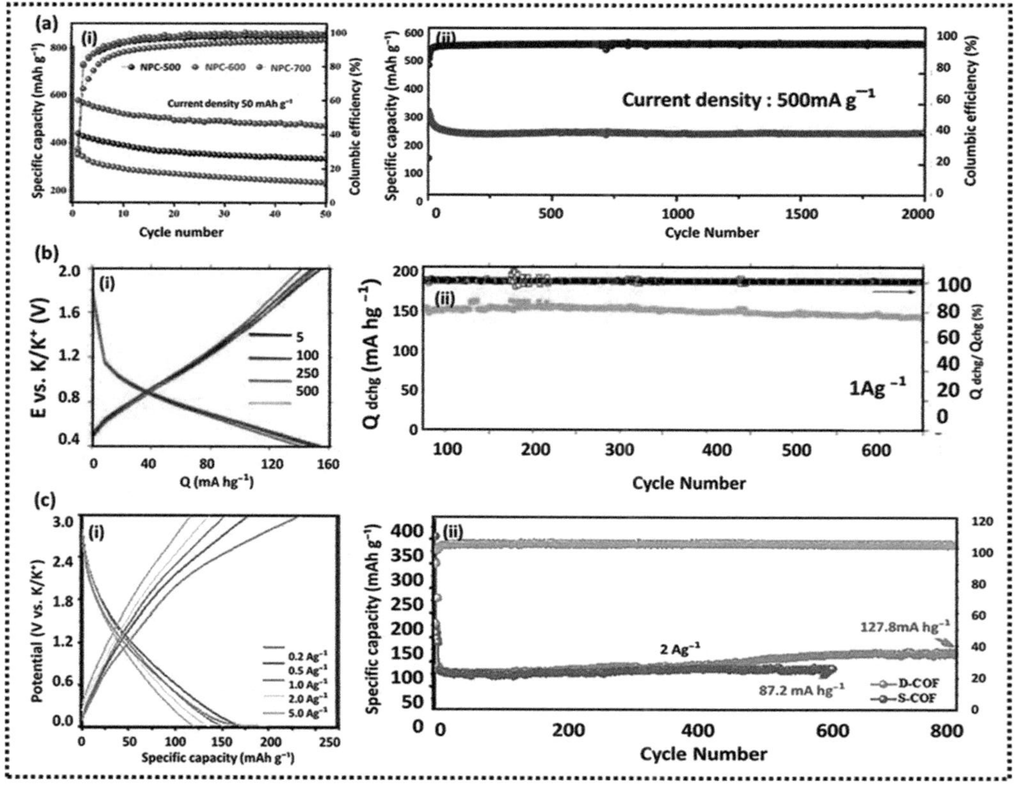

FIGURE 7.3 (a) (i) Rate ability of NPC-500, NPC-600 and NPC-700, (ii) Cycling properties at 500 mA g^{-1} of NPC-600. Reproduced with permission from reference [61] copyright claim from WILEY_VCH. (b) (i) Charge/discharge curves for cycles at 5, 100, 250, and 500, (ii) Discharge capacity and coulombic efficiency at 1 A g^{-1}. Reproduced with permission from reference [62] copyright claim from Royal Society of Chemistry. (c) (i) Galvanostatic charge/discharge capacities of the $_D$-COF, (ii) Cycling performances of $_D$-COF and S-COF at 2.0 A g^{-1}. Reproduced with permission from reference [63] copyright claim from Elsevier.

7.3.4 Coordination Nanomaterials for Zinc-Ion Batteries

Aqueous electrolyte batteries are attractive for large-scale energy storage because of their nonflammable nature, abundant supply of zinc, and low cost. Also, AZIBs are easy to produce and have two orders of magnitude more ionic conductivity than non-aqueous electrolytes [65]. A larger population can adopt them if they are improved in performance and durability. Table 7.4 summarizes ZIB physiochemical properties. Buyukcakir et al. [66] reported a new quinone-functionalized redox-active porous organic polymer (rPOP) as a cathode material by using 1,4,5,8-anthracenetetrone in aqueous ZIBs, rPOP performance, and charge/discharge profiles represented in Figure 7.4a (i). At a current density of 0.1 A g^{-1}, rPOP exhibits a specific capacity of 120 mA h g^{-1} during extended cycling. The cells' cycling stability was further assessed at a current density of 2.0 A g^{-1}. The cell maintains about 66% of its initial capacity after 30,000 cycles while conserving high CE (Figure 7.4a (ii)). In fact, redox-active quinones are effectively utilized due to the high porosity of the oxide during both charging and discharging cycles. A fused aromatic conjugated skeleton also enhances intrinsic conductivity and physical stability. In a recent study, Wu et al. [67] developed an AZIBs system using porous metal-organic coordination polymers (MOCP) nanosheets as cathode and Zn foil as anode that utilized MOCPs during the manufacturing process. The zinc cell demonstrated outstanding capacity retention, maintaining close to 100% over 1,500 cycles at a rate of 600 mA g^{-1}. Additionally, it exhibited a notable reversible capacity of 240.3 mA h g^{-1} at a lower rate of 150 mA g^{-1}. An AZIB cathode based on manganese (MOF-73) was presented by Gou et al. [29]. The MOF-73 cathode achieved discharge-specific capacities of 142.6, 1000.8, 77.2, 54.9, and 46.9 mA h g^{-1} (corresponding to 0.15, 0.11, 0.08, 0.06 and 0.05 mA h cm^{-2}) at current densities of 0.1, 0.3, 0.5, 0.8, and 1 A g^{-1} respectively (corresponding to 0.15, 0.11, 0.08, 0.06, and 0.05 mA h cm^{-2}) Figure 7.4b (i). After 1,000 cycles at 0.3 A g^{-1}, the specific capacity reached 137 mA h g^{-1}. As a result, capacity keeps increasing throughout the entire cycle testing.

TABLE 7.4
ZIBs Coordination Nanomaterials Physiochemistry

Coordination Nanomaterials	Morphology	Current Density (mA g^{-1})	Cycle Performance	Reversible Capacity (mA h g^{-1})	Cycles	Reference
rPOP	Tiny sheet-like structures	0.1 A g^{-1}	40 mA h g^{-1}	120 mAh g^{-1}	30,000	[66]
MOF-73	Rod-like micro-particles	50 mA g^{-1}	137 mA h g^{-1}	815 mA h g^{-1}	1,000	[29]
COF; TpPa-SO_3H	–	0.5 A g^{-1}	–	5 mA h cm^{-2}	100	[70]
Tp-PTO-CO	Lamellar stacking	5 A g^{-1}	300 mA h g^{-1}	192.8 mA h g^{-1}	1,000	[71]

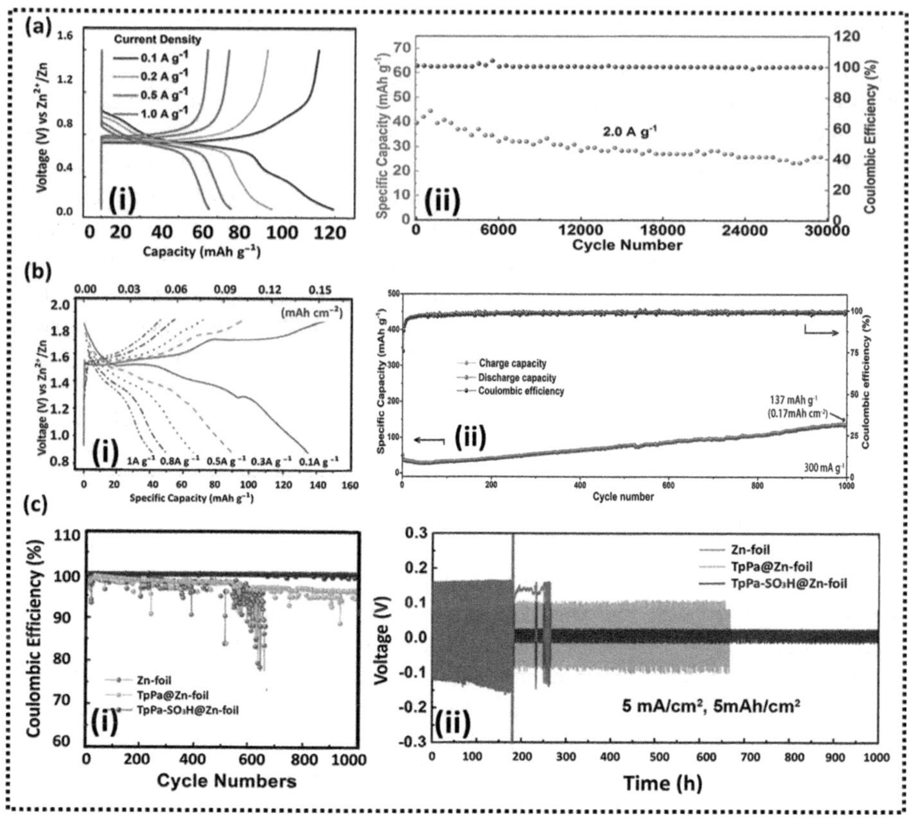

FIGURE 7.4 a) (i) Galvanostatic (dis-)charge profiles at various current densities, (ii) long-term cycling stability of rPOP at a current density of 2.0 A g^{-1} up to 30 000 cycles. Reproduced with permission from reference [66] copyright claim from American Chemical Society. (b) (i) Rate performance of the MOF-73 cathode, (ii) long cycle performance of the MOF-73 cathode. Reproduced with permission from reference [29] copyright claim from Royal Society of Chemistry. (c) (i) Coulombic efficiencies of the half cells assembled with bare Zn-foil, TpPa@Zn-foil, and TpPa-SO_3H@Zn-foil electrodes at 1 mA cm^{-2}, (ii) long-term galvanostatic discharge/charge curves of various cells at a high current density of 5 mA cm^{-2}. Reproduced with permission from reference [68] copyright claim from Elsevier.

Yagang Yao et al. [69] constructed conductive 3D vanadium-based MOF arrays in CNT fibers using a self-scarified pathway. The CNTF surface was uniformly grown with V-MOFs by a simple solvothermal reaction so that they could serve as a binder-free flexible cathode for ZIB. The TpPa-SO_3H film (COF; TpPa-SO_3H@ Zn-foil) was synthesized by Zhao et al. [70] using an interfacial reaction. As Zn^{+2} ions are uniformly deposited on the TpPa-SO_3H film, sulfonic acid groups inhibit dendrite formation. A short circuit caused by dendrites or corrosion resulted in a decline in coulombic efficiency after 600 cycles. After 1,000 cycles, the CEs of the TpPa@Zn-foil electrode and the TpPa-SO_3H@Zn-foil electrode remained around 95% and 99%, respectively. Several studies have shown that there is a possibility of reversing Zn plating on TpPa-SO_3H@Zn foil due to the high and stable CE of the foil Figure 7.4c (i). As a result of TpPaSO_3H being used in the electrolyte, it was possible to achieve high CE and facilitate the plating and stripping of Zn. In the Cu/Zn-foil cell, an overpotential of 160 mV was observed after only 170 hours of cycling at five milliamps per square foot Figure 7.4c (ii). By reducing dendrite growth and by-product formation, TpPa-SO_3H@Zn-foil can allow homogeneous deposition of Zn ions. In aqueous rechargeable ZIBs, Ma et al. [71] reported the successful introduction of a covalent organic framework with multiple carbonyl active sites (Tp-PTO-COF) achieving substantial specific capacities of 301.4 mA h g^{-1} at a current density of 0.2 A g^{-1} and 192.8 mA h g^{-1} at 5 A g^{-1}.

Furthermore, these studies demonstrate the use of PCP/MOF/COF-based electrodes in ZIBs in addition to those covered in the section on ZIBs. Performance is enhanced by catalytic metal centers, organic linkers, and intrinsic crystalline porosity. Despite significant progress, some deficiencies remain. Currently, only classical electrodes with a limited structural range are available for exploratory PCP/MOF/COF. As ZIB tests are typically conducted in extremely concentrated reactive/corrosive electrolytes, electrodes must also be extremely tolerant and stable.

7.3.5 Coordination Nanomaterials for Magnesium-Ion Batteries

In recent years, researchers have paid tremendous attention to MIBs for their potential due to their abundance of magnesium, reversibility, low reduction potential, and operational safety in addressing the growing energy crisis resulting from the increase in human population and an increase in living standards around the world [72]. The physiochemical properties of MIBs are presented in Table 7.4. According to Wang et al. [73], poly(hexaazatrinaphthalene sulfide) (PSHATN) has been reported. In a two-step procedure involving the condensation reaction between 4-fluoro-1,2-phenylenediamine and hexaketocyclohexane followed by the polymerization step using Na_2S, a sulfur-linked porous polymer was obtained. Figure 7.5a (i) shows

TABLE 7.5
MIBs Coordination Nanomaterials Physiochemistry

Coordination Nanomaterials	Morphology	Current Density (mA g^{-1})	Cycle Performance	Reversible Capacity (mA h g^{-1})	Cycles	Reference
PSHATN	Porous structure	–	142 mA h g^{-1}	196 mA h g^{-1}	500	[73]
α-$Mg_3(HCOO)_6$	Spongy-crumbled	0.1 μAcm^{-2}	–	–	20	[75]
COF	Lamellar structure	5C	–	74 mA h g^{-1}	1000	[76]
V-MOFs, MIL-47	Hierarchical structure	0.1 A cm^{-3}	85.4 mA h cm^{-3}	64.3%	400	[69]

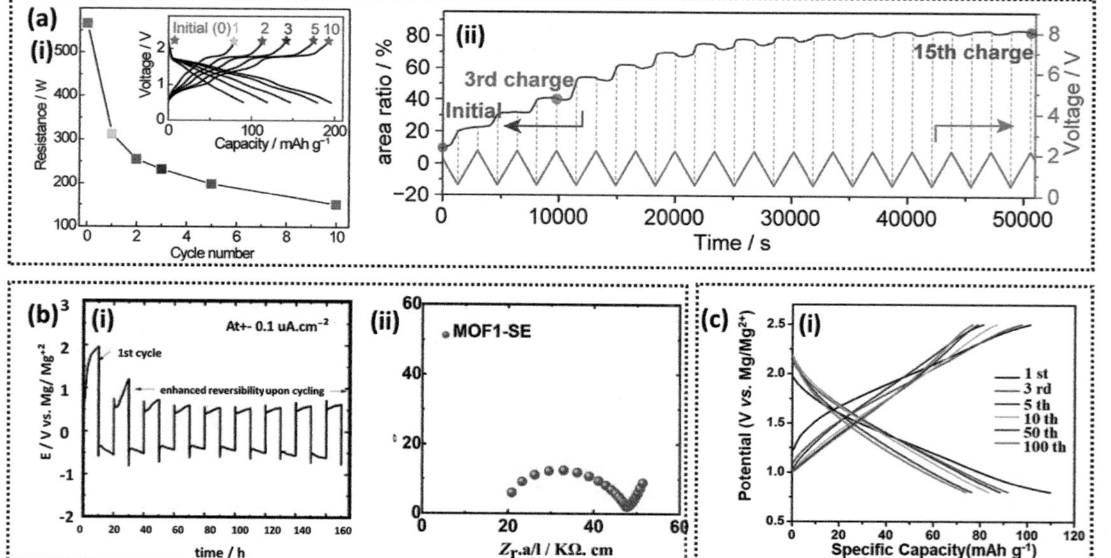

FIGURE 7.5 (a) (i) Interfacial resistance vs. cycle number plot obtained from electrochemical impedance spectroscopy (EIS) of PSHATN electrodes in Mg electrolyte (ii) voltage profiles and electrode area ratio (%) of PSHATN electrode in Mg electrolytes during cycle voltammetry measurements at a scan rate of 1 mV s^{-1}. Reproduced with permission from reference [73] copyright claim from Elsevier. (b) (i) Galvanostatic cycling of Mg/a-MOF1-$Mg(TFSI)_2$-$MgCl_2$-G_4-10 SE/Mg at 0.1 μAcm,(ii) Nyquist plots at 30°C of mixed MOFs-based SE. Reproduced with permission from reference [75] copyright claim from European Chemical Societies. (c) (i) Discharge-charge curves of COF at 0.5 C, (ii) Long-term cycling performance of COF at 5 C. Reproduced with permission from reference [76] copyright claim from American Chemical Society.

that the interfacial impedance decreases after the first cycle. With electrolyte swelling of the polymer, the diffusion of bulky $MgCl^+$ ions improves, which may explain this phenomenon. In the electrolyte containing magnesium, a different phenomenon was observed. It is believed that the electrode expanded continuously during the initial thirteen cycles, resulting in an increase in the electrode area ratio Figure 7.5a (ii). In the following months, there was relatively little expansion and contraction of the electrode area. Since both experiments were conducted using the same solvent (e.g., dimethoxyethane, DME), the swelling processes of polymers in magnesium electrolytes are considerably slower than those in lithium electrolytes. The microporous conjugated polymer (PMCP) based on phenanthrenequinone is a highly porous organic material suggested as an optimal MIB cathode material by Khan et al. [74]. Based on density functional theory calculations, this porous material shows a strong affinity for Mg ions at the double bond O sites in the C double bond. In one monomer of PMCP, four Mg^{+2} ions with specific capacities of 157.05 mA per gram and redox potential of 1.37 V can be accommodated. According to Hassan et al. [75] during a hydrothermal reaction in dimethylformamide at 100°C, magnesium nitrate and 2,2'-bipyridine-3,3'-dicarboxylic acid (bp_3dca) ligand produce monocrystalline $[Mg(bp_3dc)(H2O)_4]n$ (c-$Mgbp_3d$). The nodes were made of Mg^{2+} ions and the organic linkers were made of Bp_3dca. As part of the investigation, a MOF1-$Mg(TFSI)_2$-$MgCl_2$-G_4-10 SE was tested for stability and viability in the presence of Mg electrodes. In the first step, Mg was deposited for 10 hours, followed by nine cycles of repeated galvanostatic deposition/stripping at 0.1 μA cm. Figure 7.5b (i) indicates that cycling makes the magnesium deposition/stripping process more reversible, and the CV following the galvanostatic experiment also supports this conclusion. In this study, the deposition and stripping processes were carried out for 10 hours each cycle, which was 20 hours $cycle^{-1}$. It can be seen from Figure 7.5b (ii) that mixed MOFs exhibited the highest ionic conductivity at 3.8x10–5 cm^{-2}. As shown by the ionic conductivity data, combining two distinct MOFs, one of which is amorphized, increases the degree of disorder in MOF-based SEs.

Sun et al. report a COF cathode with high rate and stability, an Mg anode without dendrites, and an electrolyte that is chloride-free and Mg-compatible [76]. The COF electrode shows sloping magnetization and demagnetization curves from cycle 1 to 100 Figure 7.5c (i). The reversible power capacity can be retained after 100 cycles, demonstrating exceptional cycling stability.

To summarize, the recent studies on MIBs, based on PCP/MOF/COF electrodes, proved that these are still in the research and development phase, and further work is required to address technical challenges and enhance their commercial viability. The implementation of ample active species inside PCP/MOF/COF skeletons is favorable and contributes to elevating ZAB performance.

7.3.6 Coordination Nanomaterials for Aluminum-Ion Batteries

On account of their high theoretical specific capacity, plentiful Al, less flammability of ionic liquid-based electrolytes, and functionality, AIBs have become members of the advanced generation of energy storage technologies [77]. The physicochemical properties of AIBs are presented in Table 7.6. By using the self-template method, a

TABLE 7.6
AIBs Coordination Nanomaterials Physiochemistry

Coordination Nanomaterials	Morphology	Current Density (mA g^{-1})	Cycle Performance	Reversible Capacity (mA h g^{-1})	Cycles	Reference
Al/HPCO	Hierarchical porous architecture	100 mA g^{-1}	60 mA h g^{-1}	35.6 mA h g^{-1}	1000	[78]
TpBpy-COF	Hierarchical structure	2Ag^{-1}	–	150 mA h g^{-1}	13000	[80]
COF-JLU$_2$	Porous structure	–	–	83 mAhg^{-1}	30000	[64]
3DOM CoSe$_2$@C	Macro–microporous	5.0 A g^{-1}	125 mA h g^{-1}	86 mA h g^{-1}	1000	[81]

MOF derived hierarchical porous carbon octahedrons (HPCO) were synthesized by Wang et al. [78], which was utilized in urea-based Al/graphite battery as the cathode denoted as Al/HPCO. The Al/HPCO batteries' galvanostatic charge-discharge curves are shown in Figure 7.6a (i). During discharge, two plates centered at ~1.50 and ~1.95V were observed indicating the de-intercalation of HPCO intercalations. Al/HPCO batteries' cycling performances are shown in Figure 7.6a (ii). At 100 mA g^{-1} the reversible specific capacity of ~60.8 mA h g^{-1} was obtained, and the average CE could reach ~98%. This may offer widespread solutions to future research on AIBs and commercial utilization of AIBs.

A crystalline framework comprising 2,2-bipyridine moieties denoted as TpBpy-COF was developed by Lu et al. [80] which exhibits a two-dimensional hexagonal layered network. The good performance exhibited by the cell showed CE of approximately 100% and a reversible discharge capacity of 150 mA h g^{-1} was maintained after 13,000 cycles Figure 7.6b (i). At the current densities of 0.1, 0.5, 1, 2, 5, 0.5, A g^{-1}, the COFs-Al battery exhibits reversible discharge capacities of 307, 245, 162, and 125, 113, 230 mA h g^{-1}, along with an increase in CE of COFs-Al battery showing good rate capabilities Figure 7.6b (ii). This COF material has a high chemical stability, making it unaffected by Lewis acid electrolyte corrosion.

Overall, MOF and COF materials may perhaps have been optimistic competitors for current AIBs. Nevertheless, considerable effort should have been made to adapt PCP electrodes in AIBs. In contrast, pristine MOFs and COFs have received comparatively little attention in AIBs. This would encourage the ongoing research of MOF- and COF-based electrodes in AIBs.

7.4 SUMMARY AND PROSPECTS

Throughout this chapter, coordination nanomaterials, such as PCPs, MOFs, and COFs, as well as their derivatives, have been discussed extensively. The specific contributions of coordination nanomaterials to several categories of metal-ion batteries

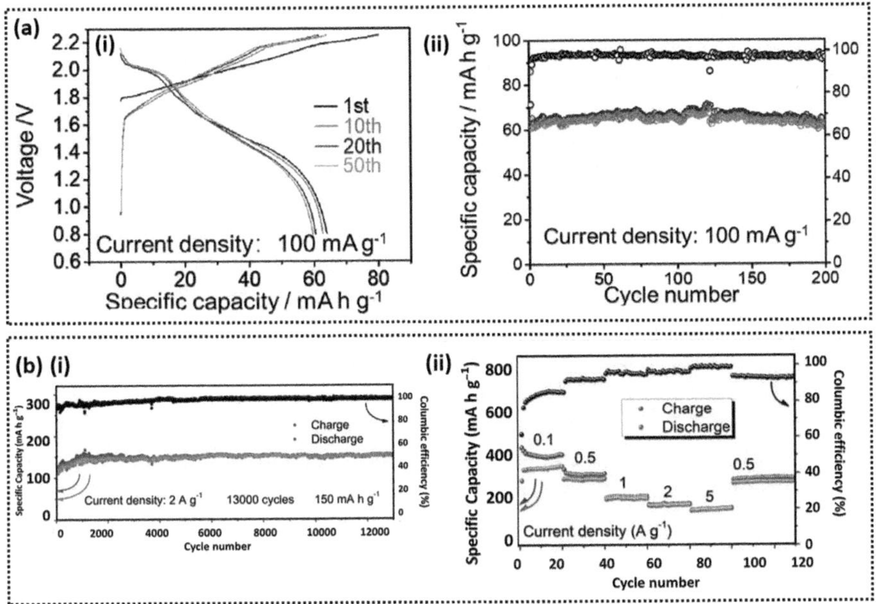

FIGURE 7.6 (a) (i) Galvanostatic charge/discharge capacities of HPCO in urea-based AIBs (ii) the Al/HPCO battery cyclic performances at 100 mA g^{-1}. Reproduced with permission from reference [78] copyright claim from Elsevier. (b) (i) The COFs-Al battery galvanostatic charge-discharge curves (ii) Galvanostatic discharge/charge capacities of COFs-Al battery at different current densities. Reproduced with permission from reference [79] copyright claim from European Chemical Societies.

have been explored, highlighting their unique advantages and potential for enhancing energy storage performance in LIBs, SIBs, KIBs, ZIBs, MIBs, and AIBs. Finally, we suggest some directions for future research. A wide range of coordination nanomaterials, including PCPs, MOFs, and COFs has been studied in EES, including metal-ion batteries, which utilize coordination sites to their full potential. Coordination nanomaterials have high surface areas and abundant pores, making mass transfer easy. Moreover, the open coordination sites reveal active sites to make certain speedy electrochemical reactions, resulting in elevating capacities under elevated current densities. Additionally, the adaptable components empower coordination nanomaterials and their derivatives to perform better electrochemically than mono-component materials. It is likely that coordination nanomaterials will exhibit previously unidentified properties as a result of the complementary actions of various metal ions and functional ligands. A variety of electrochemical conditions can also be supported by coordination nanoparticles and composites. Coordinated nanomaterials and composites could be intended rationally via a sort of topology and morphology The use of PCPs, COFs, and MOFs in energy storage and conversion systems has been extensively studied. There are, however, still some obstacles to overcome. For LIBs and SIBs low electronic conductance is the main disadvantage. Poor conductivity requires more conductive material, ensuing in less energy density and low rate performance.

To sustain a framework structure, PCPs, MOFs, and COFs contain many structural segments that are electrochemically inactive, hence the limited specific capacity of PCPs, COFs, and MOF electrodes. Moreover, MOFs, PCPs, and COFs need to be electrochemically stable. After numerous charge-discharge cycles, the coordinated structure may be destroyed. However, the cause of instability is rarely investigated. In MOFs, PCPs, COFs, and their derivatives, the porosity structure contributes to their low volumetric energy density and low coulombic efficiency. As active materials in KIB and MIB batteries, PCPs, MOFs, and COFs serve high-performance electrodes. In addition to facilitating rapid electrolyte permeability, the large inner surface allows active materials to be loaded rapidly. Further, PCPs, MOFs, and COFs are unstable in acidic and alkaline solutions, which limits their application in various batteries. Due to their elevated stability and stronger interactivity with LIBs, PCPs-, MOFs-, and COF-derived materials have advantages over others. The implementation of PCPs, COFs, and MOFs in metal-ion electrodes still presents a number of difficulties, but ongoing study into these three material types and a better comprehension of the reaction mechanism will yield more knowledge that will help researchers develop more effective and focused materials, which should encourage widespread industrial use in the future.

ACKNOWLEDGMENTS

This work was supported by the Research Fund for International Scientists (RFIS-Grant number: 52150410410) National Natural Science Foundation of China.

REFERENCES

[1] Yasin, G., et al., Chapter 1 – Nanoelectrocatalysis: An introduction, in *Nanomaterials for Electrocatalysis*, T. Maiyalagan, et al., Editors. 2022, Elsevier. p. 3–10.
[2] Yasin, G., et al., Chapter 1 – Introduction to electrochemical energy storage technologies, in *Lithium-Sulfur Batteries*, R.K. Gupta, et al., Editors. 2022, Elsevier. p. 3–10.
[3] Wang, M., et al., A high-performance tin phosphide/carbon composite anode for lithium-ion batteries. *Dalton Transactions*, 2020. **49**(46): p. 17026–17032.
[4] Wang, H., et al., A green and scalable synthesis of Na3Fe2 (PO4) P2O7/rGO cathode for high-rate and long-life sodium-ion batteries. *Small Methods*, 2021. **5**(8): p. 2100372.
[5] Liang, Y., et al., Current status and future directions of multivalent metal-ion batteries. *Nature Energy*, 2020. **5**(9): p. 646–656.
[6] Yasin, G., et al., Chapter 5 – Battery-nanogenerator hybrid systems, in *Nanobatteries and Nanogenerators*, H. Song, et al., Editors. 2021, Elsevier. p. 61–68.
[7] Zhang, C., et al., Two-dimensional organic cathode materials for alkali-metal-ion batteries. *Journal of Energy Chemistry*, 2018. **27**(1): p. 86–98.
[8] Ali, H.H., et al., Rationally designed Mo-based advanced nanostructured materials for energy storage technologies: Advances and prospects. *Sustainable Materials and Technologies*, 2023. **38**: p. e00738.
[9] Yasin, G., et al., Chapter 1 – Nanobattery: An introduction, in *Nanobatteries and Nanogenerators*, H. Song, et al., Editors. 2021, Elsevier. p. 3–9.
[10] Yuan, C., et al., Mixed transition-metal oxides: design, synthesis, and energy-related applications. *Angewandte Chemie International Edition*, 2014. **53**(6): p. 1488–1504.
[11] Kim, C.R., T. Uemura, and S. Kitagawa, Inorganic nanoparticles in porous coordination polymers. *Chemical Society Reviews*, 2016. **45**(14): p. 3828–3845.

[12] Yasin, G., et al., Chapter 1 – MOF-based nanostructures and nanomaterials for next-generation energy storage: An introduction, in *Metal-Organic Framework-Based Nanomaterials for Energy Conversion and Storage*, R.K. Gupta, T.A. Nguyen, and G. Yasin, Editors. 2022, Elsevier. p. 3–10.
[13] Zhang, L., et al., Synthesis strategies and potential applications of metal-organic frameworks for electrode materials for rechargeable lithium ion batteries. *Coordination Chemistry Reviews*, 2019. **388**: p. 293–309.
[14] Iqbal, R., et al., State of the art two-dimensional covalent organic frameworks: Prospects from rational design and reactions to applications for advanced energy storage technologies. *Coordination Chemistry Reviews*, 2021. **447**: p. 214152.
[15] Yasin, G., et al., Chapter 9 – Nanostructured anode materials in rechargeable batteries, in *Nanobatteries and Nanogenerators*, H. Song, et al., Editors. 2021, Elsevier. p. 187–219.
[16] Yasin, G., et al., Chapter 11 – Nanostructured cathode materials in rechargeable batteries, in *Nanobatteries and Nanogenerators*, H. Song, et al., Editors. 2021, Elsevier. p. 293–319.
[17] Puttaswamy, R., S.K. Nataraj, and D. Ghosh, MOFs-based nanomaterials for metal-ion batteries, in *Metal-Organic Framework-Based Nanomaterials for Energy Conversion and Storage*. 2022, Elsevier. p. 293–313.
[18] Asif, H.M., et al., Chapter 12 – Polyoxometalate-based metal organic frameworks (POMOFs) for lithium-ion batteries, in *Metal-Organic Framework-Based Nanomaterials for Energy Conversion and Storage*, R.K. Gupta, T.A. Nguyen, and G. Yasin, Editors. 2022, Elsevier. p. 245–268.
[19] Luo, Z., et al., A microporous covalent–organic framework with abundant accessible carbonyl groups for lithium-ion batteries. *Angewandte Chemie International Edition*, 2018. **57**(30): p. 9443–9446.
[20] Wang, H., et al., Reviewing the current status and development of polymer electrolytes for solid-state lithium batteries. *Energy Storage Materials*, 2020. **33**: p. 188–215.
[21] Yasin, G., et al., A novel strategy for the synthesis of hard carbon spheres encapsulated with graphene networks as a low-cost and large-scalable anode material for fast sodium storage with an ultralong cycle life. *Inorganic Chemistry Frontiers*, 2020. **7**(2): p. 402–410.
[22] Pang, X., et al., Cathode materials of metal-ion batteries for low-temperature applications. *Journal of Alloys and Compounds*, 2022. **912**: p. 165142.
[23] Park, J., et al., Stabilization of hexaaminobenzene in a 2D conductive metal–organic framework for high power sodium storage. *Journal of the American Chemical Society*, 2018. **140**(32): p. 10315–10323.
[24] Gupta, R., et al., Lithium-sulfur batteries: Materials, challenges and applications. 2022. Book, ISBN 978-0-323-91934-0, Elsevier Publisher, USA.
[25] Rajagopalan, R., et al., Advancements and challenges in potassium ion batteries: A comprehensive review. *Advanced Functional Materials*, 2020. **30**(12): p. 1909486.
[26] Deng, Q., et al., Exploration of low-cost microporous Fe (III)-based organic framework as anode material for potassium-ion batteries. *Journal of Alloys and Compounds*, 2020. **830**: p. 154714.
[27] Qiu, C., et al., Thiophene-sulfur doping in nitrogen-rich porous carbon enabling high-ICE/rate anode materials for potassium-ion storage. *Journal of Materials Chemistry A*, 2023. **11**(41): p. 22187–22197.
[28] Yasin, G., et al., Defective/graphitic synergy in a heteroatom-interlinked-triggered metal-free electrocatalyst for high-performance rechargeable zinc–air batteries. *Journal of Materials Chemistry A*, 2021. **9**(34): p. 18222–18230.
[29] Gou, W., et al., High specific capacity and mechanism of a metal–organic framework based cathode for aqueous zinc-ion batteries. *Energy Advances*, 2022. **1**(12): p. 1065–1070.

[30] Yu, D., et al., High-voltage and ultrastable aqueous zinc–iodine battery enabled by N-doped carbon materials: Revealing the contributions of nitrogen configurations. *ACS Sustainable Chemistry & Engineering*, 2020. **8**(36): p. 13769–13776.
[31] Yasin, G., et al., Chapter 21 – Lithium metal anode: An introduction, in *Lithium-Sulfur Batteries*, R.K. Gupta, et al., Editors. 2022, Elsevier. p. 489–497.
[32] Yasin, G., et al., Understanding and suppression strategies toward stable Li metal anode for safe lithium batteries. *Energy Storage Materials*, 2020. **25**: p. 644–678.
[33] Wang, X., et al., Electrochemical swelling induced high material utilization of porous polymers in magnesium electrolytes. *Materials Today*, 2022. **55**: p. 29–36.
[34] Iqbal, R., et al., Chapter 2 – 2D hybrid nanoarchitecture electrocatalysts, in *Nanomaterials for Electrocatalysis*, T. Maiyalagan, et al., Editors. 2022, Elsevier. p. 11–23.
[35] Liu, X., et al., Facile synthesis of MPS3/C (M = Ni and Sn) hybrid materials and their application in lithium-ion batteries. *ACS Omega*, 2021. **6**(27): p. 17247–17254.
[36] Ru, Y., et al., Different positive electrode materials in organic and aqueous systems for aluminium ion batteries. *Journal of Materials Chemistry A*, 2019. **7**(24): p. 14391–14418.
[37] Guo, Y., et al., Alternate storage of opposite charges in multisites for high-energy-density Al–MOF batteries. *Advanced Materials*, 2022. **34**(13): p. 2110109.
[38] Yasin, G., et al., Self-templating synthesis of heteroatom-doped large-scalable carbon anodes for high-performance lithium-ion batteries. *Inorganic Chemistry Frontiers*, 2022. **9**(6): p. 1058–1069.
[39] Muhammad, N., et al., Volumetric buffering of manganese dioxide nanotubes by employing 'as is' graphene oxide: An approach towards stable metal oxide anode material in lithium-ion batteries. *Journal of Alloys and Compounds*, 2020. **842**: p. 155803.
[40] Luo, Y., et al., Cobalt (II) coordination polymers as anodes for lithium-ion batteries with enhanced capacity and cycling stability. *Journal of Materials Science & Technology*, 2018. **34**(8): p. 1412–1418.
[41] Luo, Y., J. Liu, and L. Zhang, A monocrystalline coordination polymer with multiple redox centers as a high-performance cathode for lithium-ion batteries. *Angewandte Chemie International Edition*, 2022. **61**(38): p. e202209458.
[42] Luo, Y., et al., Cobalt(II) coordination polymers as anodes for lithium-ion batteries with enhanced capacity and cycling stability. *Journal of Materials Science & Technology*, 2018. **34**(8): p. 1412–1418.
[43] Li, Q., et al., Vanadium metaphosphate V(PO3)3 derived from V-MOF as a novel anode for lithium-ion batteries. *ChemistrySelect*, 2021. **6**(31): p. 8150–8157.
[44] Bai, L., Q. Gao, and Y. Zhao, Two fully conjugated covalent organic frameworks as anode materials for lithium ion batteries. *Journal of Materials Chemistry A*, 2016. **4**(37): p. 14106–14110.
[45] Ullah, S., et al., Construction of well-designed 1D selenium–tellurium nanorods anchored on graphene sheets as a high storage capacity anode material for lithium-ion batteries. *Inorganic Chemistry Frontiers*, 2020. **7**(8): p. 1750–1761.
[46] Li, Q., et al., Vanadium metaphosphate V (PO3) 3 derived from V-MOF as a novel anode for lithium-ion batteries. *ChemistrySelect*, 2021. **6**(31): p. 8150–8157.
[47] Chen, L., et al., Hierarchical cobalt-based metal–organic framework for high-performance lithium-ion batteries. *Chemistry–A European Journal*, 2018. **24**(50): p. 13362–13367.
[48] Luo, D., et al., Rational design of covalent organic frameworks with high capacity and stability as a lithium-ion battery cathode. *Chemical Communications*, 2023. **59**(45): p. 6853–6856.
[49] Yasin, G., et al., Facile and large-scalable synthesis of low cost hard carbon anode for sodium-ion batteries. *Results in Physics*, 2019. **14**: p. 102404.
[50] Yasin, G., et al., Chapter 8 – Advanced carbon nanomaterial–based anodes for sodium-ion batteries, in *Advanced Nanomaterials and Their Applications in Renewable Energy* (Second Edition), J.L. Liu, T.-H. Yan, and S. Bashir, Editors. 2022, Elsevier. p. 251–272.

[51] Shimizu, T., et al., Application of porous coordination polymer containing aromatic azo linkers as cathode-active materials in sodium-ion batteries. *ACS Applied Energy Materials*, 2022. **5**(4): p. 5191–5198.
[52] Yin, X., et al., Designing cobalt-based coordination polymers for high-performance sodium and lithium storage: From controllable synthesis to mechanism detection. *Materials Today Energy*, 2020. **17**: p. 100478.
[53] Zhou, L., et al., Structural and chemical synergistic effect of CoS nanoparticles and porous carbon nanorods for high-performance sodium storage. *Nano Energy*, 2017. **35**: p. 281–289.
[54] Patra, B.C., et al., Covalent organic framework based microspheres as an anode material for rechargeable sodium batteries. *Journal of Materials Chemistry A*, 2018. **6**(34): p. 16655–16663.
[55] Zhang, Y., et al., MOF based on a longer linear ligand: Electrochemical performance, reaction kinetics, and use as a novel anode material for sodium-ion batteries. *Chemical Communications*, 2018. **54**(83): p. 11793–11796.
[56] Patra, B.C., et al., Covalent organic framework based microspheres as an anode material for rechargeable sodium batteries. *Journal of Materials Chemistry A*, 2018. **6**(34): p. 16655–16663.
[57] Shehab, M.K., et al., Exceptional sodium-ion storage by an Aza-covalent organic framework for high energy and power density sodium-ion batteries. *ACS Applied Materials & Interfaces*, 2021. **13**(13): p. 15083–15091.
[58] Mehtab, T., et al., Metal-organic frameworks for energy storage devices: Batteries and supercapacitors. *Journal of Energy Storage*, 2019. **21**: p. 632–646.
[59] Kraevaya, O.A., E.V. Shchurik, and P.A. Troshin, Ni-based coordination polymer as a promising anode material for potassium batteries. *Physica Status Solidi (A)*, 2020. **217**(12): p. 1901050.
[60] Kapaev, R.R., et al., A nickel coordination polymer derived from 1, 2, 4, 5-tetraaminobenzene for fast and stable potassium battery anodes. *Chemical Communications*, 2020. **56**(10): p. 1541–1544.
[61] Li, Y., et al., High pyridine N-doped porous carbon derived from metal–organic frameworks for boosting potassium-ion storage. *Journal of Materials Chemistry A*, 2018. **6**(37): p. 17959–17966.
[62] Kraevaya, O.A., E.V. Shchurik, and P.A. Troshin, Ni-based coordination polymer as a promising anode material for potassium batteries. *Physica Status Solidi (A)*, 2020. **217**(12): p. 1901050.
[63] Su, Z., et al., Multilayer structure covalent organic frameworks (COFs) linking by double functional groups for advanced K+ batteries. *Journal of Colloid and Interface Science*, 2023. **639**: p. 7–13.
[64] Zhang, Q., et al., Accessible COF-based functional materials for potassium-ion batteries and aluminum batteries. *ACS Applied Materials & Interfaces*, 2019. **11**(47): p. 44352–44359.
[65] Yasin, G., et al., Simultaneously engineering the synergistic-effects and coordination-environment of dual-single-atomic iron/cobalt-sites as a bifunctional oxygen electrocatalyst for rechargeable zinc-air batteries. *ACS Catalysis*, 2023. **13**(4): p. 2313–2325.
[66] Buyukcakir, O., et al., Ultralong-life quinone-based porous organic polymer cathode for high-performance aqueous zinc-ion batteries. *ACS Applied Energy Materials*, 2023. **6**(14): p. 7672–7680.
[67] Wu, S., et al., A high-capacity and long-life aqueous rechargeable zinc battery using a porous metal–organic coordination polymer nanosheet cathode. *Inorganic Chemistry Frontiers*, 2018. **5**(12): p. 3067–3073.
[68] Zhao, J., et al., Covalent organic framework film protected zinc anode for highly stable rechargeable aqueous zinc-ion batteries. *Energy Storage Materials*, 2022. **48**: p. 82–89.

[69] He, B., et al., Self-sacrificed synthesis of conductive vanadium-based metal–organic framework nanowire-bundle arrays as binder-free cathodes for high-rate and high-energy-density wearable Zn-ion batteries. *Nano Energy*, 2019. **64**: p. 103935.
[70] Zhao, J., et al., Covalent organic framework film protected zinc anode for highly stable rechargeable aqueous zinc-ion batteries. *Energy Storage Materials*, 2022. **48**: p. 82–89.
[71] Ma, D., et al., A carbonyl-rich covalent organic framework as a high-performance cathode material for aqueous rechargeable zinc-ion batteries. *Chemical Science*, 2022. **13**(8): p. 2385–2390.
[72] Liu, X., et al., One-pot synthesis of high-performance tin chalcogenides/C anodes for Li-Ion batteries. *ACS Omega*, 2021. **6**(27): p. 17391–17399.
[73] Wang, X., et al., Electrochemical swelling induced high material utilization of porous polymers in magnesium electrolytes. *Materials Today*, 2022. **55**: p. 29–36.
[74] Khan, A.A., et al., Use of a porous phenanthrenequinone polymer-based cathode in Mg-ion batteries. *The Journal of Physical Chemistry C*, 2023. **127**(29): p. 14075–14085.
[75] Hassan, H.K., et al., Mixed metal-organic frameworks as efficient semi-solid electrolytes for magnesium-ion batteries. *Batteries & Supercaps*, 2022. **5**(10): p. e202200260.
[76] Sun, R., et al., A covalent organic framework for fast-charge and durable rechargeable Mg storage. *Nano Letters*, 2020. **20**(5): p. 3880–3888.
[77] Kumar, A., et al., Chapter 21 – Battery-supercapacitor hybrid systems: An introduction, in *Nanotechnology in the Automotive Industry*, H. Song, et al., Editors. 2022, Elsevier. p. 453–458.
[78] Wang, L., et al., MOF-derived hierarchical porous carbon octahedrons for aluminum-ion batteries. *Carbon*, 2023. **202**: p. 305–313.
[79] Lu, H., et al., Two-dimensional covalent organic frameworks with enhanced aluminum storage properties. *ChemSusChem*, 2020. **13**(13): p. 3447–3454.
[80] Lu, H., et al., Two-dimensional covalent organic frameworks with enhanced aluminum storage properties. *ChemSusChem*, 2020. **13**(13): p. 3447–3454.
[81] Hong, H., et al., Ordered macro–microporous metal–organic framework single crystals and their derivatives for rechargeable aluminum-ion batteries. *Journal of the American Chemical Society*, 2019. **141**(37): p. 14764–14771.

8 Coordination Nanomaterials for Metal-Air Batteries

Khalid Aziz, Waseem Raza, Nadeem Raza, Andleeb Mehmood, Arshad Hussain, Muhammad Asim Mushtaq and Ghulam Yasin

8.1 INTRODUCTION TO METAL-AIR BATTERIES

One of the key elements of a society operating properly is the equitable distribution and use of the many sources of energy, particularly electricity. To lessen reliance on non-renewable sources, it appears important to establish a reliable and sustainable method of energy harvesting, storing, and delivery [1]. Conservation of energy is becoming more and more well liked, as a power storage facility's performance is determined by how quickly it can respond to adjustments in demand, how much energy it can hold overall, how quickly energy is lost during its storage period, and how simple it is to recharge [2].

Batteries provide a number of benefits over other energy storage options including excellent productivity and a variety of size options. One form of battery that is both more secure and has a better energy density than other varieties is metal-air batteries (MABs) [Figure. 8.1 (a)]. Metal-air batteries, especially lithium-air and zinc-air batteries, are gaining popularity as viable forms of energy preservation due to their affordable price and exceptionally large energy density [3]. Additionally, using airborne oxygen as the primary reactant in metal-air batteries may substantially lower the battery's weight [Figure. 8.1 (b)]. The employment of oxygen from the air around us as a cathode supply has the extra benefit of cutting the expenses and weight of the MAB substantially [4].

Despite decades of development, Li-air batteries (LABs) have only recently attracted a lot of attention. In the development of LABs, the electrolyte has shown to be a limiting element (Figure. 8.1 (c)). Aprotic (non-aqueous) and aqueous electrolyte compositions are the two that Li-air systems use the most. There have also been ideas for mixed electrolyte systems. The initial wave of LAB research concentrated on nonaqueous electrolytes. Abraham et al. [8] published the first research on the nonaqueous Li-air system ($LiPF_6$ in ethylene carbonate (EC)) and suggested a general reaction that would produce Li_2O_2 or Li_2O. Early aprotic Li-air batteries relied on a carbonate solvent, but research has since revealed that carbonate solvents are fragile, releasing carbon dioxide while charging and creating lithium-carbonates

DOI: 10.1201/9781003345886-10

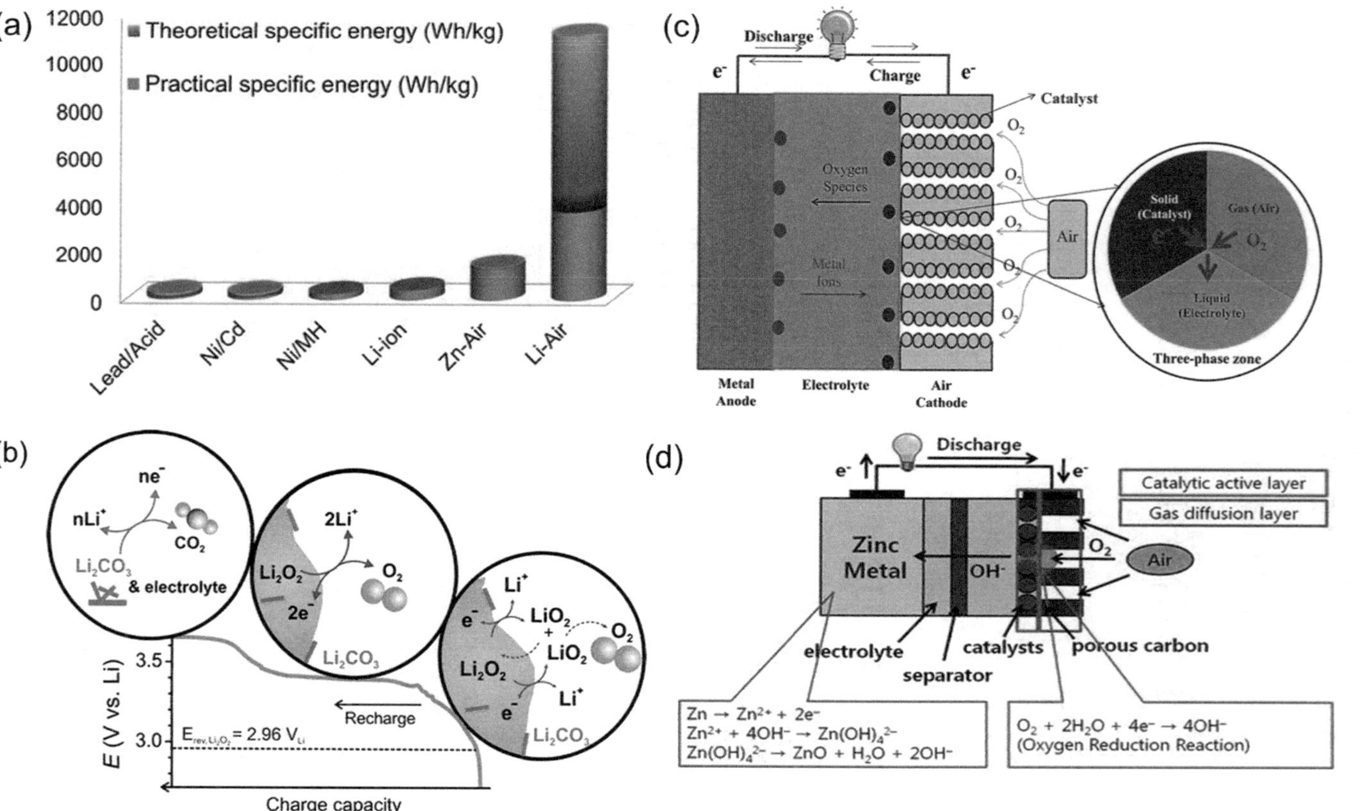

FIGURE 8.1 (a) theoretical vs practical energy densities for MABs [5]; (b) general structure of MABs, [6]; (c) proposed reaction mechanism for Li batteries [7]; (d) Zn batteries [5].

while discharging [9]. In recent times, ether and ester solvents with lithium salts have replaced carbonate electrolytes as the preferred electrolyte [10]. The efficacy of LAB can be hard due to the mobility of oxygen in aprotic electrolytes, particularly at greater electrical densities. It has encouraged researchers to study LABs with aqueous electrolytes and draw lessons from the effectiveness of the gas diffusion electrode in fuel cells.

Zinc anode, zinc surface, and zinc smear or powder combined with additive substance are the typical components of a zinc-oxygen battery [Figure 8.1(d)]. Furthermore, they consist of an alkaline electrolyte-integrated membrane divider and an air-cathode composed of metal, metallic oxide, or carbon-based substances [11]. Zinc-O_2 batteries have an air-based manufacturing process and all of their constituents are moisture stable, in contrast to Li-O_2 batteries. Zn–O_2 batteries are far easier to manage than lithium–oxygen batteries. This is a result of the latter ingredients' propensity to become fragile when exposed to water. Zn-O_2 batteries have an extra benefit over Li-O_2 devices in that their assembly can be performed under normal air ailments whereas Li-O_2 cells need to be constructed in an inert setting. Compared to Li–O_2 batteries, Zn–O_2 batteries are far easier to build. As zinc and oxygen batteries employ water-based electrolytes and zinc is less expensive than lithium, zinc-oxide cells are also a more affordable option than lithium-oxide batteries [12].

Although the design for magnesium-oxygen batteries has been described since the 1960s, no significant progress has been accomplished regarding its commercialism because of a number of unsolved technical issues. These issues include precipitate surveillance, the production of hydrogen, designing electrolytic composition, air dispersion, the separation of anode and cathode materials, and the expensive nature of the production procedures. Theoretical investigation indicates that under neutral circumstances, magnesium-oxygen cells can offer a significant power density, substantial power retrieval, and discharging voltage [13]. Moreover, magnesium metal exhibits strong biological absorption capability, and Mg^{2+} poses no toxicity to humans or the atmosphere. Nevertheless, the broad use of Mg–O_2 batteries is restricted by their significant polarizability and inadequate voltage ratings relative to the theoretical potential in the implementation phase. In the coming years, a significant advancement in magnesium-oxygen batteries will involve the robust functioning of the air electrodes ORR reaction [14].

8.2 COORDINATION NANOMATERIALS FOR MABs

MOCCs, or metal-organic coordination complexes, are made up of an endless array of metal ions connected by coordinating ligands. This generic phrase encompasses a wide range of topologies, from huge mesoporous complexes such as metal-organic frameworks (MOFs) to fundamental a single-dimensional networks with tiny constituents [15]. Metal-organic coordination complexes, among which metal-organic frameworks are particularly prominent, have garnered significant attention in recent times due to their potential uses in a cellular appliance, surface design, gas preservation, and catalytic processes [16]. Owing to its unique characteristics and greater stability over regular complexes because of the chelation or macrocyclic impact, coordination substances containing chelated components are a fascinating area of study in the field of coordination science. They pertain to a unique family of polymeric metallic chelates that have metals in their primary chain and are distinguished by the ability to rupture the polymer core following the removal of the metal [17]. Due to their attractive physicochemical qualities, outstanding resilience, enormous surface area, versatile structure, and minimal expense, substances based on metal-organic frameworks (MOFs) are being extensively researched in the field of solar cell applications. Coordination compounds, which have the following benefits, are often strong candidates for use as electrode materials for MABs: (i) The skeletal variability and versatility of the ligands can be regulated at the molecular scale, enabling the ligands to be redox-active, managed effectively, and made from renewable resources, (ii) the redox-active metal center may contain multiple electrons; (iii) redox-active metal components and natural ligands can work together to improve electrochemical efficiency, and (iv) the production of coordination substances uses only a small amount of energy. In reusable metal-air batteries, the chemical reactions of the oxygen reduction reaction (ORR) and oxygen evolution reaction (OER) generally play significant roles and have a significant impact on their charging and discharging behavior. There is evidence that expensive metals and their analogs, such as Pt and IrO_2, are effective electrodes for each ORR and OER. On the other hand, the lack of supplies for precious metals and their high price has a significant impact on their usability [18].

8.2.1 Porous Coordination Polymers

Porous coordination polymers have greater inherent benefits in the production of integrated electrocatalysts involving mesoporous carbon as a matrix [19]. The consistent organic ligand connections between the metallic ions in the coordinated polymer result in the organized dispersion of the metallic component in the mesoporous carbon matrix. The nearby organic ligands also affect the development of metals [20]. Further, metallic ions may also be produced when the coordinated polymer is calcined, which is advantageous for increasing the electrocatalytic efficiency [21]. Numerous kinds of metals and carbon have been utilized as porous coordination polymers as electrodes for MABs. An inorganic nanocarbon hybrid's construction concurrently boosts the spatial distribution of active sites and the ability to conduct electricity. Superior catalytic effectiveness is the outcome of the catalyst and substrate's complementary connection [22]. Moreover, using porous coordination composites, the nitrogen atoms produced by the calcination of a heterocyclic component incorporating nitrogen can form metal-N bonds with thriving centers. Carbon-supported transition metal/nitrogen (M-Nx/C) substances (M = Co, Fe, Ni, Mn, etc.; typically, $x = 2$ or 4) are among the foremost exciting non-precious metallic catalysts in fuel cell usage owing to their exciting catalytic properties demonstrated through ORR and the use of available affordable prelude materials. The M-Nx/C catalysts fall into two groups: non-pyrolyzed catalysts containing organic states and pyrolyzed catalysts having inorganic states, depending on the production method. During straightforward synthesis processes, non-pyrolyzed M-Nx/C catalyst materials preserve the clearly defined structure of macrocycle complexes, offering beneficial structural oversight for their activities. Through high-temperature treatment, the non-pyrolyzed

M-Nx/C catalysts serve as the foundation for the pyrolyzed M-Nx/C catalysts [23]. For instance, Li et al. [24] suggest that for MABs, a Co-integrated N-doped carbon polygon can be formed by swiftly calcining ZIF-67 at 800°C while an N_2 influx is maintained. Co-PBA was employed by Zeng et al. as an ingredient to produce substantially graphitized carbon cages on conductive Co nanoparticles [25]. A remarkable 350 mWcm_2 power density was shown for a typical Zn-air battery using a prepared electrode following an acidic etching procedure by removing the majority of the metallic Co nanoparticles using 1 M HCl. Ren et al. [26] carbonized ZIF-67 seeded externally on nickel foam to create a transparent nanowire mosaic made of Co_3O_4 precursors and porous carbon substrate. The ensuing composite substance demonstrated a substantial energy density of 118 mW cm^{-2} and good operational sustainability when utilized as an air electrode for ZABs. In the Li-air arrangement, Dong et al. [27] examined the behavior of eight different types of substances, comprising manganese and carbon-based organic polymers. The Mn_2(DOBDC)-based electrodes demonstrated maximal discharging strength of 18,022 mAh g^{-1} at 50 mA g^{-1}, owing to the substantial adsorption of carbon dioxide. Additionally, Mn $(HCOO)_2$ maintained fifty cycles of operation and displayed a reduced charge voltage (4.0 V) at an elevated current density of 200 mA g^{-1}. Leveraging a MOF antecedent (ZIF-8), Zheng et al. reported producing heavily nitrogen (N)-doped porous carbon. Increased N-doping of carbon-containing substances is necessary to improve Li-batteries' electrical functionality. The specimen's substantial amount of nitrogen (17.72 weight%) was more than that of the majority of N-doped carbon compounds that have been reported [28]. A method by Xia et al [29]. transformed MOF (Co-ZIF-67; ZIF) into an extensively graphitized vacuous scaffolding made of N-doped CNT structures adorned with Co particles. Following the calcination process of the ZIF predecessor, a hollow component exhibiting cavities and an N-doped CNT infrastructure was produced. The catalyst performed exceptionally well in terms of ORR/OER. For instance, the NCNTFs catalyst has a better half-wave potential than Pt/C electrocatalyst sold in stores. Subsequently, this study served as inspiration for the creation of many TM-N-C composites generated from MOFs for use in Zn-air batteries. Fe, Cu-coordinated ZIF-obtained carbon structure (Cu@Fe-N-C), which is effectively made by adding Fe^{2+} and Cu^{2+} prior to ZIF-8 development and then pyrolyzing it, has an established topology of the elongated rhombic dodecahedron, as reported by Wang et al [30]. Cu@Fe-N-C, which has been developed, performs exceptionally well in ORR due to its huge expanse of surface, substantial nitrogen loading level, bimetallic component active sites, and conducting carbon scaffolds. A schematic representation of the Cu@Fe-N-C production procedure is shown in Figure. 8.2(a). The ensuing Cu@Fe-N-C was uniformly distributed in Fe and Cu species and possessed the appearance of a tapered rhombic dodecahedron. Following exposure to heat, Cu@Fe-N-C maintained its original tapered rhombic dodecahedron geometry [Figure. 8.2 (a)]. Cu@Fe-N-C exhibits remarkable longevity and a greater energy density of 92 mW cm^{-2} compared to Pt/C (74 mW cm^{-2}) when used as an electrode stimulant in zinc–air batteries. Cu@Fe-N-C had a significantly higher positive onset-potential (E_{onset}) of 1.01 V and significant electrocatalytic operation, as demonstrated by the linear sweep voltammetry (LSV) curves, as illustrated in Figure 8.2 (b). Cu@Fe-N-C was found to have a smaller Tafel slope (78 mV dec-1) than Fe-N-C and

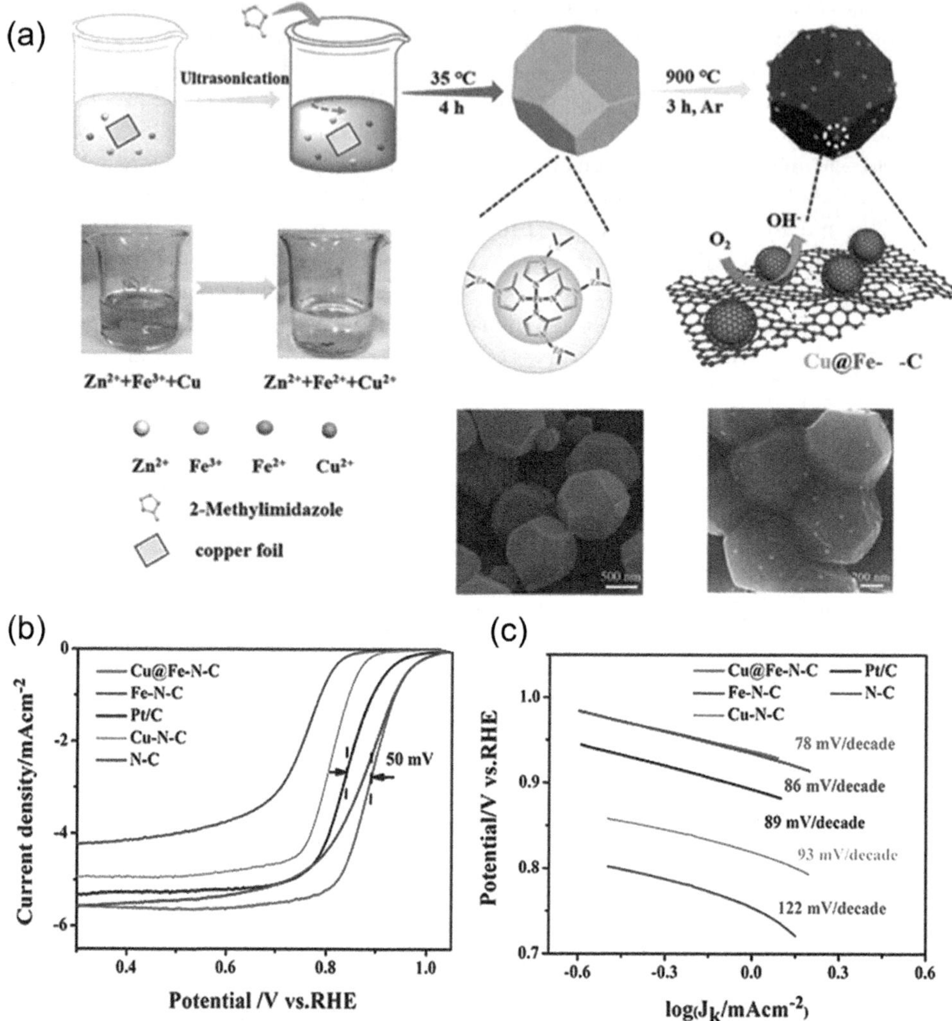

FIGURE 8.2 (a) schematic illustration for synthesis of Cu@Fe-N-C; (b) SEM image of Fe-Cu-ZIF, SEM image of Cu@Fe-N-C, LSV curves; (c) Tofel slope [30].

more favorable ORR kinetics [Figure 8.2(c)]. Further, results illustrate that following 20,000 s, the proportional current of Cu@Fe-N-C stayed at 97.8%, which is significantly higher compared to the 66.2% observed for commercial Pt/C. By dispersion of the substance in 0.1 m Fe $(NO_3)_3$ water solutions at room temperature for 24 h, the Cu particulates were largely eliminated. Analyses revealed that although Fe^{3+} leaching cannot entirely eliminate Cu particles, the alterations in catalytic behavior predicted that it might enhance ORR function. To confirm the exceptional ORR performance of Cu@Fe-N-C, a ring disk electrode (RRDE) experiment and peroxide yield to find the electron transfer rate was performed. Additionally, a domestic, all-solid-state battery provided a substantial 1.35 V open-circuit voltage and was able to adequately operate an LED when used in series with four batteries.

8.3 MOF-DERIVATIVES AS ELECTROCATALYSTS

Metal-organic coordination compounds, a class of organic material that uses organic ligands as intermediaries and metallic cations as the core atoms, offer an extensive range of options for the anodes of MABs that have tunable properties and are inexpensive. To create highly active MOF-derived electrocatalysts, four general approaches are used: (i) enhancing the inherent activity of the influential sites; (ii) raising the mass content of the engaged sites; (iii) quickening the procedure of electron dissemination; and (iv) enhancing the electrocatalysts' resilience [31]. There are two primary types of MOFs as progenitors: MOF antecedents and MOF interconnected precursors. The annealing approach can be used to create M-N-C catalysts, considered to be among the best prospects for use in the discipline of electrochemical processes. DFT calculations revealed that a significant catalysis impact on ORR was played by the M-Nx active sites [32]. Thus, selecting appropriate ligands is crucial. Furthermore, exposure to high temperatures will alter the MOF framework, porousness, and chemical arrangement; the amount of heat and duration of pyrolysis will impact their conductance, ultimately affecting the ORR performance [88–90]. Using a straightforward pyrolysis technique, Wen et al [33] synthesized a Co/N-co doped carbon-graphene electrode (Co@NC-G). They discovered that Co@NC-G, when treated at 700°C, demonstrated outstanding trifunctional electrocatalytic activity in HER, OER, and ORR [33]. Metallic atoms and ligands may interact synergistically to modify the electrical framework, accelerate mass transfer, and increase ORR catalyst efficiency. A well-organized three-dimensional structure is also helpful for displaying more vibrant sites [34]. MOFs are assembled using carbon nanotubes (CNTs) or graphene as carriers through an in-situ fabrication technique. The MOF-derived substances' pore shape, metallic site, and particle dimension all significantly affect how well they catalyze reactions. The variations in the shape and size of particles of the underlying MOFs can be reduced in the derived substances under the right pyrolysis circumstances [35]. In particular, MCCs with adjustable architectures (which incorporate one-dimensional chains, multifaceted layers, and three-dimensional arrangements), a substantial specific surface area, and abundant active sites provide rich and constant routes for the movement of ions between charging and discharging operations. By virtue of their strong oxidation-reduction reaction aptitude and multi-electron interaction mechanism, the described MCCs as anode components have demonstrated astonishing electrochemical performance. Using Zn_4O(1,3,5-benzenetribenzoate)$_2$ (MOF-177) material as an anode, Li et al. tested the lithium storage performance in 2006; this material provided a reasonably high preliminary discharge capacity of 400 mAh g^{-1} at 0.05 A g^{-1}. Despite having unsatisfactory cycle stability, MOF-177 has created a solid framework for using MCCs with LIBs [36]. The types of metal ions and organic ligands are then changed to produce a number of different MCC anode materials, such as $Zn_3(HCOO)_6$ [37], [Pb(4,4'- ocppy)$_2$], $7H_2O$ [38], and Mn-based 1,3,5-benzenetricarboxylate [39], which have shown exceptional electrochemical activity. After 150 cycles of operation at 0.1 mA cm^{-1}, Xu et al.

fabricated a heterometallic (Ni, Co) 2,2 -bipyridine-4–4 -dicarboxylic coordination material via a one-pot solvothermal method [40]. The material has an established discharge/charge capacity of 0.65/0.64 mAh cm^{-2}. However, the mechanical disintegration or fragmentation processes caused by the redox interaction of functional groups and metal cations, which results in elevated irreversible capabilities and rapid capacity decline, pose a challenge to MCCs. MOF materials – particularly MOF-based composite substances and transition-metal-based MOFs containing Ni, Co, Cu, and Fe – have grown quickly as OER and other processes, since they can supply a large number of active sites that speed up the electrocatalytic activity. However, for electrocatalysis usage, the monometallic MOF materials' limited electroconductivity continues to be the main impediment [41]. However, the one-pot approach has been effective in synthesizing a range of bimetallic MOFs. By using the one-pot reflux approach, Fan's group [42] effectively manufactured a range of customized bimetallic MOF catalysts with different molar ratios of Al^{+3} to Fe^{+3}. Trimeric Al^{+3}/Fe^{+3} octahedral clusters and 2-amine-1.4-benzenedicarboxylate ligands were used to build the resulting bimetallic MOFs for OER. Co-N-C nanomaterials have been thoroughly investigated for their potential applications as both bifunctional and monofunctional metal-air battery catalysts for rechargeable Zn-air batteries. It is necessary to incorporate adequate stiff structural components into organic ligand molecules for acceptable architectures in order to produce MCCs with robust structures and strong redox reaction activity. Compared to carboxylic acid ligands containing two oxygen atoms, phosphonic acid having three oxygen atoms possesses a better ability to coordinate. Due to the ease with which phosphonic acid may form P-O-M (M = metal ions) bonds and P-C bonds, the ensuing phosphonic acid-based MCCs have solid frameworks with adjustable voids [43]. Zn-air battery: ZIF-67 was exclusively grown on prepared ZIF-8 nanostructures by Liu et al. [44] to produce ZIF-8@ZIF-67 structures with a core-shell configuration. Following elevated temperature pyrolysis and preserving, the double-shell synthetic nanocage NC@Co-NGC DSNCs containing an N-doped microporous carbon (NC) core-shell and an outer coating of Co-N loaded graphite carbon (Co-NGC) emerged. Figure 8.3 (a) shows a schematic depiction of the synthesis procedure, as well as the epitaxial proliferation and development of the nanocage framework following the carbonization procedure. As seen in Figure. 8.3 [b (I)], an SEM image of the fabricated nanostructures revealed the creation of a polygonal geometry. By using TEM description Figure. 8.3 [b (II)], the arrangement of densely packed dual casings with a consistent thickness was perceived. The ZIF-8-derived NC core-shell enhanced dissemination kinetics with a nanotube-shaped vacuous framework, whereas the ZIF-67-derived Co-NGC exterior had strong ORR catalytic properties, an enduring framework, and good conductance. When used as bifunctional electrodes for ORR/OER, the nanomaterial outperformed Pt/C and RuO_2 in terms of electrocatalytic activity. Using a cyclic voltammogram (CV) with a scan rate of 10 mV s^{-1}, the electrocatalytic efficiency of NC@Co-NGC DSNCs for ORR was initially determined [Figure. 8.3(c)]. Linear sweep voltammograms [Figure. 8.3 (d)] were used to obtain further information about the electrocatalytic function and kinetic variables of NC@Co-NGC DSNCs for ORR. Having an onset potential of around 0.92 V and an equivalent half-wave voltage ($E_{1/2}$) of 0.82 V, the contours of NC@ Co-NGC DSNC conducting electrodes clearly exhibit a point of equilibrium.

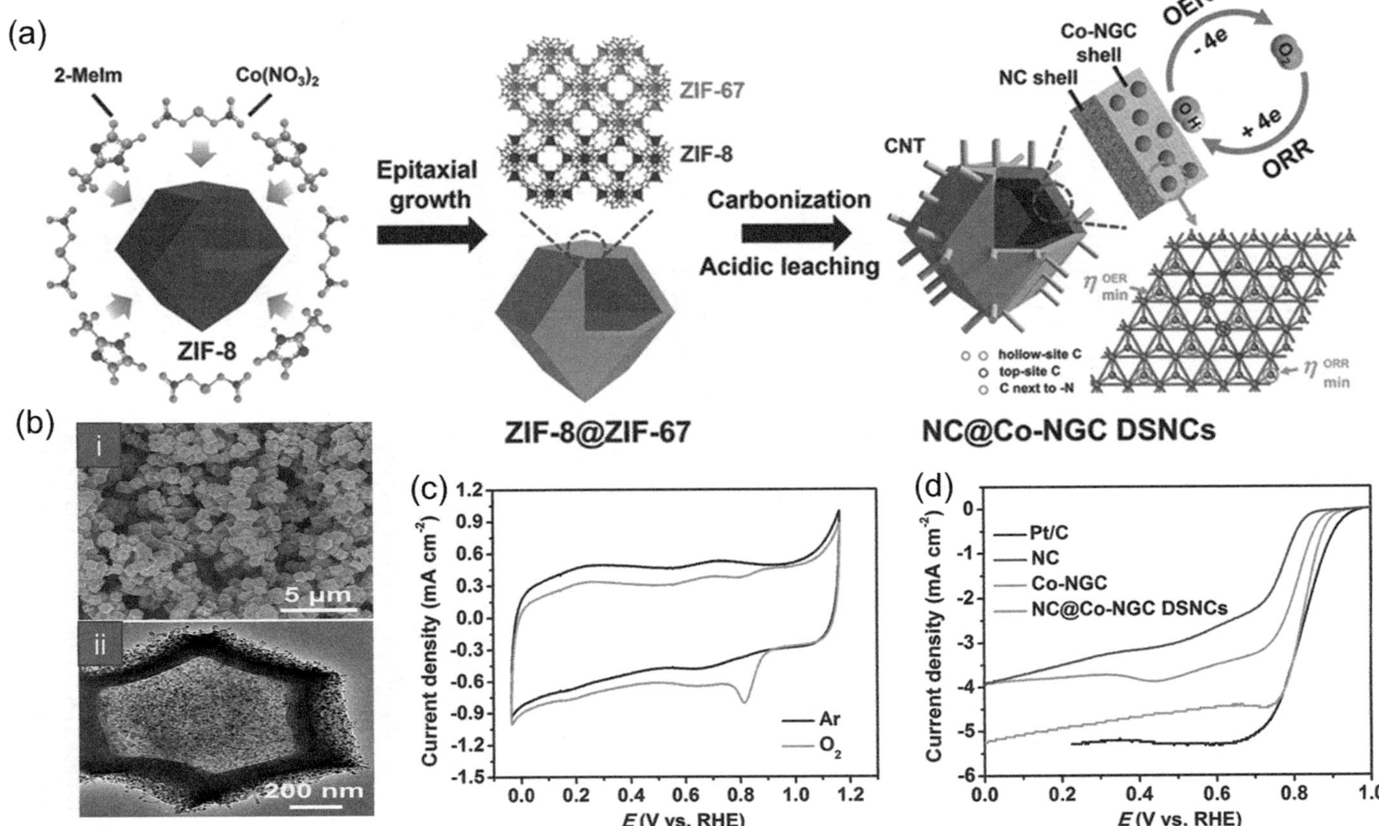

FIGURE 8.3 (a) schematic diagram for synthesis of NC@Co-NGC DSNCs; (b) SEM image [i], TEM image, [ii], of NC@Co-NGC; (c) CV curves for NC@Co-NGC DSNC; (d) ORR polarization curves for NC, Co-NGC, NC@Co-NGC DSNC, and Pt/C [44].

This is similar to the values of Pt/C (0.96 V, $E_{1/2}$ = 0.82 V), and they operate superior to the majority of earlier studied carbon-based ORR electrodes.

In the recent past, creating nanoporous metal/carbon synthetic materials at elevated temperatures without the aid of an additional structure has become one of the increasingly notable uses for metal-organic coordination substances. These composites execute exceptionally well in electrochemical energy storage and transformation devices like supercapacitors, fuel cells, and batteries. Using cobalt ions and N-rich imidazole ligands to build the zeolitic imidazolate framework 67 (ZIF-67), for instance, can result in Co/N-doped nonporous carbon composites after carbonization. The generated carbon composites have the potential to serve as inexpensive catalysts for the oxygen reduction reaction (ORR) and oxygen evaluation reaction (OER), two crucial processes for fuel cells, metal-air batteries, and water splitting. These responses call for noble metal-based catalysts like Pt and RuO_2 [45, 46]. Choi and colleagues [47] created multi-layered MOFs that alternate between disintegrating and anchoring phases in order to generate and sustain nanostructures in a modular sequentially loading process. Crucially, π-antibonding over many shell layers is what stabilizes nanostructures, and the cross-layer switching process can greatly reduce the propagation susceptibility of empty interspaces to attain excellent conductivity. $Zn_3[Fe(CN)_6]_2$ cubes were annealed by Hou et al. to create a multi-layer mesoporous bimetallic component $ZnO/ZnFe_2O_4$, (ZZFO). Following 200 cycles of operation, the ZZFO's discharging specific capacity held steady and even increased to approximately 837 mA h g^{-1} at 1000 mA g^{-1} [48]. A Co/CoOx embellished 1D N-decorated carbon nanotube framework (Co@CoO_x/NCNTs) was described by Chao et al. [49]. ZnO nanowires were used to provide extra oxygen atoms for oxidizing Co. Compared with alternative composite materials, a primary Zn–air battery constructed using the aforementioned material showed significantly improved charging efficiency. Leveraging MOF-derived material, one of the finest high-rate outcomes in primary Zn-air batteries was the outstanding discharging capacity of 0.69 V at 500 $mAcm^{-2}$.

In order to develop M-FeCo-ZIFs-X nanostructures with uniformly dispersed macropores, polystyrene tiny particles stencils were utilized as precursors. Stacked M-FeCo-ZIFs-X also functioned as progenitors. After 192 rounds of continuous operation, the M-FeCo-N-C-0 based cathode demonstrated an excellent potential of 7900 mAh/g at a current density of 0.5 A g^{-1}, in addition to a substantial starting capacity of 18,750 mAh/g at a current density of 0.1 A g^{-1}. The remarkable outcome is attributed to the combined action of N co-doping, Fe, and Co nanocrystals on carbon scaffolds with unique micro-meso structures [50]. An autonomous array-structured Co_3O_4@$NiCo_2O_4$ electrode was created by growing cobalt-based metal-organic framework (Co-MOF) layers existing on carbon fiber, impregnating the material with nickel nitrate remedy, and then pyrolyzing the mixture during relocation. A schematic representation for the fabrication of a composite material using a mixture of aerosol decomposition and electrodeposition is shown in Figure 8.4a (i). The special composition improves electronic conduction, electrolytic proximity, Li^+/O_2 transportation and dissemination, and simultaneous ORR/OER kinetics when employed as a cathode stimulant in Li-O_2 cells [Figure. 8.4 a (ii)]. SEM and TEM images of the Fe_2O_3 @C prelude received following aerosol spray pyrolysis are presented in Figure 8.4 b (i, ii). The precursor has a cylindrical shape having a diameter

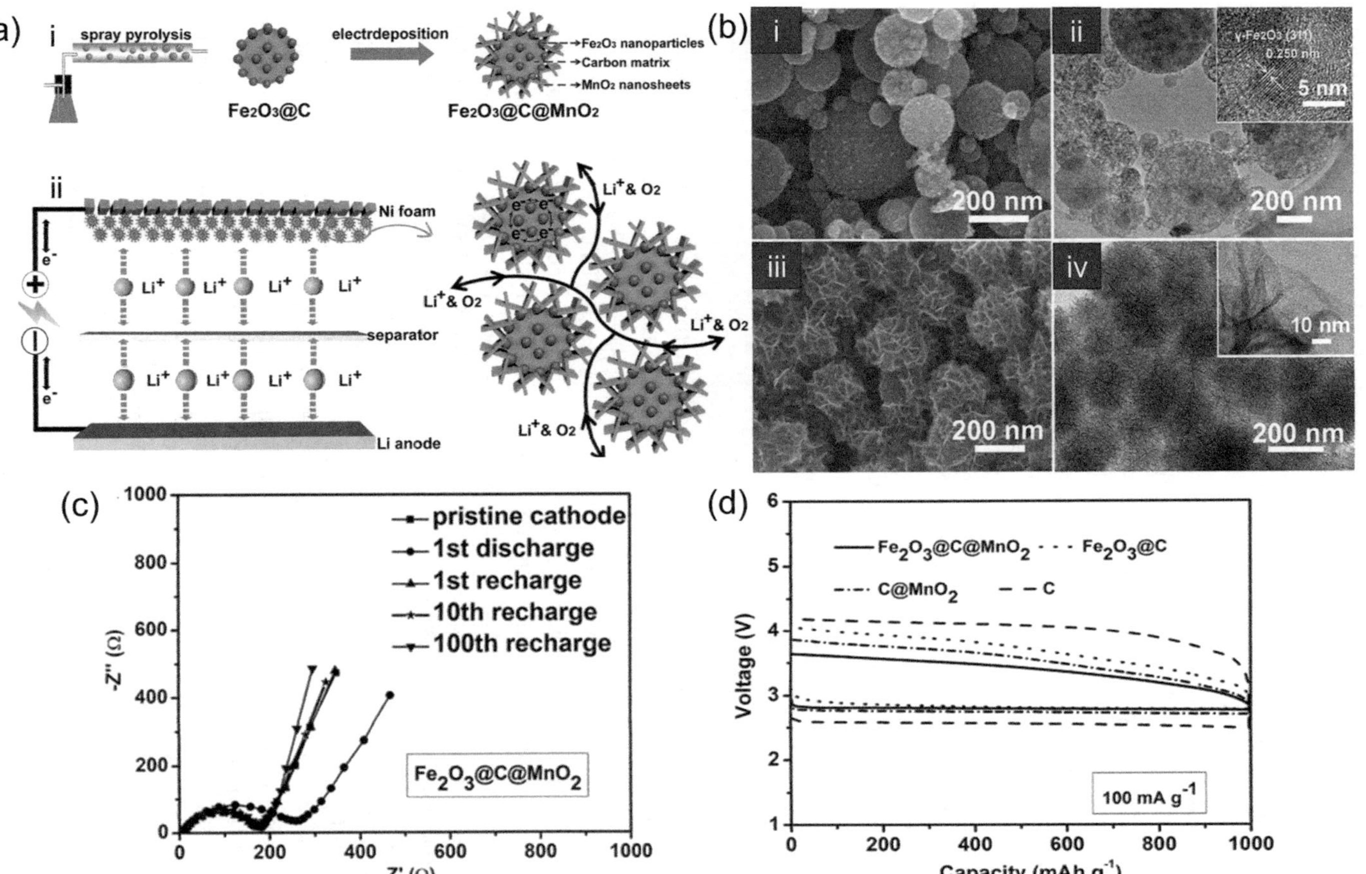

FIGURE 8.4 (ai) schematic diagram for synthesis; (aii) applications in Li-O_2 batteries; (b) SEM and TEM images, [i, ii, iii and iv)], (c) EIS curves; (d) charging-discharging curves of Fe_2O_3@C@MnO_2 [51].

of 50–400 nm, and each of its spheres contains uniformly dispersed nanoparticles that display fringe-like lattices of crystallized γ-Fe_2O_3. [Figure 8.4 b (ii), inset]. After the ferric oxide was etched with HCl, the spherical shape was retained, but there were many pits, which showed that the nanostructures of ferric oxide had been embedded in the framework of carbon. The SEM and TEM images of the synthesized Fe O_3 @C@MnO_2 are displayed in Figure 8.4b (iii, iv). A striking textural contrast between the pre- and post-electrodeposition of MnO_2 shows that the cylindrical Fe_2O_3 @C progenitor is consistently enveloped in fragile, shredded, and linked nanosheets. Charge-transfer resistance (R_{ct}) jumped dramatically from 175 to 270 Ω following the initial discharge, according to the EIS analysis [Figure 8.4(c)]. This may be explained by the interface between the electrode and the electrolyte producing a substantial amount of insulating accumulates or discharging substances. The usual charge/discharge curves at a maintained ultimate capacity of 1,000 mAh g^{-1} are displayed in Figure 8.4(d). Fe_2O_3 @C has a voltage gradient that is substantially greater (2.81 V) compared to that of C (2.56 V) and slightly greater (2.75 V) compared to that of C@MnO_2. Featuring a specific capacity of 10645 mAh g^{-1} as opposed to 6773 mAh g^{-1} for Co_3O_4/CC and 9354 mAh g^{-1} for $NiCo_2O_4$/CC, the Co_3O_4@$NiCo_2O_4$/CC demonstrated superior charging and discharging behavior. It also attained a specific capacity of 10,645 mAh g^{-1} and robustness for 225 sessions without any discernible decline [51].

8.3.1 Covalent Organic Frameworks

A possible platform for achieving sequentially metrical manipulation of organic molecular constituents with strong chemical and constructional durability is covalent organic frameworks, or COFs. The configurable chemistry and customizable architectures of COFs are a result of their organic origin. The energy density of the electrode substance can be changed by predesigning the redox-active domains into COF progenitors with exact densities and locations. The slow ion transport issues in batteries can be effectively solved by carefully anchoring organic functionalities that enhance metallic ion movement in the framework of COFs. Since they have the potential to interact with more engaged sites in addition to having innate exertion, COFs with spacious characteristics make excellent heterogeneous electrocatalysts. For the Zn-air battery, Zhao et al. produced a Fe-N doped mesoporous carbon (mC-TpBpy-Fe) as an ORR catalyst. [52] Following carbonization and template erasure, the mC-TpBpy-Fe was synthesized by synthesizing a COF encapsulating bipyridine using a p-toluene sulfonic acid-supported mechanochemical technique. In mesoporous carbon-based substances, the bipyridine group in the COF helped for coordinating the movement of Fe metal ions. Furthermore, mesoporous carbon's wide pores and copious permeability efficiently aided in the passage of O_2 and electrolyte to the Fe-N_x active sites, resulting in superior ORR performance. Peng and colleagues [53] developed an operational COF (COFBTC) to serve as a cathode material for Zn-air flow batteries. These batteries were interconnected by a stiff conjugated framework and a plentiful Fe-N_4 active core. Solvent molecules in solution are susceptible to in-situ incorporation and abrasion due to the strong interactions that the Fe single-atom core can generate with hydroxide alongside other solvent molecules. Crucially, the Fe-N-C single-atom framework, which is enriched in nitrogen coordination, has exceptional electrocatalytic efficiency for ORR. These findings offer a fresh concept, a new approach, and a point of reference for the growth of soluble COFs in Zn-air batteries. Porphyrin covalent organic framework was created by Li et al [54] to serve as a cathode component in zinc air batteries. In particular, the liquid Zn air battery using the CNT@POF cathode performs substantially better compared to the noble metal cathode, displaying an exceptional robustness for 200 cycles and a narrow voltage gap of 0.71 V. When bent to various degrees, adaptable all-solid-state Zn air batteries show an excellent energy capacity of 61.6% at 1.0 mA cm > 2 and the ability to illuminate a red light-emitting diode (LED, 2.0 V) steadily. To synthesize a very stable COF, Peng et al [55] suggested a novel in-situ charge exfoliating in an alkaline environment. They used the solution straight away as a non-noble Pt electrode, which allowed them to achieve excellent efficiency in Zn-air flow batteries. The COF demonstrated great potential as the oxygen reduction catalyst due to its conjugated network and clearly defined N-coordinate Fe single-atom cores. Even in terms of stability, the COF solution outperformed conventional Pt/C-coated air electrodes. A COF with atomically meticulously planned positively charged centers restricted in a planar configuration was synthesized by Peng et al [56] and showed remarkable mobility via an in-situ charge exfoliating mechanism. With its conjugated architectures and regular N-coordinated Fe single-atom centers, as-designed soluble COF exhibits outstanding catalytic efficiency for oxygen reduction, resulting in a modest work function of 4.84 eV and an excellent half-wave potential of around 900 mV. With zinc-air flow batteries, the resulting COF solution was utilized directly as an incredibly efficient Pt-replaced catalyst, producing notable efficiency and exceptional stability.

8.4 CHALLENGES

Even while coordination-based nanomaterials for energy preservation systems have advanced significantly thus far, there are still a number of issues that need to be resolved. For pertinent energy storage systems, particularly electrochemical ones in which repressive potentials and electrolytes are employed, the inadequate cyclical resilience and impoverished electrical resistance of coordination substances and MOF-based materials must be addressed. Evaluation of the adaptability, interactions, and development of the synergistic impact involving the coordination ingredient and the added constituents are still difficult tasks for coordination-based nanocomposites. In order to improve resilience and electrocatalytic efficiency, COFs commonly develop insoluble, cross-linked flakes or films, which limits their processabilities and potential uses in MABs. The complexity of the transformation process from coordination antecedents to coordination-derived substances in the realm of coordination-derived materials causes unregulated compositional variations and architectural progression deviating from the specified patterns and formulations.

8.5 CONCLUSION

Even though MABs emerged as the subject of extensive research as a renewable and environment-friendly source of energy over the past years, there are still several obstacles in the way of their actual deployment. It is difficult to formulate an enduring, vibrant, affordable, and bi-functional air coordination catalyst that is suitable for both OER and ORR. Current studies demonstrate the advancements made recently in the advancement of bifunctional ORR/OER electrocatalysts as MAB air cathodes. The benefits and drawbacks of the coordination materials that include pristine MOFs, MOF-derived single-atom catalysts, metal-based catalysts, nitrogen-based catalysts, and oxygen-based catalysts for MABs, have been compiled. These include exertion, endurance, conductance, selective nature, affordability, and ecological friendliness. The primary objective of this study is to develop bifunctional ORR/OER catalysts that may substitute noble metal nanomaterials in rechargeable MAB applications, allowing for the efficient and cost-effective use of energy.

Although there are still many obstacles to overcome, it is anticipated that coordination-based components with appealing electro/photochemical features could one day be achieved for use in batteries thanks to the development of more sophisticated methods for characterization and a fundamentally profound comprehension of the structure-performance attachment of these substances for storage of energy by electrochemical operations. Furthermore, it is thought that further study could lead to a substantially improved design for metal-air batteries in the foreseeable future. To improve the electrode's catalytic efficacy for MABs, coordination materials' electrical conductivity needs to be significantly increased. By skillfully adjusting the framework of coordination substances, such as MOFs and COFs, one can improve their conductive properties and cyclic resilience. This can be achieved through the use of alloy-based nanomaterials, single-atom catalysts, metallic catalysts, oxygen, carbon, and nitrogen. To maintain extended cycling endurance as a cathode component for MABs, coordination substances' pore and transmit configurations should be rationally designed by controlling their composition, and their durability should be improved.

REFERENCES

[1] M. Rehan, M.A. Raza, A.G. Abro, M. Aman, I.M.I. Ismail, A.S. Nizami, M.I. Rashid, A. Summan, K. Shahzad, N. Ali, A sustainable use of biomass for electrical energy harvesting using distributed generation systems, *Energy* (2023) 128036.

[2] C.S. Li, Y. Sun, F. Gebert, S.L. Chou, Current progress on rechargeable magnesium–air battery, *Advanced Energy Materials* 7(24) (2017) 1700869.

[3] K.F. Blurton, A.F. Sammells, Metal/air batteries: their status and potential – a review, *Journal of Power Sources* 4(4) (1979) 263–279.

[4] D.U. Lee, P. Xu, Z.P. Cano, A.G. Kashkooli, M.G. Park, Z. Chen, Recent progress and perspectives on bi-functional oxygen electrocatalysts for advanced rechargeable metal–air batteries, *Journal of Materials Chemistry A* 4(19) (2016) 7107–7134.

[5] J.S. Lee, S. Tai Kim, R. Cao, N.S. Choi, M. Liu, K.T. Lee, J. Cho, Metal–air batteries with high energy density: Li–air versus Zn–air, *Advanced Energy Materials* 1(1) (2011) 34–50.

[6] Y.-J. Wang, B. Fang, D. Zhang, A. Li, D.P. Wilkinson, A. Ignaszak, L. Zhang, J. Zhang, A review of carbon-composited materials as air-electrode bifunctional electrocatalysts for metal–air batteries, *Electrochemical Energy Reviews* 1 (2018) 1–34.

[7] Z.-L. Wang, D. Xu, J.-J. Xu, X.-B. Zhang, Oxygen electrocatalysts in metal–air batteries: from aqueous to nonaqueous electrolytes, *Chemical Society Reviews* 43(22) (2014) 7746–7786.

[8] K. Abraham, Z. Jiang, A polymer electrolyte-based rechargeable lithium/oxygen battery, *Journal of the Electrochemical Society* 143(1) (1996) 1.
[9] X. Wei, D. Desai, G.G. Yadav, D.E. Turney, A. Couzis, S. Banerjee, Impact of anode substrates on electrodeposited zinc over cycling in zinc-anode rechargeable alkaline batteries, *Electrochimica Acta* 212 (2016) 603–613.
[10] N. Mahne, O. Fontaine, M.O. Thotiyl, M. Wilkening, S.A. Freunberger, Mechanism and performance of lithium–oxygen batteries–a perspective, *Chemical Science* 8(10) (2017) 6716–6729.
[11] J. Zhang, Q. Zhou, Y. Tang, L. Zhang, Y. Li, Zinc–air batteries: are they ready for prime time?, *Chemical Science* 10(39) (2019) 8924–8929.
[12] Y. Li, H. Dai, Recent advances in zinc–air batteries, *Chemical Society Reviews* 43(15) (2014) 5257–5275.
[13] T. Zhang, Z. Tao, J. Chen, Magnesium–air batteries: from principle to application, *Materials Horizons* 1(2) (2014) 196–206.
[14] L. Zhang, Q. Shao, J. Zhang, An overview of non-noble metal electrocatalysts and their associated air cathodes for Mg-air batteries, *Materials Reports: Energy* 1(1) (2021) 100002.
[15] B. Zhu, Z. Liang, D. Xia, R. Zou, Metal-organic frameworks and their derivatives for metal-air batteries, *Energy Storage Materials* 23 (2019) 757–771.
[16] X. Wen, Q. Zhang, J. Guan, Applications of metal–organic framework-derived materials in fuel cells and metal-air batteries, *Coordination Chemistry Reviews* 409 (2020) 213214.
[17] M.E. Hilal, A. Aboulouard, A.R. Akbar, H.A. Younus, N. Horzum, F. Verpoort, Progress of MOF-derived functional materials toward industrialization in solar cells and metal-air batteries, *Catalysts* 10(8) (2020) 897.
[18] L. Gong, D. Zhang, C.Y. Lin, Y. Zhu, Y. Shen, J. Zhang, X. Han, L. Zhang, Z. Xia, Catalytic mechanisms and design principles for single-atom catalysts in highly efficient CO_2 conversion, *Advanced Energy Materials* 9(44) (2019) 1902625.
[19] Y. Tao, D.-H. Yang, H.-Y. Kong, T.-X. Wang, Z. Li, X. Ding, B.-H. Han, Covalent triazine polymer derived porous carbon with high porosity and nitrogen content for bifunctional oxygen catalysis in zinc–air battery, *Applied Catalysis B: Environmental* 339 (2023) 123088.
[20] Q. Niu, B. Chen, J. Guo, J. Nie, X. Guo, G. Ma, Flexible, porous, and metal–heteroatom-doped carbon nanofibers as efficient ORR electrocatalysts for Zn–air battery, *Nano-Micro Letters* 11 (2019) 1–17.
[21] Y. Deng, J. Zheng, B. Liu, H. Li, M. Yang, Z. Wang, Schiff-base polymer derived FeCo-N-doped porous carbon flowers as bifunctional oxygen electrocatalyst for long-life rechargeable zinc-air batteries, *Journal of Energy Chemistry* 76 (2023) 470–478.
[22] X.-X. Xing, H.-L. Guo, T. Feng, T.-M. He, W.-S. Zhu, H.-M. Li, J.-Y. Pang, Y. Bai, D.-B. Dang, Design and synthesis of amphiphilic catalyst [C16mim] 5VW12O40Br and its application in deep desulfurization with superior cyclability at room temperature, *Inorganic Chemistry* 62(14) (2023) 5780–5790.
[23] J. Lilloja, E. Kibena-Põldsepp, A. Sarapuu, M. Käärik, J. Kozlova, P. Paiste, A. Kikas, A. Treshchalov, J. Leis, A. Tamm, Transition metal and nitrogen-doped mesoporous carbons as cathode catalysts for anion-exchange membrane fuel cells, *Applied Catalysis B: Environmental* 306 (2022) 121113.
[24] J.-C. Li, X.-T. Wu, L.-J. Chen, N. Li, Z.-Q. Liu, Bifunctional MOF-derived Co-N-doped carbon electrocatalysts for high-performance zinc-air batteries and MFCs, *Energy* 156 (2018) 95–102.
[25] M. Zeng, Y. Liu, F. Zhao, K. Nie, N. Han, X. Wang, W. Huang, X. Song, J. Zhong, Y. Li, Metallic cobalt nanoparticles encapsulated in nitrogen-enriched graphene shells: its bifunctional electrocatalysis and application in zinc–air batteries, *Advanced Functional Materials* 26(24) (2016) 4397–4404.

[26] J.-T. Ren, G.-G. Yuan, C.-C. Weng, Z.-Y. Yuan, Rationally designed Co3O4–C nanowire arrays on Ni foam derived from metal organic framework as reversible oxygen evolution electrodes with enhanced performance for Zn–air batteries, *ACS Sustainable Chemistry & Engineering* 6(1) (2018) 707–718.
[27] S. Li, Y. Dong, J. Zhou, Y. Liu, J. Wang, X. Gao, Y. Han, P. Qi, B. Wang, Carbon dioxide in the cage: manganese metal–organic frameworks for high performance CO 2 electrodes in Li–CO 2 batteries, *Energy & Environmental Science* 11(5) (2018) 1318–1325.
[28] F. Zheng, Y. Yang, Q. Chen, High lithium anodic performance of highly nitrogen-doped porous carbon prepared from a metal-organic framework, *Nature Communications* 5(1) (2014) 5261.
[29] B.Y. Xia, Y. Yan, N. Li, H.B. Wu, X.W.D. Lou, X. Wang, A metal–organic framework-derived bifunctional oxygen electrocatalyst, *Nature Energy* 1(1) (2016) 1–8.
[30] Z. Wang, H. Jin, T. Meng, K. Liao, W. Meng, J. Yang, D. He, Y. Xiong, S. Mu, Fe, Cu-coordinated ZIF-derived carbon framework for efficient oxygen reduction reaction and zinc–air batteries, *Advanced Functional Materials* 28(39) (2018) 1802596.
[31] U.I. Kramm, J. Herranz, N. Larouche, T.M. Arruda, M. Lefèvre, F. Jaouen, P. Bogdanoff, S. Fiechter, I. Abs-Wurmbach, S. Mukerjee, Structure of the catalytic sites in Fe/N/C-catalysts for O 2-reduction in PEM fuel cells, *Physical Chemistry Chemical Physics* 14(33) (2012) 11673–11688.
[32] X. Wen, Z. Duan, L. Bai, J. Guan, Atomic scandium and nitrogen-codoped graphene for oxygen reduction reaction, *Journal of Power Sources* 431 (2019) 265–273.
[33] X. Wen, X. Yang, M. Li, L. Bai, J. Guan, Co/CoOx nanoparticles inlaid onto nitrogen-doped carbon-graphene as a trifunctional electrocatalyst, *Electrochimica Acta* 296 (2019) 830–841.
[34] T. Schuler, J.M. Ciccone, B. Krentscher, F. Marone, C. Peter, T.J. Schmidt, F.N. Büchi, Hierarchically structured porous transport layers for polymer electrolyte water electrolysis, *Advanced Energy Materials* 10(2) (2020) 1903216.
[35] C.C. Hou, L. Zou, Q. Xu, A hydrangea-like superstructure of open carbon cages with hierarchical porosity and highly active metal sites, *Advanced Materials* 31(46) (2019) 1904689.
[36] X. Li, F. Cheng, S. Zhang, J. Chen, Shape-controlled synthesis and lithium-storage study of metal-organic frameworks Zn4O (1, 3, 5-benzenetribenzoate) 2, *Journal of Power Sources* 160(1) (2006) 542–547.
[37] K. Saravanan, M. Nagarathinam, P. Balaya, J.J. Vittal, Lithium storage in a metal organic framework with diamondoid topology–a case study on metal formates, *Journal of Materials Chemistry* 20(38) (2010) 8329–8335.
[38] L. Hu, X.-M. Lin, J.-T. Mo, J. Lin, H.-L. Gan, X.-L. Yang, Y.-P. Cai, Lead-based metal–organic framework with stable lithium anodic performance, *Inorganic Chemistry* 56(8) (2017) 4289–4295.
[39] S. Maiti, A. Pramanik, U. Manju, S. Mahanty, Reversible lithium storage in manganese 1, 3, 5-benzenetricarboxylate metal–organic framework with high capacity and rate performance, *ACS Applied Materials & Interfaces* 7(30) (2015) 16357–16363.
[40] N. Xu, Q. Han, L. Zhu, L. Xie, J. Xu, W. Zhang, X. Yang, X. Cao, Design and synthesis of heterometallic Ni–Co organic frameworks as anode materials for high-performance lithium storage, *Journal of The Electrochemical Society* 169(3) (2022) 030526.
[41] S. Li, Y. Gao, N. Li, L. Ge, X. Bu, P. Feng, Transition metal-based bimetallic MOFs and MOF-derived catalysts for electrochemical oxygen evolution reaction, *Energy & Environmental Science* 14(4) (2021) 1897–1927.
[42] Y. Hu, J. Zhang, H. Huo, Z. Wang, X. Xu, Y. Yang, K. Lin, R. Fan, One-pot synthesis of bimetallic metal–organic frameworks (MOFs) as acid–base bifunctional catalysts for tandem reaction, *Catalysis Science & Technology* 10(2) (2020) 315–322.

[43] Y. Lu, L. Wang, Z. Lou, L. Wang, Y. Zhao, W. Sun, L. Lv, Y. Wang, S. Chen, A Nickel-Based coordination compound with tunable morphology for high-performance anode and the lithium storage mechanism, *Batteries* 9(6) (2023) 313.
[44] S. Liu, Z. Wang, S. Zhou, F. Yu, M. Yu, C.Y. Chiang, W. Zhou, J. Zhao, J. Qiu, Metal–organic-framework-derived hybrid carbon nanocages as a bifunctional electrocatalyst for oxygen reduction and evolution, *Advanced Materials* 29(31) (2017) 1700874.
[45] Y. Deng, Y. Dong, G. Wang, K. Sun, X. Shi, L. Zheng, X. Li, S. Liao, Well-defined ZIF-derived Fe–N codoped carbon nanoframes as efficient oxygen reduction catalysts, *ACS Applied Materials & Interfaces* 9(11) (2017) 9699–9709.
[46] C.Y. Su, H. Cheng, W. Li, Z.Q. Liu, N. Li, Z. Hou, F.Q. Bai, H.X. Zhang, T.Y. Ma, Atomic modulation of FeCo–nitrogen–carbon bifunctional oxygen electrodes for rechargeable and flexible all-solid-state zinc–air battery, *Advanced Energy Materials* 7(13) (2017) 1602420.
[47] W.H. Choi, B.C. Moon, D.G. Park, J.W. Choi, K.H. Kim, J.S. Shin, M.G. Kim, K.M. Choi, J.K. Kang, Autogenous production and stabilization of highly loaded sub-nanometric particles within multishell hollow metal–organic frameworks and their utilization for high performance in Li–O2 batteries, *Advanced Science* 7(9) (2020) 2000283.
[48] L. Hou, L. Lian, L. Zhang, G. Pang, C. Yuan, X. Zhang, Self-sacrifice template fabrication of hierarchical mesoporous Bi-component-active ZnO/ZnFe2O4 sub-microcubes as superior anode towards high-performance lithium-ion battery, *Advanced Functional Materials* 25(2) (2015) 238–246.
[49] C. Lin, S.S. Shinde, Z. Jiang, X. Song, Y. Sun, L. Guo, H. Zhang, J.-Y. Jung, X. Li, J.-H. Lee, In situ directional formation of Co@ CoO x-embedded 1D carbon nanotubes as an efficient oxygen electrocatalyst for ultra-high rate Zn–air batteries, *Journal of Materials Chemistry A* 5(27) (2017) 13994–14002.
[50] J. Li, Y. Deng, L. Leng, M. Liu, L. Huang, X. Tian, H. Song, X. Lu, S. Liao, MOF-templated sword-like Co3O4@ NiCo2O4 sheet arrays on carbon cloth as highly efficient Li–O2 battery cathode, *Journal of Power Sources* 450 (2020) 227725.
[51] X. Hu, F. Cheng, N. Zhang, X. Han, J. Chen, Nanocomposite of Fe2O3@ C@ MnO2 as an efficient cathode catalyst for rechargeable lithium–oxygen batteries, *Small* 11(41) (2015) 5545–5550.
[52] X. Zhao, P. Pachfule, S. Li, T. Langenhahn, M. Ye, G. Tian, J. Schmidt, A. Thomas, Silica-templated covalent organic framework-derived Fe–N-doped mesoporous carbon as oxygen reduction electrocatalyst, *Chemistry of Materials* 31(9) (2019) 3274–3280.
[53] P. Peng, L. Shi, F. Huo, S. Zhang, C. Mi, Y. Cheng, Z. Xiang, In situ charge exfoliated soluble covalent organic framework directly used for Zn–air flow battery, *ACS Nano* 13(1) (2019) 878–884.
[54] B.-Q. Li, S.-Y. Zhang, B. Wang, Z.-J. Xia, C. Tang, Q. Zhang, A porphyrin covalent organic framework cathode for flexible Zn–air batteries, *Energy & Environmental Science* 11(7) (2018) 1723–1729.
[55] P. Peng, L. Shi, F. Huo, C. Mi, X. Wu, S. Zhang, Z. Xiang, A pyrolysis-free path toward superiorly catalytic nitrogen-coordinated single atom, *Science Advances* 5(8) (2019) eaaw2322.
[56] S. Wei, Y. Wang, W. Chen, Z. Li, W.-C. Cheong, Q. Zhang, Y. Gong, L. Gu, C. Chen, D. Wang, Atomically dispersed Fe atoms anchored on COF-derived N-doped carbon nanospheres as efficient multi-functional catalysts, *Chemical Science* 11(3) (2020) 786–790.

9 Coordination Materials for Supercapacitors

Jai Kumar, Rana R. Neiber, Mohammad Tabish and Ghulam Yasin

9.1 INTRODUCTION

Non-renewable energy sources are being depleted at an ever-increasing rate as a direct consequence of the rapid expansion of the economy. The usage of various energy sources simultaneously results in a greenhouse impact in addition to other environmental issues, which need to be addressed. For the purpose of addressing these pressing environmental issues, it is absolutely necessary to encourage the development of energy sources that are both environmentally friendly and sustainable [1, 2]. It has been demonstrated that electrochemical energy storage devices, such as supercapacitors (SCs), have a significant potential to facilitate the utilization of energy that is generated from new renewable technology. This potential has been established through research and development [3, 4]. Consequently, it is of paramount significance to research the technologies that convert and store energy using electrochemical processes [5]. Electrochemical devices known as SCs have demonstrated significant potential for application in various industries, such as electric vehicles, backup power storage, portable electronic gadgets, and other settings [6]. Their achievement can be attributed to the high-power density, excellent electrochemical reversibility, and remarkable cycle stability of the materials. SCs have been widely employed in various sectors of the business, science, technology, and everyday life across the country. Empirical data has demonstrated that these applications provide a substantial contribution to alleviating the energy and environmental crisis, while concurrently enhancing the overall quality of life for the general populace. Furthermore, stem cells have become a new subject of interest for both the academic community and the general public [7]. The investigation of high-performance materials is the central topic of interest in this specific field. The act of SCs is mostly influenced through the materials used in their electrode fabrication.

9.1.1 Overview of Supercapacitors

The first electrochemical capacitor (EC) was introduced by NEC company in 1978 under the brand name "Supercapacitor" [8]. The development of EC began back in the early 1950s by engineers working for General Electric. The decade of the 1980s saw a gradual expansion of the market, during which time a number of businesses started producing ECs. The underlying behavior of ruthenium oxide as a base

 DOI: 10.1201/9781003345886-11

electrochemical capacitor was investigated by Brian Evans Conway in the year 1999. This was done before the capacitance was revealed by the surface redox reaction with faradaic charge allocation between electrodes and ions [8]. Our understanding of electrochemical capacitors and the consequent mechanics of pseudocapacitors was greatly improved as a result of Conway's study [9]. TESLA, Maxwell Technologies, Panasonic, Nesscap, Tecate, EPCOS, NEC, Murata, ELNA, and TOKIN are among the many firms that are making significant investments in the research and development of ECs. The automotive industry, hybrid mobility systems, grid stabilizations, and commercial vehicles are some of the application areas that are now being considered for electric vehicles. An electrolyte that ionically connects both electrodes and an ion-permeable membrane (separator) that separates the two electrodes are the two components that make up the fundamental design of an electrochemical cell. As shown in Figure 9.1a, the key components of the supercapacitor device included electrode material, electrolytes, a current collector, and separators. There are three distinct configurations that can be applied to supercapacitors, and these configurations are determined by the working mechanism (Figure 9.1b):

Electric double-layer capacitors (EDLCs): An electrochemical process called non-faradaic charge storage occurs when an applied voltage causes the separation of charges in the electrolyte. This results in the formation of a thin layer of ions that adhere to the surface of the electrode material, known as electric double layer capacitance.

Pseudocapacitors: Ions are absorbed onto the surface of the electrode as a consequence of redox reactions, electrosorption, or intercalation activities that take place on the electrode surface. Therefore, this results in the production of a charge storage mechanism that is referred to as faradaic pseudocapacitance. This mechanism is characterized by the occurrence of reversible charge transfer on the electrode.

Hybrid supercapacitors: The construction of hybrid supercapacitors involves merging the electrical characteristics of EDLCs and pseudocapacitors in order to reap the benefits of both types of storage mechanisms.

Battery-type supercapacitors: Hybrid capacitors or lithium-ion capacitors (LICs) merge the substantial energy storage capability of batteries with the formidable power discharge of supercapacitors. Regular batteries and supercapacitors exhibit varying levels of energy and power density. The design of these batteries aims to effectively balance the trade-offs inherent in both types. The battery-type supercapacitors are constructed utilizing a hybrid technology for their structural composition. The electrode acts as an anode in a similar way to a battery. The functioning of this electrode relies on a faradaic mechanism, specifically the intercalation or alloying of ions, with lithium ions being the most typically used. This mechanism is similar to the reactions observed in batteries. The anode can consist of several materials, including graphite, lithium titanate ($Li_4Ti_5O_{12}$), or other materials with lithium intercalation capabilities. The utilization of activated carbon in the production of the cathode electrode is a prevalent practice due to its advantageous electric EDLC features. The energy storage mechanism of this electrode is due to the electrostatic accumulation of charges in both the electrolyte and the electrode. SCs have garnered considerable interest for their remarkable attributes, such as their superior electrochemical properties [10]. However, their application in real-time systems has

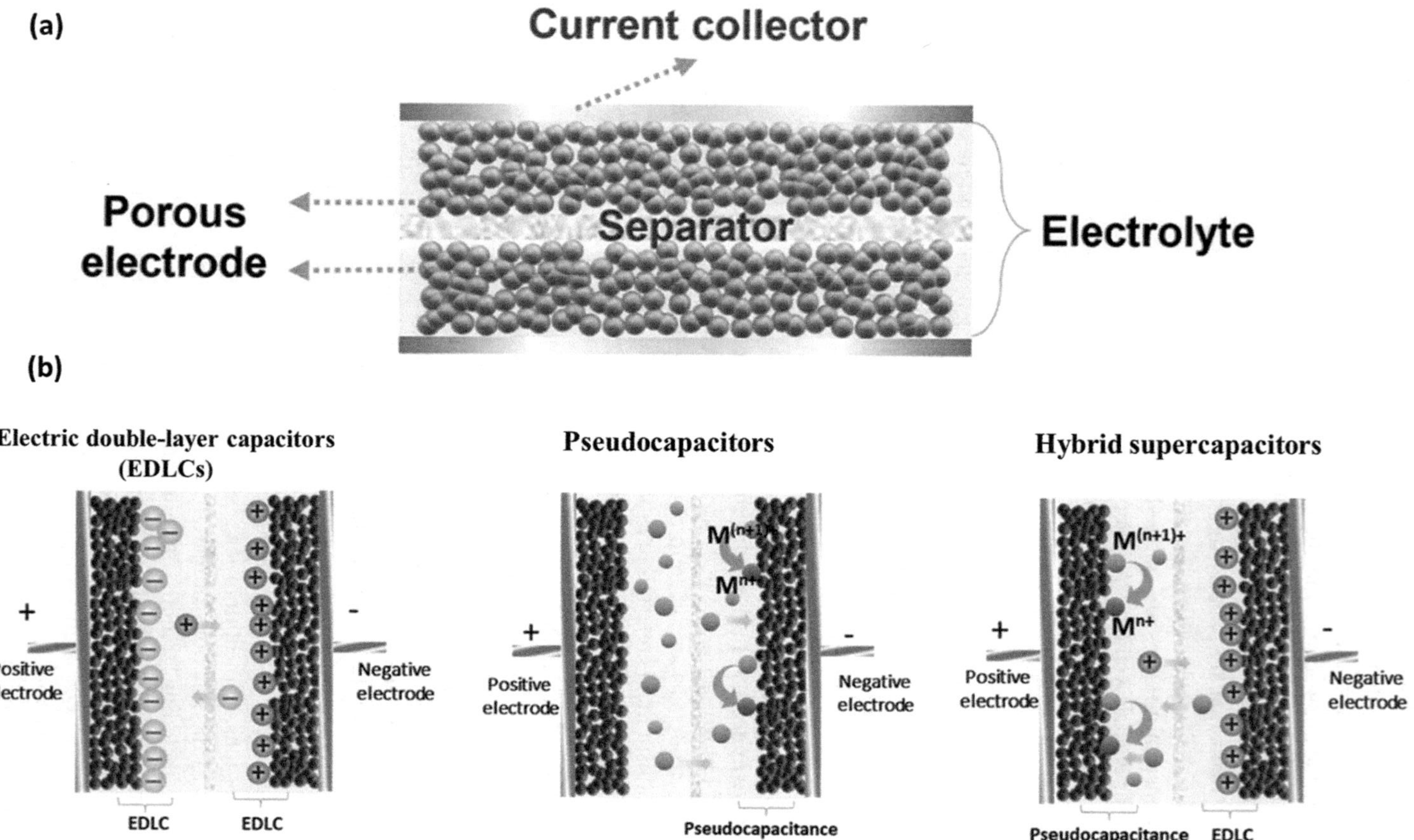

FIGURE 9.1 (a) Component of electrochemical supercapacitor device (b) and their different types.

been limited due to their lower energy density than LIBs at 180 Wh kg^{-1} [11] and impacted electrochemical performance of SCs. Therefore, the scientific community has mainly focused on investigating new electrode materials for supercapacitors [12].

This chapter specifically addressed the most recent progress in the study and creation of coordination materials, including porous coordination polymers (PCPs), metal-organic frameworks (MOFs), and covalent organic frameworks (COFs), with the purpose of enhancing supercapacitors. The aim is to facilitate future advancements and innovations in SCs. The synthesis methodologies and correlations between physical and electrochemical characteristics of coordination materials are summarized.

9.2 COORDINATION MATERIALS FOR SUPERCAPACITORS

SCs commonly employ coordination materials such as PCPs, MOFs, and COFs. Coordination compounds arise from the establishment of coordination bonds between ligands and/or metal atoms or ions. The remarkable functionalities of PCPs, MOFs, and COFs have sparked considerable interest and financial support to advance the creation of state-of-the-art materials for supercapacitors. Prior evaluations have furnished exceptional synopses of coordination materials, encompassing MOFs, COFs, PCPs, and energy-storage apparatus.

9.2.1 PCPs in Supercapacitors

The phrase "coordination polymer" was first coined in the early 1960s. Liu and his colleagues introduced the notion of PCPs in 1964 [13]. The study of PCPs did not see substantial expansion until the 1990s due to limitations in scientific research equipment during that time. It was during this period that PCPs began to attract attention as excellent coordination materials [14]. PCPs can be categorized according to their framework structures as one, two, and three dimensional (1-, 2-, 3D) infinite network systems [15]. Coordination polymers can be classified according to the metal center they contain, which may include parodic metals. Furthermore, depending on the particular organic-ligand, they can be classified as N-containing heterocycles, O-containing organic ligands, and so on with mixed ligands [16, 17].

The order and dimensions of PCPs exert a substantial influence on their properties. 2D-PCPs provide distinct structural benefits in comparison to bulk PCPs, including exceptional flexibility and a significant abundance of surface-active sites. The primary approach employed to produce 2D-PCPs involves constraining the expansion in SCs, which exhibit a significantly expansive surface area, making them more attractive in comparison to their bulk counterparts with increased surface area, which minimizes the diffusion distance for reactants and products [18]. In contrast, 2D-PCPs increase the effectiveness of active sites, resulting in a significant enhancement in the performance of applications such as catalysis and sensing. The polymer framed membrane shown a very high flexibilities as SC [19]. The Young's modulus of 2D PCP nanosheets is high enough to enable the production of a strong and self-regulating PCP a thin layer with exceptional photoluminescent characteristics. The exceptional malleability and physical characteristics of two-dimensional polymer-coated polymer films provide significant prospects for practical applications. Furthermore, the flexibility of two-dimensional photonic crystal polymers (2D PCPs) also exhibits specific morphological traits and structural alterations. An illustrative work showcased the mechanism through which a Mo-polydopamine coordination complex undergoes a shape metamorphosis from 2D single nanosheets to hierarchical micro-flowers, while simultaneously exerting effective control over this change [20, 21]. Furthermore, 2D-PCPs can augment electrical conductivity by reducing the length of the channel through which electrons/charges are transported in SCs. Recent investigations have demonstrated that precise arrangement of secondary building units (SBUs) and organic ligands can yield porous coordination polymers (PCPs) with inherent electrical conductivity. Further decreasing the dimensions of electrically conductive PCPs resulted in a significant increase in their electrical conductivity. These PCPs are highly appropriate for use in SCs [22, 23].

9.2.2 MOFs in Supercapacitors

MOFs are a specific type of porous coordination polymers that were initially described by Yaghi et al. in the late 1990s. Since then, they have garnered significant attention throughout the past two decades [24]. The materials comprise crystalline structures comprising metal-containing and organic ligands. More than twenty thousand unique MOFs have been synthesized, and the quantity of MOFs is still increasing through the modification of secondary building units (SBUs) and the functionalization of organic ligands [25]. Moreover, MOFs can be synthesized using cost-effective raw materials. Usually, metal-ion precursors consist of inorganic salts, such as sulphates, chlorides, and nitrates. Conversely, organic ligands, typically having many binding sites, are composed of carboxylates, nitriles – or azoles [26–28]. MOFs have superior characteristics compared to conventional inorganic porous materials, including higher specific surface areas, greater porosity, and a wider range of structural and functional diversity. In addition, other functional elements have been included into MOFs to enhance their properties. Many MOF composites have been effectively synthesized by combining MOFs with various functional species, such as carbon materials, conducting polymers, metal doping, hydroxide metal salts, and nickel oxalate [29]. During heat treatment, MOFs serve as precursors for the production of metal moieties. The size and shape of these materials can be controlled [30]. Hence, there has been considerable enthusiasm in employing MOFs, MOF composites, and MOF derivatives as electrode materials for supercapacitors.

The use of pure MOFs for SCs has been recorded in preliminary studies carried out by Díaz and colleagues [33]. The scientists manufactured Co8-MOF-5 to serve as an electrode substance in supercapacitors. However, the electrochemical property is restricted to the specific MOF and type of electrolyte. Furthermore, Lee et al. [34] examined Co-MOF as a viable substance for supercapacitors (SCs) and established a foundation for future research on various MOFs as electrode materials. The Co-MOF established a performance around 200 F g^{-1} and exhibited favorable energy density of 7.18 Wh kg^{-1}. Moreover, Lee and his colleagues have created Co-MOFs with unique shapes and pore sizes [35] with longer linkers demonstrate greater surface area and wider pores among the three. The electrochemical properties of SCs were recorded as: 179 F g^{-1}, 5.64 kW kg^{-1}, and 31.4 Wh kg^{-1}, respectively. These measurements were based on the surface area and pore diameter of the electrodes. Liu and co-workers reported the syntheses of a two-dimensional (2D) Co-MOF (Co-LMOF. The electrochemical performance of Co-LMOF as electrode material for supercapacitors was investigated. It was observed that Co-LMOF exhibited a significantly high specific capacitance of around 1900 F g^{-1} and maintained outstanding initial capacitance after 2,000 cycles [36]. Liao et al. effectively synthesized a range of Ni-MOF by combining iso-nicotinic acid and nickel nitrate in N, N-dimethylformamide (DMF) at a temperature of 140°C for a duration of 96 hours in an autoclave. These Ni-MOF materials were intended for use in electrochemical supercapacitors [37]. The electrochemical experiments demonstrated excellent capacitive behavior, with a capacitance of around 600 F g^{-1} at a scan rate of 5 mV s^{-1}. Additionally, the material exhibited remarkable cycling stability, retaining 84% of its capacity across 2,000 cycles. Kang and coauthors [38] conducted an investigation on a novel pseudo-capacitive substance called Ni-MOF ($Ni_3(btc)_{212}H_2O$). The test findings demonstrate that the Ni-MOF has remarkable pseudocapacitive characteristics and possesses outstanding cycle stability. Yang et al. effectively created a multi-layer nanostructured Ni-MOF by mixing p-benzenedicarboxylic acid and nickel chloride in DMF at a temperature of 120°C for different lengths of time [9]. The material demonstrated a notable capacitance and sustained a cycling stability of 90% after 3,000 cycles. The exceptional electrochemical property can be attributed to the intrinsic characteristics of the Ni-MOF, such as its layered structure and selectively exposed facets. Examples of these characteristics exist. The utilization of carboxylate ligands allowed Qu and his colleagues to successfully construct novel nickel-based MOFs that had a structure that was comparable to the original. The electrochemical properties of these DMOF capacitors were excellent when they were used as electrode materials for SCs. When compared to the other two, the Ni-DMOF-ADC demonstrated higher cycle stability, with a retention rate of 98% after 16,000 cycles. In a recent study, Sheberla et al. discovered that $Ni_3(HITP)_2$ has a remarkably high electrical conductivity. This makes it an ideal candidate for use as the only electrode material in electric EDLCs [31]. Figure 9.2a illustrates that $Ni_3(HITP)_2$ consists of stacked *p*-conjugated two-dimensional layers, which are traversed by one-dimensional cylindrical channels with a diameter of 1.5 nm. $Ni_3(HITP)_2$ has an exceptionally high areal capacitance. Figure 9.2b illustrates synthesis of $Ni_3(HITP)_2$ MOF with exceptional flexibility and electrochemical performance. Hence, the porous MOF ($Ni_3(HITP)_2$) with its excellent electrical conductivity shows great potential as an active electrode material for EDLCs (Figure 9.2 c-d) [32]. Furthermore, there have been reports of additional Ni-MOFs being used as electrode materials for SCs [39]. Campagnol and colleagues created three iron-based metal-organic frameworks (MOFs) named MIL-88(Fe), MIL-53(Fe), and MIL-100(Fe). These MOFs were employed as electrode materials for SCs [40]. They exhibited the impact of pellet thickness, pore diameters, and cations of varying sizes on performance. The experimental results indicate that MIL-100 (Fe) outperforms a combination of carbon and nanotubes in the same fluid. However, the disruption of the reduction process greatly impairs the electrode's long-term durability. Furthermore, Choi et al. constructed an SC apparatus utilizing coin-type electrodes composed of nMOFs [41]. Twenty-three unique MOF compounds were synthesized and analyzed. The compounds were composed of several organic ligands and central metal ions.

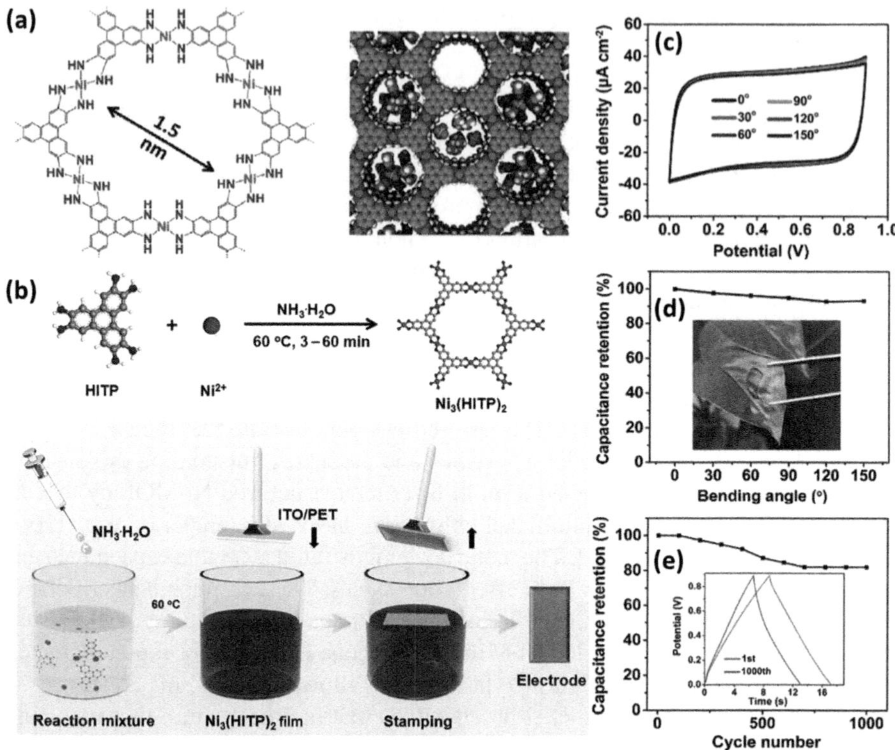

FIGURE 9.2 The $Ni_3(HITP)_2$ MOF preparation (reference [31], extracted with authorization, Nature, 2016), (b) The $Ni_3(HITP)_2$/ITO/PET electrode is prepared using the Langmuir-Schäfer method at the air/liquid interface. (c) The CV curves and (e) capacitance retention and (d) bending were assessed by conducting measurements at different bending angles, spanning from 0° to 150° (Reference [32], extracted with authorization, Elsevier, 2020).

The compounds comprised the MOF-74 structure, characterized by the presence of mixed multi-metallic metal oxide units and one-dimensional pores. Moreover, there existed zirconium (IV) metal-organic frameworks (MOFs) that exhibited differences in the dimensions and configuration of their connections, as well as the dimensions of their nanocrystals. The MOF-5 structure exhibited diverse mixed functions and possessed three-dimensional pores. Finally, there were MOFs with varying numbers of metal-containing units. Recently, n-MOF-867, which is composed of zirconium ($Zr_6O_4(OH)_4$(2,20-bipyridine-5,50-dicarboxylate)$_6$), exhibits exceptional capacitances in both volume and area, measuring 0.64 F cm^{-3} and 5.09 mF cm^{-2}, respectively. Moreover, these characteristic stays stable even after experiencing a minimum of 10,000 cycles (Figure 9.3f). Jiao et al. successfully synthesized mixed-MOFs (M-MOFs) by partially substituting Ni^{2+} ions in the Ni-MOF with either Co^{2+} or Zn^{2+} ions [42], with the aim of improving the conductivity of MOFs (Figure 9.3a). The scanning electron microscope (SEM) images of Co/Ni-MOF and Zn/Ni-MOF (Figure 9.3b–e) clearly demonstrate that the Co/Ni-MOF flower-like structures consist of several nanosheets, each with an average thickness of 30 nm. Moreover, the

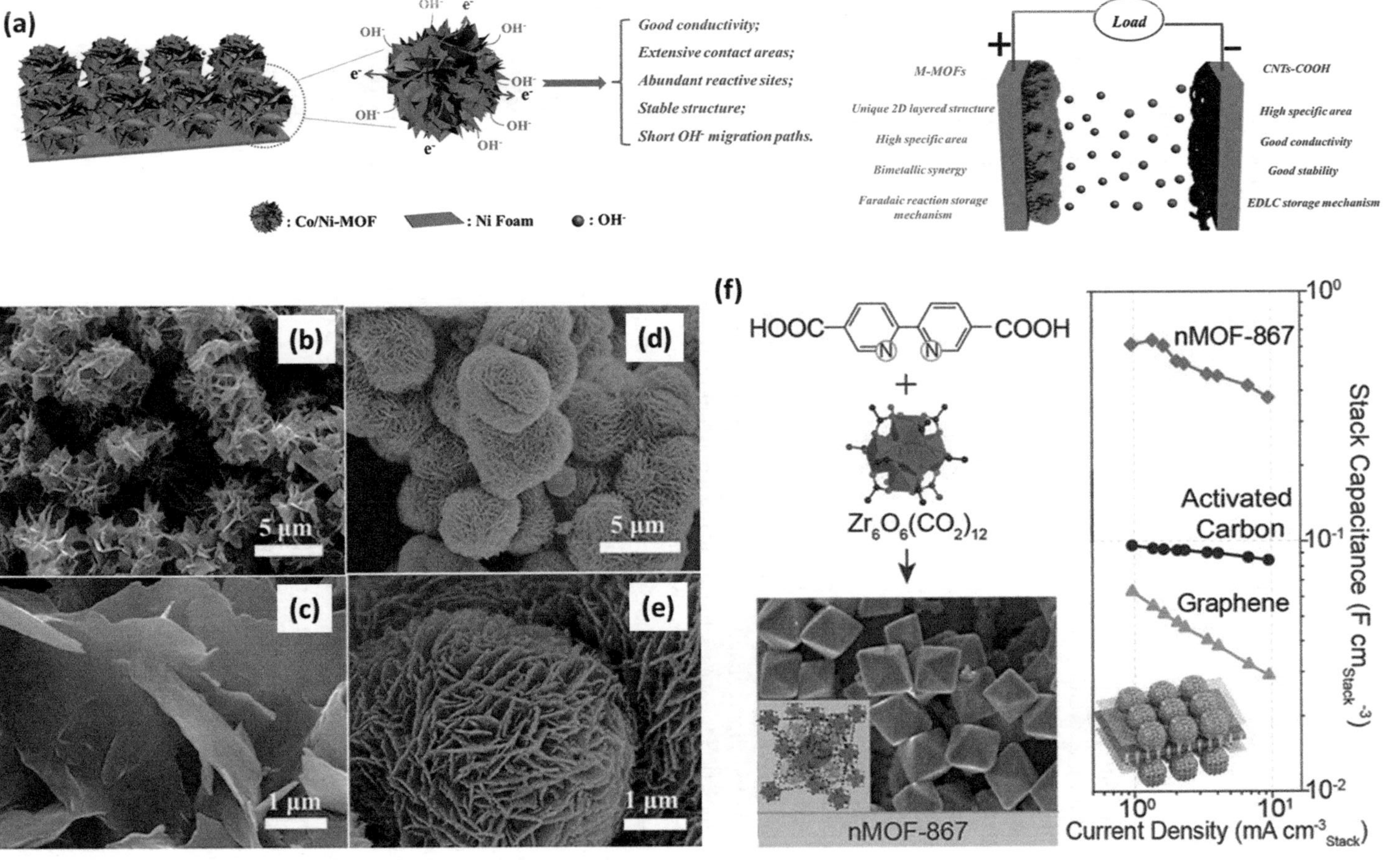

FIGURE 9.3 (a) Illustration of Co/Ni-MOF electrode charge deposition. (b) Graphical display of hybrid supercapacitors. (c, d) SEM images of the resulting Co/Ni-MOF and (e, f) Zn/Ni-MOF (Reference [42]. Extracted with permission, The Royal Society of Chemistry, 2016). (f) morphology of MOF-867 and its comparison performance with n-MOF SCs (Reference [41]. Extracted with permission, American Chemical Society, 2014).

structures of Zn/Ni-MOFs closely resemble those of Co/Ni-MOF. M-MOFs, acting as electrode materials similar to batteries, have a unique 2D layered structure that provides sufficient space for the insertion and removal of OH ions during faradaic reactions. Hybrid supercapacitors have the capacity to take advantage of both capacitive behavior and faradaic reactivity, hence increasing power density and energy density. The 2D M-MOFs displayed higher electrochemical properties in comparison to the Ni-MOF. The remarkable electrochemical behavior of M-MOF can be ascribed to the subsequent considerations. At first, the M-MOF structures, which resemble flowers, as shown in Figure 9.3b-e, serve as both the conductive channel and backbone. They provide robust support for the structural integrity and enable efficient charge transfer for the faradic reaction. Moreover, the study of MOFs not only broadens their possible applications but also provides a chance to tackle the performance discrepancy between batteries and supercapacitors.

MOF composites: Combining MOFs with various functional materials is a highly effective and practical method to further improve the characteristics of MOFs [43]. MOF composites have been utilized as electrodes for SCs, including combinations of MOF with carbon materials [44], conducting polymers [45], metal doping [46], hydroxide metal salts [47], nickel oxalate [48], and several other substances. Zhang and coauthors successfully fabricated a composite material by merging Mn-MOF with carbon nanotubes (CNTs@Mn-MOF) by a cost effective and simple hydrothermal technique. Subsequently, this composite material was employed as a novel electrode for supercapacitors, based on manganese [49]. A remarkable capacitance of around 200 F g^{-1} was achieved using a conventional three-electrode setup. In addition, the symmetric supercapacitors (SSCs) were fabricated using a two-electrode configuration in a 1 M Na_2SO_4 electrolyte. The SSC exhibited exceptional characteristics, with an energy density of 6.9 Wh kg^{-1} and a power density of 2240 W kg^{-1}. In general, carbon nanotubes (CNTs) were introduced into metal-organic frameworks (MOFs) as a composite in order to enhance the conductivity of the materials. Wen et al. developed a new type of asymmetric supercapacitor (ASC) using Ni-MOF/CNT as positive electrodes and rGO/C_3N_4 (rGO, reduced graphene oxide; C_3N_4, graphitic carbon nitride) as negative electrodes [44]. This ASC showed a significant energy density and gave a superb cycle stability (95% retention after 5,000 cycles) with remarkable capacitance conventional three-electrode setup. In addition, Zhou and his colleagues documented a collection of innovative nickel-based MOF (Ni-MOFs) using a simple solvothermal method [50]. When the acquired Ni-MOF was utilized as an electrode for SCs, it exhibited a notable specific capacitance. To enhance the electrochemical characteristics, Ni-MOF was applied onto GO nanosheets using an in-situ technique. The Ni-MOFs@GO composite, with a GO composition of 3 wt.%, demonstrated an excellent capacitance.

9.2.3 COFs in Supercapacitors

COFs, also known as covalent organic frameworks, are crystalline polymers that are created through reversible mechanisms and display a systematic arrangement of organic building blocks. Reversible reactions are employed to generate thermodynamically stable networks with extensive organization by means of "error corrections." COFs have been widely employed in several domains including gas separation and storage, optoelectronic devices, energy conversion, energy storage, and heterogeneous catalysis. These qualities are a result of their remarkable porosity, capacity to regulate and manipulate pore size, precise and well-defined structures, and varied frameworks with distinct functional capabilities.

Covalent organic frameworks (COFs) in both 2D and 3D forms can be created by selecting suitable organic building pieces and employing reversible methods. COFs have been synthesized using various methods. The synthetic techniques encompass methods, as illustrated in Figure 9.4 [51, 52]. COFs are extremely exciting materials for use in electrochemistry because of their exact chemical structures, the ability to customize pore sizes, and the variety of pore layouts that they offer. The electrochemical characteristics of these materials have been improved by the application of a variety of techniques. The purposeful selection of suitable connecting motifs and building components allows for the synthesis of redox-active and stable COFs. Additionally, in order to improve mass movement and achieve higher electrical conductivity, researchers have developed hybrid materials that are composed of carbon fiber or carbon monoxide [52].

Increasing the number of active sites is considered to be a crucial approach for enhancing characteristics of redox-active COFs. COFs, namely 1KT-Tp COF, 2KT-Tp COF, and 4KT-Tp COF, were synthesized using the solvothermal Schiff-base condensation method and these COFs have varying quantities of carbonyl groups (Figure 9.5a) [53]. At 0.27/0.22 V (against Ag/AgCl), the cyclic voltammetry (CV) profile of the 2KT-Tp COF demonstrated a substantial degree of reversibility, exhibiting a characteristic symmetrical structure. The CV profile also featured a pair of redox peaks. On the other hand, the cyclic voltammograms (CVs) of the 4KT-Tp COF displayed two separate sets of redox peaks at 0.28/0.24 and 0.37/0.33 V, as it can be seen in Figure 9.5 b-c. Based on the data, it can be concluded that the presence of redox-active orthoquinone groups in 2KT-Tp COF and 4KT-Tp COF results in a significant number of redox-active sites. As a result, the involvement of carbonyl groups in neighboring oxidation-reduction reactions is promoted, leading to an enhanced level of conjugation after reduction. The electrodes utilizing 2KT-Tp COF and 4KT-Tp COF demonstrated capacitances of 256 and 583 F g^{-1}, correspondingly, when discharged at a rate of 0.2 A g^{-1}. The values exceeded the capacitance of 1KT-Tp COF, which was measured at 61 F g^{-1}. This study indicated that orthoquinone moieties outperformed COFs with separated carbonyl groups.

Redox capabilities can be attained by choosing reaction precursors with redox characteristics and by utilizing post-synthetic functionalization on unaltered COFs. The synthesis of two redox-active COFs, with different percentages, was carried out utilizing 4-azido-2,2,6,6-tetramethy-l-1-piperidinyloxy (TEMPO) as the primary component (Figure 9.6a) [54]. The x-ray diffraction (XRD) signals of [TEMPO]100%-NiP-COF exhibited considerable enlargement due to the presence of pliable chains inside the COFs' pores. This reduced the diffractions, as observed in Figure 9.6b. TEMPO immobilization led to a decrease in surface areas and pore diameters of the resulting COFs, as depicted in Figure 9.6c-d. The CV performance indicates the [TEMPO]100%-NiP-COF demonstrated enhanced performance due to the abundant distribution of TEMPO radicals on its surfaces (Figure 9.6e-f).

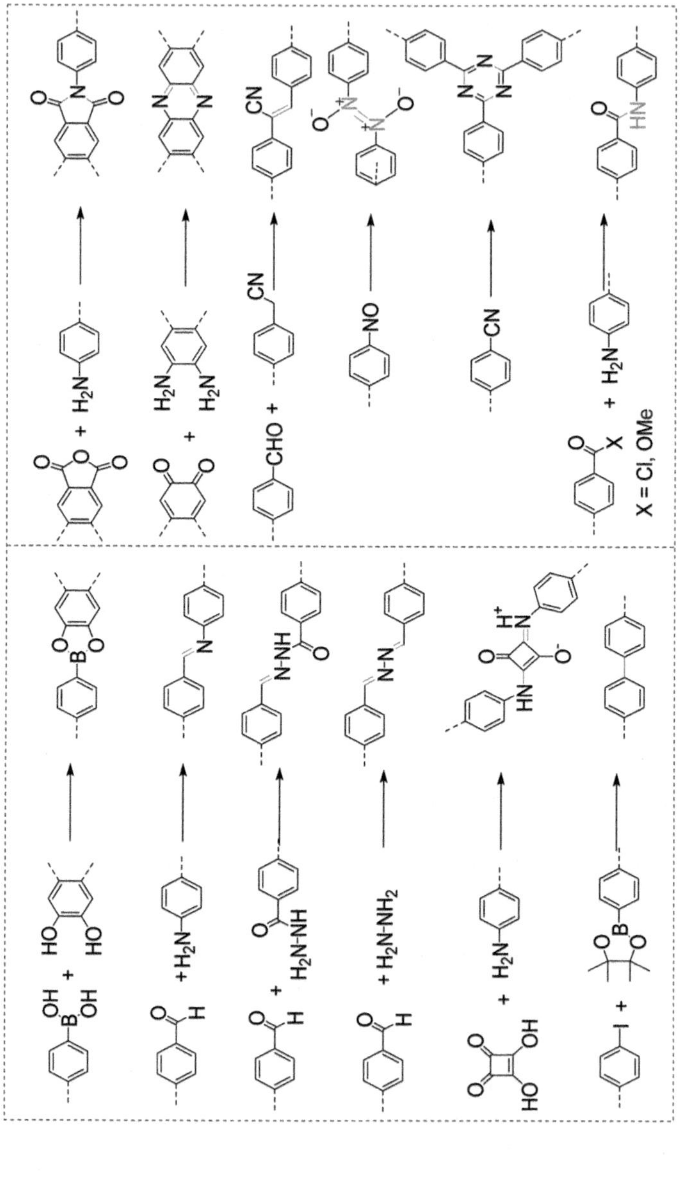

FIGURE 9.4 The development of different types of COFs.

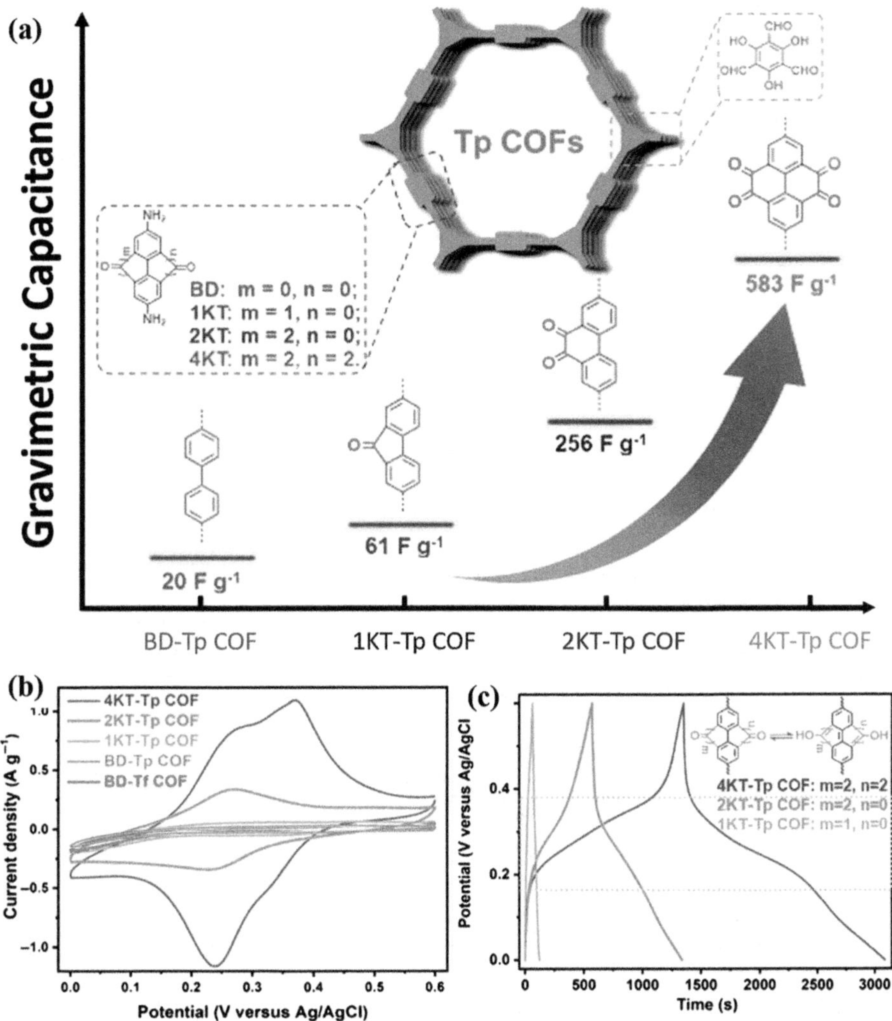

FIGURE 9.5 (a) The chemical structure representation of Tp COFs. (b-c) Electrochemical performance (CV and GCD) of prepared COF. (Reference [53]. Extracted with authorization, Chinese Chemical Society, 2020.)

Conversely, the [TEMPO]50%-NiP-COF demonstrated rapid electrode kinetics to enable effective charging and discharging. This was attributed to its enlarged surface area and larger pore size, which enhanced the movement of ions. Ultimately, the [TEMPO]100%-NiP-COF and [TEMPO]50%-NiP-COF demonstrated a capacitance of 167 and 124 F g^{-1} at 100 mA g^{-1}, respectively. The [TEMPO]50%-NiP-COF exhibited superior capacitance retention compared to the [TEMPO]100%-NiP-COF due to its elevated porosity. Significantly, the devices fabricated using the COFs derived from this process exhibited exceptional cycling stability due to the limited solubility of the TEMPO radical included inside the COFs. In contrast to physical doping,

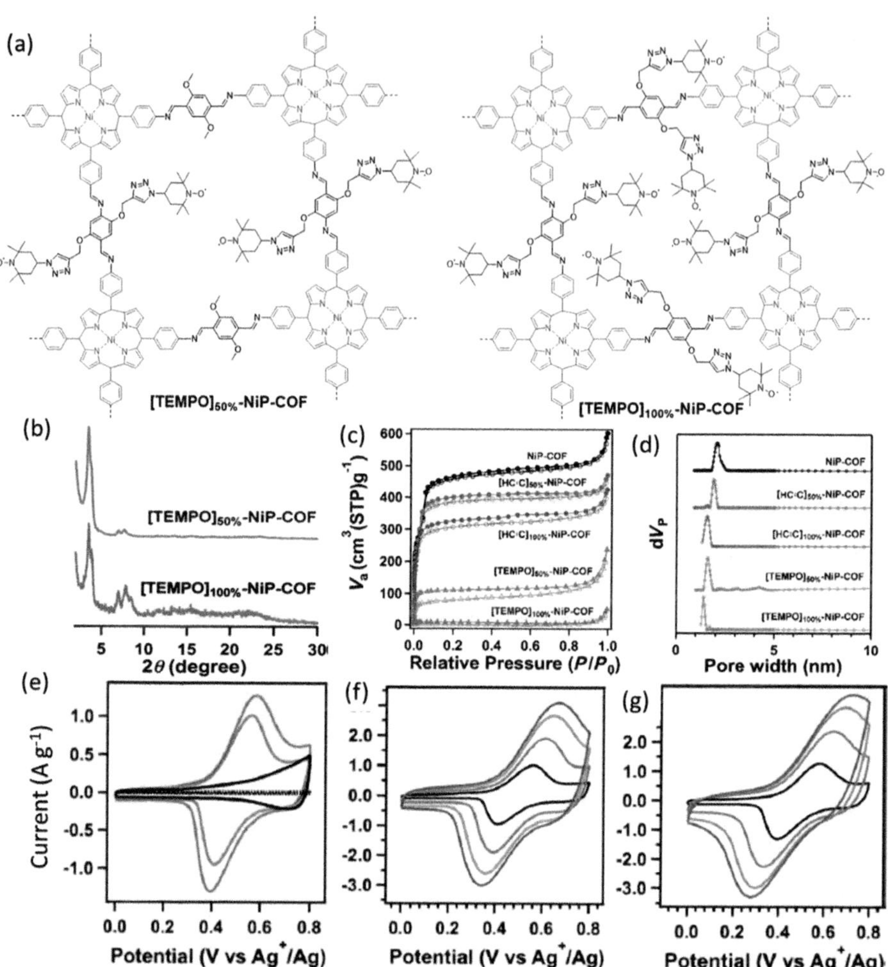

FIGURE 9.6 (a-d) Preparation, XRD, BET, and pore size of COFs, respectively. (e-g) Electrochemical performance of different prepared COF materials 100% (red curve), 50% (blue curve), [Ph]100% (black curve). (Reference [54]. Extracted with authorization, Wiley-VCH, 2015.)

post-synthetic functionalization approaches have the advantage of achieving both high stabilities, but this comes at the cost of sacrificing the surface functionality of COFs. Therefore, including redox-active groups into precursors during the creation of pseudocapacitive COFs can be an effective approach. This strategy allows COFs to possess pseudo capacitance properties while preserving their surface capacitance to a great extent.

9.3 CONCLUSION

This chapter provides a concise overview of the most recent uses of coordination materials, including POP, MOF, and COF, along with their nano-composites. Initially, we compiled an all-encompassing depiction of the supercapacitor, encompassing its several often employed types. Subsequently, a thorough analysis of the mechanism, along with the merits and drawbacks of these synthetic approaches of coordinating material, is conducted. Subsequently, an investigation was conducted on the utilization of two-dimensional polycyclic aromatic hydrocarbons (PCPs) and their electrochemical capabilities specifically about supercapacitors. This chapter serves as a comprehensive platform for presenting an overview of the key characteristics, material coordination in supercapacitors, and the future potential of these materials.

ACKNOWLEDGMENTS

This work was supported by the Research Fund for International Scientists (RFIS-Grant number: 52150410410) National Natural Science Foundation of China.

REFERENCES

[1] Deng, Q.; R. Alvarado; E. Toledo; L. Caraguay, 2020. Greenhouse gas emissions, non-renewable energy consumption, and output in South America: the role of the productive structure. *Environmental Science and Pollution Research*. 27, 14477–14491.

[2] Katzin, D.; L. F. M. Marcelis; S. van Mourik, 2021. Energy savings in greenhouses by transition from high-pressure sodium to LED lighting. *Applied Energy*. 281, 116019.

[3] Poizot, P.; J. Gaubicher; S. Renault; L. Dubois; Y. Liang; Y. Yao, 2020. Opportunities and challenges for organic electrodes in electrochemical energy storage. *Chemical Reviews*. 120, 14, 6490–6557.

[4] Wu, N.; X. Bai; D. Pan; B. Dong; R. Wei; N. Naik; R. R. Patil; Z. Guo, 2021. Recent advances of asymmetric supercapacitors. *Advanced Materials Interfaces*. 8, 1, 2001710.

[5] Zhang, Q.; E. Uchaker; S. L. Candelaria; G. Cao, 2013. Nanomaterials for energy conversion and storage. *Chemical Society Reviews*. 42, 7, 3127–3171.

[6] Zhang, S.; N. Pan, 2015. Supercapacitors performance evaluation. *Advanced Energy Materials*. 5, 6, 1401401.

[7] Ye, W.; H. Wang; J. Ning; Y. Zhong; Y. Hu, 2021. New types of hybrid electrolytes for supercapacitors. *Journal of Energy Chemistry*. 57, 219–232.

[8] Kasprzak, D.; J. Liu, Toward green energy storage: chitin and cellulose as constituents of efficient, sustainable, and flexible zinc-ion hybrid supercapacitors. *Sustainable, and Flexible Zinc-Ion Hybrid Supercapacitors*. Sustainable Materials and Technologies, 38 (2023) e00726.

[9] Yang, J.; P. Xiong; C. Zheng; H. Qiu; M. Wei, 2014. Metal–organic frameworks: a new promising class of materials for a high performance supercapacitor electrode. *Journal of Materials Chemistry A*. 2, 39, 16640–16644.

[10] Simon, P.; Y. Gogotsi, 2020. Perspectives for electrochemical capacitors and related devices. *Nature Materials*. 19, 11, 1151–1163.

[11] Lin, L.; J. Chen; D. Liu; X. Li; G. G. Wallace; S. Zhang, 2020. Engineering 2D materials: a viable pathway for improved electrochemical energy storage. *Advanced Energy Materials*. 10, 45, 2002621.

[12] Ipadeola, A. K.; K. Eid; A. M. Abdullah, 2023. Porous transition metal-based nanostructures as efficient cathodes for aluminium-air batteries. *Current Opinion in Electrochemistry*. 37, 101198.

[13] Huang, Y.; M. Zhao; S. Han; Z. Lai; J. Yang; C. Tan; Q. Ma; Q. Lu; J. Chen; X. Zhang, 2017. Growth of au nanoparticles on 2D metalloporphyrinic metal-organic framework nanosheets used as biomimetic catalysts for cascade reactions. *Advanced Materials*. 29, 32, 1700102.

[14] Liu, C. F.; N. C. Liu; J. C. Bailar Jr, 1964. The stereochemistry of complex inorganic compounds. XXVII. Asymmetric syntheses of tris (bipyridine) complexes of ruthenium (II) and osmium (II). *Inorganic Chemistry*. 3, 8, 1085–1087.
[15] Batten, S. R.; R. Robson, 1998. Interpenetrating nets: ordered, periodic entanglement. *Angewandte Chemie International Edition*. 37, 11, 1460–1494.
[16] Zhao, Y. S.; H. Fu; A. Peng; Y. Ma; D. Xiao; J. Yao, 2008. Low-dimensional nanomaterials based on small organic molecules: preparation and optoelectronic properties. *Advanced Materials*. 20, 15, 2859–2876.
[17] Deng, D., 2016. K. S. Novoselov, Q. Fu, N. Zheng, Z. Tian; X. Bao, Catalysis with two-dimensional materials and their heterostructures. *Nature Nanotechnology* 11, 3, 218–230.
[18] Hu, M. L.; M. Abbasi-Azad; B. Habibi; F. Rouhani; H. Moghanni-Bavil-Olyaei; K. G. Liu; A. Morsali, 2020. Electrochemical applications of ferrocene-based coordination polymers. *ChemPlusChem*. 85, 11, 2397–2418.
[19] Xu, W.; L.-H. Wang; Y. Chen; Y. Liu, 2022. Flexible carbon membrane supercapacitor based on γ-cyclodextrin-MOF. *Materials Today Chemistry*. 24, 100896.
[20] Sun, L.; C. Wang; X. Wang; L. Wang, 2018. Morphology evolution and control of Mo-polydopamine coordination complex from 2D single nanopetal to hierarchical micro-flowers. *Small*. 14, 27, 1800090.
[21] Wang, C.; L. Sun; F. Zhang; X. Wang; Q. Sun; Y. Cheng; L. Wang, 2017. Formation of Mo–polydopamine hollow spheres and their conversions to MoO2/C and Mo2C/C for efficient electrochemical energy storage and catalyst. *Small*. 13, 32, 1701246.
[22] Sheberla, D.; L. Sun; M. A. Blood-Forsythe; S. L. Er; C. R. Wade; C. K. Brozek; A. Aspuru-Guzik; M. Dincă, 2014. High electrical conductivity in Ni3 (2, 3, 6, 7, 10, 11-hexaiminotriphenylene) 2, a semiconducting metal–organic graphene analogue. *Journal of the American Chemical Society*. 136, 25, 8859–8862.
[23] Li, X.; L. Tao; Z. Chen; H. Fang; X. Li; X. Wang; J.-B. Xu; H. Zhu, 2017. Graphene and related two-dimensional materials: structure-property relationships for electronics and optoelectronics. *Applied Physics Reviews*. 4, 2,.
[24] Yaghi, O. M.; H. Li, 1995. Hydrothermal synthesis of a metal-organic framework containing large rectangular channels. *Journal of the American Chemical Society*. 117, 41, 10401–10402.
[25] Wang, L.; Y. Han; X. Feng; J. Zhou; P. Qi; B. Wang, 2016. Metal–organic frameworks for energy storage: batteries and supercapacitors. *Coordination Chemistry Reviews*. 307, 361–381.
[26] Liu, W.; J. Huang; Q. Yang; S. Wang; X. Sun; W. Zhang; J. Liu; F. Huo, 2017. Multi-shelled hollow metal–organic frameworks. *Angewandte Chemie International Edition*. 56, 20, 5512–5516.
[27] Yaghi, O. M.; M. O'Keeffe; N. W. Ockwig; H. K. Chae; M. Eddaoudi; J. Kim, 2003. Reticular synthesis and the design of new materials. *Nature*. 423, 6941, 705–714.
[28] Wiktor, C.; M. Meledina; S. Turner; O. I. Lebedev; R. A. Fischer, 2017. Transmission electron microscopy on metal–organic frameworks–a review. *Journal of Materials Chemistry A*. 5, 29, 14969–14989.
[29] Zheng, S.; H. Xue; H. Pang, 2018. Supercapacitors based on metal coordination materials. *Coordination Chemistry Reviews*. 373, 2–21.
[30] Sun, J.-K.; Q. Xu, 2014. Functional materials derived from open framework templates/precursors: synthesis and applications. *Energy & Environmental Science*. 7, 7, 2071–2100.
[31] Sheberla, D.; J. C. Bachman; J. S. Elias; C.-J. Sun; Y. Shao-Horn; M. Dincă, 2017. Conductive MOF electrodes for stable supercapacitors with high areal capacitance. *Nature Materials*. 16, 2, 220–224.

[32] Zhao, W.; T. Chen; W. Wang; B. Jin; J. Peng; S. Bi; M. Jiang; S. Liu; Q. Zhao; W. Huang, 2020. Conductive Ni3(HITP)2 MOFs thin films for flexible transparent supercapacitors with high rate capability. *Science Bulletin*. 65, 21, 1803–1811.

[33] Díaz, R.; M. G. Orcajo; J. A. Botas; G. Calleja; J. Palma, 2012. Co8-MOF-5 as electrode for supercapacitors. *Materials Letters*. 68, 126–128.

[34] Lee, D. Y.; S. J. Yoon; N. K. Shrestha; S.-H. Lee; H. Ahn; S.-H. Han, 2012. Unusual energy storage and charge retention in Co-based metal–organic-frameworks. *Microporous and Mesoporous Materials*. 153, 163–165.

[35] Lee, D. Y.; D. V. Shinde; E.-K. Kim; W. Lee; I.-W. Oh; N. K. Shrestha; J. K. Lee; S.-H. Han, 2013. Supercapacitive property of metal–organic-frameworks with different pore dimensions and morphology. *Microporous and Mesoporous Materials*. 171, 53–57.

[36] Liu, X.; C. Shi; C. Zhai; M. Cheng; Q. Liu; G. Wang, 2016. Cobalt-based layered metal–organic framework as an ultrahigh capacity supercapacitor electrode material. *ACS Applied Materials & Interfaces*. 8, 7, 4585–4591.

[37] Liao, C.; Y. Zuo; W. Zhang; J. Zhao; B. Tang; A. Tang; Y. Sun; J. Xu, 2013. Electrochemical performance of metal-organic framework synthesized by a solvothermal method for supercapacitors. *Russian Journal of Electrochemistry*. 49, 983–986.

[38] Kang, L.; S.-X. Sun; L.-B. Kong; J.-W. Lang; Y.-C. Luo, 2014. Investigating metal-organic framework as a new pseudo-capacitive material for supercapacitors. *Chinese Chemical Letters*. 25, 6, 957–961.

[39] Yan, Y.; P. Gu; S. Zheng; M. Zheng; H. Pang; H. Xue, 2016. Facile synthesis of an accordion-like Ni-MOF superstructure for high-performance flexible supercapacitors. *Journal of Materials Chemistry A*. 4, 48, 19078–19085.

[40] Campagnol, N.; R. Romero-Vara; W. Deleu; L. Stappers; K. Binnemans; D. E. De Vos; J. Fransaer, 2014. A hybrid supercapacitor based on porous carbon and the metal-organic framework MIL-100 (Fe). *ChemElectroChem*. 1, 7, 1182–1188.

[41] Choi, K. M.; H. M. Jeong; J. H. Park; Y.-B. Zhang; J. K. Kang; O. M. Yaghi, 2014. Supercapacitors of nanocrystalline metal–organic frameworks. *ACS Nano*. 8, 7, 7451–7457.

[42] Jiao, Y.; J. Pei; D. Chen; C. Yan; Y. Hu; Q. Zhang; G. Chen, 2017. Mixed-metallic MOF based electrode materials for high performance hybrid supercapacitors. *Journal of Materials Chemistry A*. 5, 3, 1094–1102.

[43] Zhang, L.; S. Zheng; L. Wang; H. Tang; H. Xue; G. Wang; H. Pang, 2017. Fabrication of metal molybdate micro/nanomaterials for electrochemical energy storage. *Small*. 13, 33, 1700917.

[44] Wen, P.; P. Gong; J. Sun; J. Wang; S. Yang, 2015. Design and synthesis of Ni-MOF/CNT composites and rGO/carbon nitride composites for an asymmetric supercapacitor with high energy and power density. *Journal of Materials Chemistry A*. 3, 26, 13874–13883.

[45] Ehsani, A.; J. Khodayari; M. Hadi; H. M. Shiri; H. Mostaanzadeh, 2017. Nanocomposite of p-type conductive polymer/Cu (II)-based metal-organic frameworks as a novel and hybrid electrode material for highly capacitive pseudocapacitors. *Ionics*. 23, 131–138.

[46] Banerjee, P. C.; D. E. Lobo; R. Middag; W. K. Ng; M. E. Shaibani; M. Majumder, 2015. Electrochemical capacitance of Ni-doped metal organic framework and reduced graphene oxide composites: more than the sum of its parts. *ACS Applied Materials & Interfaces*. 7, 6, 3655–3664.

[47] Gao, Y.; J. Wu; W. Zhang; Y. Tan; J. Gao; B. Tang; J. Zhao, 2014. Synthesis of nickel carbonate hydroxide/zeolitic imidazolate framework-8 as a supercapacitors electrode. *RSC Advances*. 4, 68, 36366–36371.

[48] Gao, Y.; J. Wu; W. Zhang; Y. Tan; J. Gao; J. Zhao; B. Tang, 2015. Synthesis of nickel oxalate/zeolitic imidazolate framework-67 (NiC 2 O 4/ZIF-67) as a supercapacitor electrode. *New Journal of Chemistry*. 39, 1, 94–97.

[49] Zhang, Y.; B. Lin; Y. Sun; X. Zhang; H. Yang; J. Wang, 2015. Carbon nanotubes@ metal–organic frameworks as Mn-based symmetrical supercapacitor electrodes for enhanced charge storage. *RSC Advances*. 5, 72, 58100–58106.
[50] Zhou, Y.; Z. Mao; W. Wang; Z. Yang; X. Liu, 2016. In-situ fabrication of graphene oxide hybrid Ni-based metal–organic framework (Ni–MOFs@ GO) with ultrahigh capacitance as electrochemical pseudocapacitor materials. *ACS Applied Materials & Interfaces*. 8, 42, 28904–28916.
[51] Dogru, M.; A. Sonnauer; A. Gavryushin; P. Knochel; T. Bein, 2011. A covalent organic framework with 4 nm open pores. *Chemical Communications*. 47, 6, 1707–1709.
[52] Uribe-Romo, F. J.; J. R. Hunt; H. Furukawa; C. Klock; M. O'Keeffe; O. M. Yaghi, 2009. A crystalline imine-linked 3-D porous covalent organic framework. *Journal of the American Chemical Society*. 131, 13, 4570–4571.
[53] Li, M.; J. Liu; Y. Li; G. Xing; X. Yu; C. Peng; L. Chen, 2021. Skeleton engineering of isostructural 2D covalent organic frameworks: orthoquinone redox-active sites enhanced energy storage. *CCS Chemistry*. 3, 2, 696–706.
[54] Xu, F.; H. Xu; X. Chen; D. Wu; Y. Wu; H. Liu; C. Gu; R. Fu; D. Jiang, 2015. Radical covalent organic frameworks: a general strategy to immobilize open-accessible polyradicals for high-performance capacitive energy storage. *Angewandte Chemie*. 127, 23, 6918–6922.

10 Multifunctional Coordination Materials for Green Energy Production

An Introduction

Saira Ajmal, Mohammad Tabish, Safana Haqani, Tuan Anh Nguyen and Ghulam Yasin

10.1 COORDINATION MATERIALS FOR GREEN ENERGY PRODUCTION

Researchers are driven by the increasing demand for energy, the limited availability of fossil fuels, and rising environmental concerns to explore new sustainable energy production methods that reduce greenhouse gas emissions and offer higher efficiency and lower costs [1]. Innovative technologies designed for the efficient conversion and storage of energy, like water splitting, fuel cells, and rechargeable batteries, have garnered significant attention. These technologies for converting energy involve different catalytic reactions. For instance, the oxygen evolution reaction (OER) produces oxygen through a chemical reaction [2], the hydrogen evolution reaction (HER) generates H_2 [3], oxygen reduction reaction (ORR) involves the reduction of O_2 to water or hydrogen peroxide [4], nitrogen reduction reaction (NRR) is an electrochemical process for producing ammonia (NH_3) from N_2 and a proton source [5], and carbon dioxide reduction reaction (CO_2RR) reduces carbon dioxide to carbon monoxide or hydrocarbons [6]. The utilization of efficient catalysts, such as noble metals like IrO_2, RuO_2, and Pt/C, has shown outstanding catalytic activities with reduced overpotential. Nevertheless, their widespread use is hindered by their high cost and limited availability of resources. Hence, there is a significant focus on developing non-precious metal catalysts that are efficient, exhibit higher activity, and possess practical durability.

A novel family of coordination materials, including covalent organic frameworks (COFs), porous coordination polymers (PCPs), and metal-organic frameworks (MOFs) is gaining increased recognition nowadays. Self-assembly of organic ligands can create these materials through the clusters or metal ions via coordination interaction. They are becoming increasingly popular for various catalytic applications such as CO_2RR, ORR, HER, NRR, and OER because of their impressive adjustable

DOI: 10.1201/9781003345886-12

porous structure, active metal sites, and specific surface area [7]. Various coordination materials with different morphologies and diverse compositions have been constructed and utilized in numerous catalytic applications. Due to their ability to be produced on a large scale through a facile synthesis either indirectly or directly through high-temperature calcination or chemical conversion, these materials are extensively employed as efficient catalysts and are promising candidates for next-generation green energy production systems.

Therefore, the direct utilization of coordination materials as catalysts is realized in electrochemical energy conversion applications. Additionally, their derivatives, which can be easily transformed into nitrides, oxides, or carbon materials are also employed as effective catalysts [8]. The catalytic mechanism and active sites in coordination materials are the topics of today that are being explored further in more systematic studies. Similarly, MOFs, COFs, and PCPs can yield carbon materials simply through high-temperature carbonization, which are excellent choices as catalysts with larger specific areas, improved stability, and outstanding electrical conductivity, along with enhanced dispersion [9]. Precisely, coordination materials and their derivates have porous structures that act as catalysts offering a valuable opportunity for innovation and effectiveness in green energy production (Figure 10.1). Furthermore, bearing in mind the further progress in developing high-performance renewable energy devices, we can develop novel coordination material-based multifunctional catalysts with enhanced electrochemical performance and outstanding stability by recognizing and maximizing the positive aspects while addressing impeding drawbacks.

Moreover, the catalytically active metal centers in multifunctional coordination materials are employed as catalysts due to their essential role in promoting and

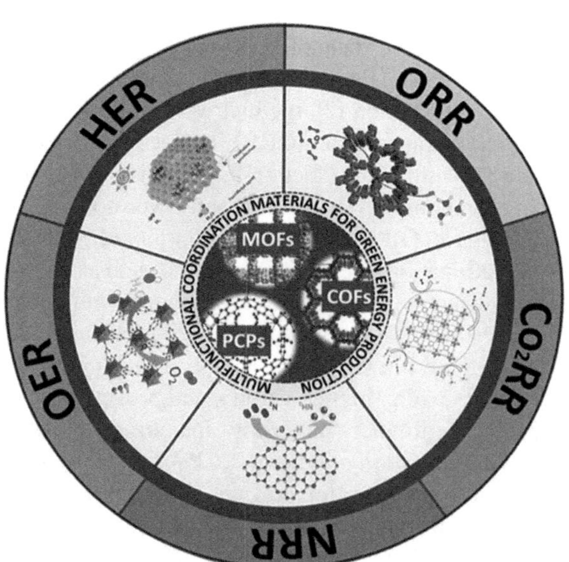

FIGURE 10.1 Schematic illustration of MOFs, COFs, and PCPs for green energy production.

facilitating diverse chemical reactions, enhancing the effectiveness of the materials in driving desired processes. Various methods, like carbonization MOFs and/or creating heterointerfaces on dissimilar host surfaces, have been employed to boost the catalytic performance of these porous materials [10]. These strategies have demonstrated combined effectiveness in different morphologies, utilizing isotropic/anisotropic or fully covered structures.

The resulting hybrids based on coordination materials possess their inherent properties, enabling multifunctionality and broader applications, while coordination materials with unique characteristics still suffer their inadequate stability in alkaline or acidic environments. So incorporating transition metal compounds (like Ni, Fe, Co, etc.) into porous carbon nanomaterials or mesoporous materials results in outstanding electrical conductivity and electrochemical performance [11]. This process takes advantage of the enhanced chemical and environmental stability offered by these frameworks [12]. The introduction of heteroatom doping, functional metal/metal oxide components, and hybridization contribute to stabilizing and creating multiple active sites and desired compositions, promoting enhancement and long-term durability of the catalysts [13]. Furthermore, diverse catalyst systems based on heterostructures of MOFs/COFs/PCPs can be produced by introducing various heteroatoms such as P, N, Se, or S atoms under specific atmospheres [14]. The heteroatoms have a double impact, significantly changing the electronic characteristics of the carbon structure to make more active sites and forming connections with other essential components [15]. Ultimately, designing and fabricating hybrid materials of MOFs/COFs/PCPs and heterostructures with their derivatives is crucial, as the interface interactions play an important role in determining material and/or catalyst performance. In this chapter, we have therefore overviewed the various applications of multifunctional coordination materials as effective catalysts in different electrochemical energy conversion reactions (HER, OER, ORR, CO_2RR, and NRR). These applications hold significance for the advancement of next-generation renewable energy technologies, contributing to sustainable social and economic development.

10.2 CONCLUSION AND OUTLOOK

Multifunctional coordination materials exhibit attractive catalytic performances, showcasing significant potential and distinct advantages in energy conversion systems when compared to traditional inorganic catalysts. The majority of pure MOFs, PCPs, or COFs, even their hybrids, exhibit excellent performance as catalysts for energy transfer or energy production. Multifunctional coordination materials and their heterostructures can effectively combine the features of each component, leading to growing and widespread applications. Researchers are keenly interested in utilizing multifunctional coordination materials in catalysis because of their notable features, such as adjustable structures, porosity, and high surface areas. However, there are still some challenges that need to be solved. First, developing innovative MOFs, COFs, and/or PCPs is crucial to address the inadequate conductivity and stability issues associated with these materials. Second, coordination materials often undergo irreversible transformations, and the actual active sites are easily overlooked at a higher current rate or voltage. Hence, advanced in-situ characterization techniques are essential for understanding the catalytic mechanism. Third, there is a need to deal with unpredictable things that happen while turning these materials into carbon, like structures breaking and metal particles sticking together. Last, the high cost of coordination materials ligands poses a limitation on their industrial application. Therefore, the exploration of cost-effective ligands for designing coordination materials catalysts with enhanced efficiency and durability holds significant potential for practical applications. Furthermore, enhancing the stability of these materials in acidic electrolytes is crucial. In summary, the rapid progress of coordination materials presents numerous new opportunities for catalysis, significantly advancing energy conversion technologies.

ACKNOWLEDGMENTS

This work was supported by the Research Fund for International Scientists (RFIS-Grant number: 52150410410) National Natural Science Foundation of China.

REFERENCES

[1] X. Yin, R. Yang, G. Tan, S. Fan, Terrestrial radiative cooling: Using the cold universe as a renewable and sustainable energy source, *Science*, 370 (2020) 786–791.
[2] G. Yasin, S. Ibraheem, S. Ali, M. Arif, S. Ibrahim, R. Iqbal, A. Kumar, M. Tabish, M.A. Mushtaq, A. Saad, H. Xu, W. Zhao, Defects-engineered tailoring of tri-doped interlinked metal-free bifunctional catalyst with lower Gibbs free energy of OER/HER intermediates for overall water splitting, *Materials Today Chemistry*, 23 (2022) 100634.
[3] N. Dubouis, A. Grimaud, The hydrogen evolution reaction: From material to interfacial descriptors, *Chemical Science*, 10 (2019) 9165–9181.
[4] M. Shao, Q. Chang, J.-P. Dodelet, R. Chenitz, Recent advances in electrocatalysts for oxygen reduction reaction, *Chemical Reviews*, 116 (2016) 3594–3657.
[5] X. Zhao, G. Hu, G.-F. Chen, H. Zhang, S. Zhang, H. Wang, Comprehensive understanding of the thriving ambient electrochemical nitrogen reduction reaction, *Advanced Materials*, 33 (2021) 2007650.
[6] Q. Lu, F. Jiao, Electrochemical CO2 reduction: Electrocatalyst, reaction mechanism, and process engineering, *Nano Energy*, 29 (2016) 439–456.
[7] K. Wang, Q. Li, Z. Ren, C. Li, Y. Chu, Z. Wang, M. Zhang, H. Wu, Q. Zhang, 2D metal–organic frameworks (MOFs) for high-performance BatCap hybrid devices, *Small*, 16 (2020) 2001987.
[8] S. Ajmal, A. Kumar, M. Selvaraj, M.M. Alam, Y. Yang, D.K. Das, R.K. Gupta, G. Yasin, MXenes and their interfaces for the taming of carbon dioxide & nitrate: A critical review, *Coordination Chemistry Reviews*, 483 (2023) 215094.
[9] L. Yang, X. Zeng, W. Wang, D. Cao, Recent progress in MOF-derived, heteroatom-doped porous carbons as highly efficient electrocatalysts for oxygen reduction reaction in fuel cells, *Advanced Functional Materials*, 28 (2018) 1704537.
[10] T. Mehtab, G. Yasin, M. Arif, M. Shakeel, R.M. Korai, M. Nadeem, N. Muhammad, X. Lu, Metal-organic frameworks for energy storage devices: Batteries and supercapacitors, *Journal of Energy Storage*, 21 (2019) 632–646.
[11] G. Yasin, S. Ali, S. Ibraheem, A. Kumar, M. Tabish, M.A. Mushtaq, S. Ajmal, M. Arif, M.A. Khan, A. Saad, L. Qiao, W. Zhao, Simultaneously engineering the synergistic-effects and coordination-environment of dual-single-atomic iron/cobalt-sites as a bifunctional oxygen electrocatalyst for rechargeable zinc-air batteries, *ACS Catalysis*, 13 (2023) 2313–2325.

[12] F. Yu, W. Liu, S.-W. Ke, M. Kurmoo, J.-L. Zuo, Q. Zhang, Electrochromic two-dimensional covalent organic framework with a reversible dark-to-transparent switch, *Nature Communications*, 11 (2020) 5534.
[13] S. Ajmal, A. Kumar, M. Tabish, M. Selvaraj, M.M. Alam, M.A. Mushtaq, J. Zhao, K.A. Owusu, A. Saad, M. Tariq Nazir, G. Yasin, Modulating the microenvironment of single atom catalysts with tailored activity to benchmark the CO2 reduction, *Materials Today*, 67 (2023) 203–228.
[14] A. Bala Musa, M. Tabish, A. Kumar, M. Selvaraj, M. Abubaker Khan, B.M. Al-Shehri, M. Arif, M. Asim Mushtaq, S. Ibraheem, Y. Slimani, S. Ajmal, T. Anh Nguyen, G. Yasin, Microenvironment engineering of Fe-single-atomic-site with nitrogen coordination anchored on carbon nanotubes for boosting oxygen electrocatalysis in alkaline and acidic media, *Chemical Engineering Journal*, 451 (2023) 138684.
[15] M. Nadeem, G. Yasin, M. Arif, M.H. Bhatti, K. Sayin, M. Mehmood, U. Yunus, S. Mehboob, I. Ahmed, U. Flörke, Pt-Ni@PC900 hybrid derived from layered-structure Cd-MOF for fuel cell ORR activity, *ACS Omega*, 5 (2020) 2123–2132.

11 Coordination Nanostructures for Green Hydrogen Generation

Muhammad Zahoor, Muhammad Ikram, Shahab Khan, Sajjad Ali and Ghulam Yasin

11.1 INTRODUCTION

Coordination nanomaterials (NMs) represent intricate structures where metals form coordinate covalent bonds with organic compounds. This diverse category encompasses various forms, including metal–organic nanocages, atomically precise metal nanoclusters, nanoscale metallocycles, metallodendrimers, and organometallic carborane nanoclusters[1]. These coordination-based nanomaterials exhibit specific molecular weights and structures, allowing precise adjustments at the molecular level in regulated manners. This capability renders them versatile for potential applications across various sectors [2]. Metal-organic frameworks (MOFs) represent a fascinating class of porous and crystalline materials distinguished by the intricate binding of organic ligands to metallic nodes, resulting in a sophisticated network of remarkable structure[3,4]. The versatility of MOFs is manifested in their extensive range of areas and porosities, an outcome of the abundant metallic nodes and organic ligands that are readily accessible. This inherent flexibility allows for precise adjustments, either in situ during the synthesis process or post-synthesis, thereby enabling the tailoring of these materials to exhibit optimal performance for a given application[5]. These permeable crystalline structures are created through the self-assembly of metal ions or clusters with organic monomers, commonly referred to as linkers. This unique synthesis process imparts MOFs with distinctive properties, making them highly adaptable and suitable for diverse applications. It is noteworthy that MOFs were first structurally characterized and formally documented in the scientific literature in 1943, marking a significant milestone in the exploration of these intriguing materials. Since then, the field of MOFs has witnessed continuous advancements, with researchers continually refining their understanding and manipulation of these structures for myriad technological and industrial purposes[5]. The subject of early research was primarily Prussian blue and its analogues, which are distinguished by Fe-CN-Fe bonds. But until Richard suggested the "node and spacer" strategy in the late 1980s, the MOF field's advancement remained constrained. This technique was

DOI: 10.1201/9781003345886-13

a major breakthrough in MOF design as it made it possible to include elongated organic ligands and transition metal ions with particular coordination geometry into the framework structure. In the 1980s, Richard proposed the "node and- spacer" method to introduce both TM-ions of well-defined coordination geometry and rod-like organic ligand into the design of the framework structure[6]. Similarly, the work presented by Hailian, Moulton, and others contributed considerably to this field, and it is now growing very rapidly. In the Richard node-and-spacer approach, the network is generally composed of metal nodes as well as organic linker spacers that could be octahedral, tetrahedral, and square, etc. In the following years, other MOFs with fascinating structures, such as HKUST-and MOF-5 were synthesized, and both of them are among the most investigated materials till date[7]. In 2002, Serre and colleagues introduced flexible and non-flexible permeable MOFs, namely, MIL-47/88 and MIL-53, respectively, pioneering the concept of isoreticular chemistry. This innovative approach, employing zinc dicarboxylates as building units, gained traction and expanded to diverse materials[3]. Metal nanomaterials, renowned for their intriguing properties and potential applications in optics, electronics, catalysis, and beyond, have since emerged as a focal point in nanomaterial research[8]. This chapter provides an in-depth exploration of the characteristics, molecular manipulation, and versatile applications of coordination nanomaterials, highlighting their significance in advancing interdisciplinary research and technology

11.2 HYDROGEN AND ITS PRODUCTION

Hydrogen emerges as a preeminent fuel source, distinguished by its environmental benignity, impressive energy yield, and a marked reduction in greenhouse gas emissions post-combustion. Current scientific investigations within the realm of hydrogen energy encompass a multifaceted exploration of diverse facets, including advanced methodologies for hydrogen generation, intricate economic considerations, the prospective integration of hydrogen in fuel cell electric vehicles, and the innovative application of hydrogen in co-combustion with diesel fuel[9]. Despite the promising strides made in hydrogen research, a notable incongruity persists, with over 80% of the global energy demand being still fulfilled by fossil fuels. This reliance on fossil fuels precipitates direct contributions to climate change, exacerbating greenhouse gas emissions and precipitating deleterious consequences such as air pollution. These environmental perturbations pose significant threats not only to ecological equilibrium but also to human health. In the imperative quest to expedite the decarbonization of the contemporary energy paradigm, there arises a pressing need to promulgate and implement sustainable and renewable energy resources as efficacious substitutes. This proactive paradigm shift, deeply rooted in scientific foresight, not only underscores a commitment to environmental conservation but also heralds a transformative stride towards establishing a resilient and eco-centric energy infrastructure poised for future generations[10]. In this perspective, hydrogen is recognized as the best candidate for a carbon-free energy vector[11,12]. Among the available energy sources; due to its power effectiveness and clean combustion byproduct, hydrogen fuel is considered as a possible contender. According to the MARKAL model, hydrogen could contribute for between 22 and 50% of Japan's electrical production while cutting CO_2

emissions to environment by 80%. Therefore, it is considered better fuel than other listed energy storage products[13]. There are two main categories of hydrogen production processes: non-renewable energy sources such as steam methane reforming, gasification, and pyrolysis and renewable energy sources, namely, electrolysis, biohydrogen, photocatalysis, thermochemical cycles, and plasmolysis[14]. Today, however, the majority of the energy needed to create hydrogen comes from non-renewable fossil fuels, and the process typically involves steam reforming of hydrocarbons at high temperatures which range from > 650 to 1,000°C[13]. Hydrogen can be produced in a variety of ways. However there are four main sources for the commercial production of H_2 with economic benefits, which are as follows: natural gas, oil, coal, and electrolysis[15]. Figure 11.1 illustrates hydrogen production using photocatalysts.

Hydrogen may also be produced by combining Ag_2S ZnO@ZnS core-shell nanorods with metal wire mesh-based immobilised photocatalysts[16]. Figure 11.2

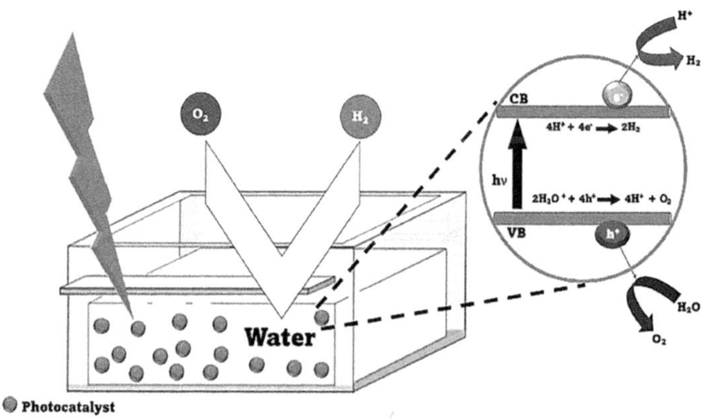

FIGURE 11.1 Hydrogen production from water using photocatalysts. Under creative commons attribution (CC BY) license (https://creativecommons.org/licenses/by/4.0/) from ref. 15.

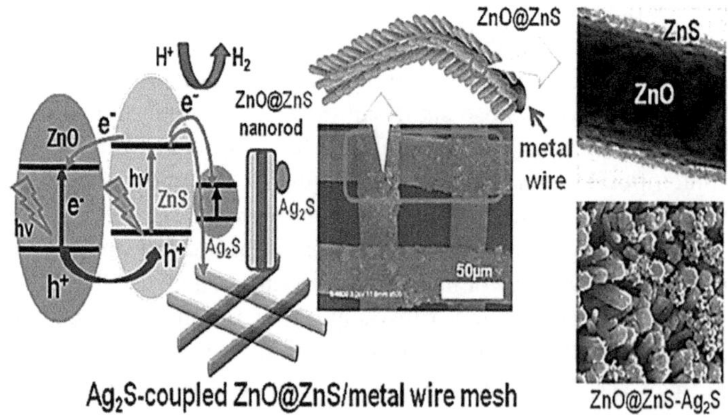

FIGURE 11.2 Photocatalytic H_2 production using metal wire mesh based immobilized photocatalyst with Ag_2S-coupled ZnO@ZnS core–shell nanorod. Copyright ACS (2016). Reproduced with permission from ref. 16.

depicts the morphology and a suggested mechanism of the Ag_2S-coupled ZnO@ZnS core-shell nanorod-based immobilised photocatalysts' photocatalytic H_2 generation[16].

David et al. reported the precise design and synthesis of a conductive copper-based MOF ultrathin layer (Cu-MOF) catalyst surface-coated with iron hydroxide [Fe $(OH)_x$] nano boxes [NBs]. Fe (OH)x@Cu-MOF was prepared and synthesized using a solvothermal reaction combined with a later redox etching strategy as shown in Figure 11.3. x-ray absorption fine structure (XAFS) spectroscopy and x-ray photoelectron spectroscopy (XPS) analysis show that there are a large number of coordination unsaturated Cu^{I}-O_2 centres in Fe (OH) x @Cu-MOF

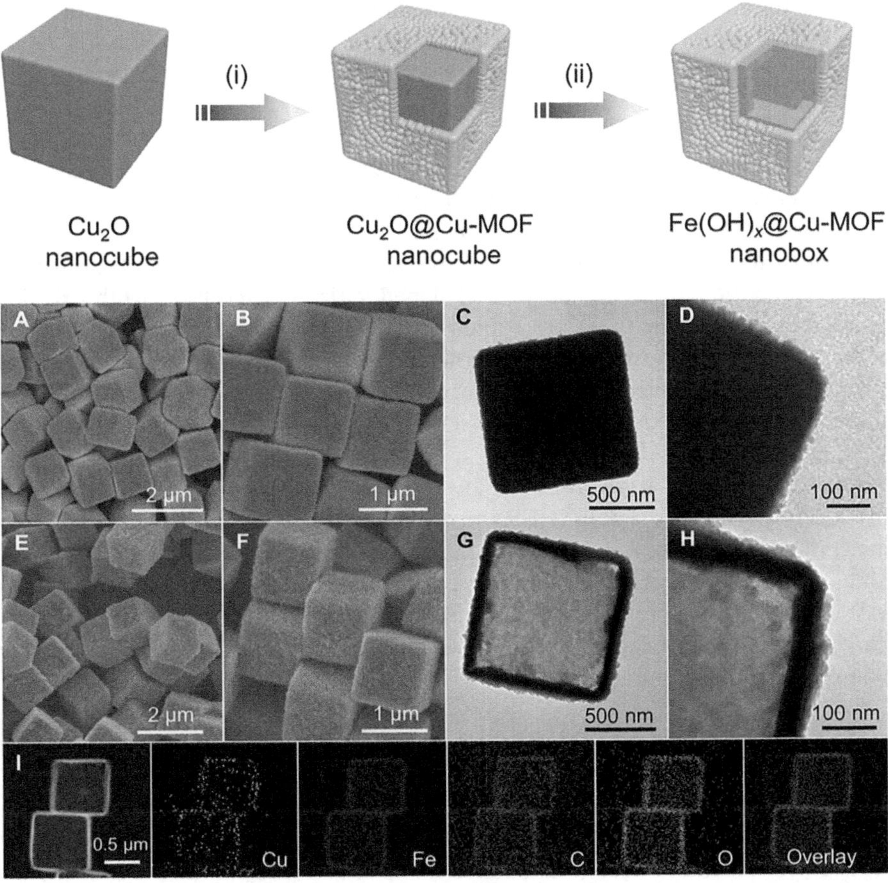

FIGURE 11.3 (i, ii) shows Schematic illustration of the synthetic process for $Fe(OH)_x$@ Cu-MOF NB. (A, B) FESEM images of Cu_2O@Cu-MOF nanocubes, (C, D) TEM images of Cu_2O@Cu-MOF nanocubes, (E, F) Fe(OH)x@ FESEM image of Cu-MOF NBs, TEM image of (G,H)Fe(OH)x @Cu-MOF NBs, HAADF-STEM image of (I)Fe(OH)x@Cu-MOF NBs and corresponding element distribution image. Science Advances (2021); under a Creative Commons Attribution NonCommercial License 4.0 (CC BY-NC), which permit reproduction in any medium. https://creativecommons.org/licenses/by-nc/4.0/ref. 18.

NB. With highly exposed active Cu centres and carefully designed material structure, the synthesized Fe (OH)x@Cu-MOF NB exhibits excellent HER activity and stability in alkaline solution, and a current density of 10 mA/cm corresponding to the overpotential is only 112 mV, and the corresponding Tafel slope is 76 mV/dec, making it one of the best MOF-based HER electrocatalysts reported so far. Density functional theory (DFT) calculations[17] show that the local electronic polarization of the coordinated unsaturated Cu^{1-} O_2 centres in the Cu-MOF layer helps promote the formation of *H reaction intermediates, thereby significantly accelerating the HER kinetics[18].

As producers of flexible substrates, the creation of infinitely coordinated polymer particles (CPPs) is of increasing importance for efficient conversion of solar energy. However, the lack of advanced organic ligands and modifications has largely hindered their applications. Controlling the morphology of CPPs through the competition and cooperation of two linkers is an effective way to develop high-quality structures. However, research on the plasticity of zinc-CPPs has been limited due to the difficulty in selecting mutually effective working ligands. Based on this, Professor Li Yuliang studied the morphological evolution and structural pyrolysis process of Zn-based CPPs for the first time[19]. Through the cooperative reaction of salen ligand and carboxylic acid ligand, 6 Zn-CPP substructures were obtained. The advantages of hollow spherical and tubular ZnO structures prepared by CPPs in photocatalysis are highlighted. Yanlin Qin at.al presented new Hierarchical Hollow polymer coordination materials with O-doped $ZnIn_2S_4$ nanosheets with excellent hydrogen production activity were finely modified onto the obtained hollow ZnO material through a hydrothermal method for optimization. The obtained O-doped $ZnIn_2S_4$/ZnO composite photocatalyst showed excellent catalyst-free hydrogen evolution reaction (HER) activity (4041.8–4512.5 μmol h-1 g-1) and stability as shown in Figure 11.4

DFT calculations were also employed to ascertain the crucial steps of the Hydrogen Evolution Reaction (HER) on the O-$ZnIn_2S_4$/ZnO heterostructure. These calculations also serve as an activity descriptor, aiding in a more comprehensive comprehension of the HER performance within the photocatalytic system. In contrast to alternative sites such as S, Zn, and in atoms, the binding free energy (ΔGH^*) for hydrogen on the oxygen (O) site of the In-S terminated surface is particularly noteworthy as show in Figure 11.5.

11.3 NANOMATERIALS FOR HYDROGEN PRODUCTION

For photocatalytic hydrogen generation, nanomaterials such CdS, SiC, $CuInSe_2$, and TiO_2 have been employed and their improved photocatalytic qualities have been highlighted. At the moment, nanomaterials such as Nb_2O_5, Ta_2O_5, α-Fe_2O_3, ZnO, TaON, $BiVO_4$, and WO_3 are being studied. Limitation of the band gap can reduce H_2 generation in most of them. Techniques including sensitization, metal-ion implantation, and noble metal or metallic ion doping have been tried to address this problem. Ag, Ru, Pd, Ni, Cu, and Ir have all been investigated concurrently because of their cheap cost and availability. Pt is the finest noble metal doping material, but it is also the more costly[20]. The increasing interest in preparing nanocatalysts

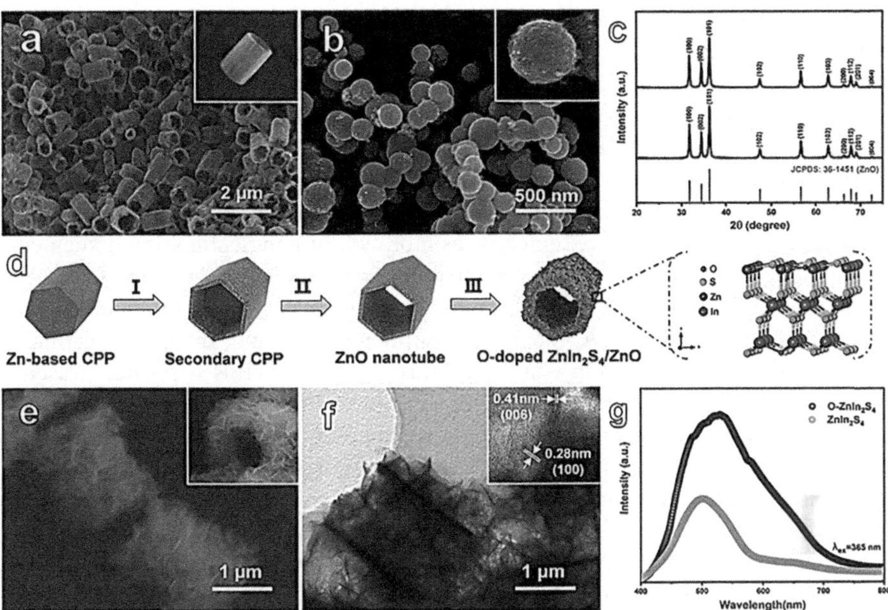

FIGURE 11.4 (a, b) FESEM and (c) XRD of CPPs-derived ZnO hollow microtubes and microspheres. (d) Schematic diagram of the formation of hollow O-doped $ZnIn_2S4$/ ZnO microtubes. (e) FESEM and (f) TEM) of hollow O-doped $ZnIn_2S_4$/ZnO microtubes (inset: HRTEM). (g) Photoluminescence spectra of O-doped $ZnIn_2S_4$ and pristine $ZnIn_2S_4$ nanosheets. Ref. 19.

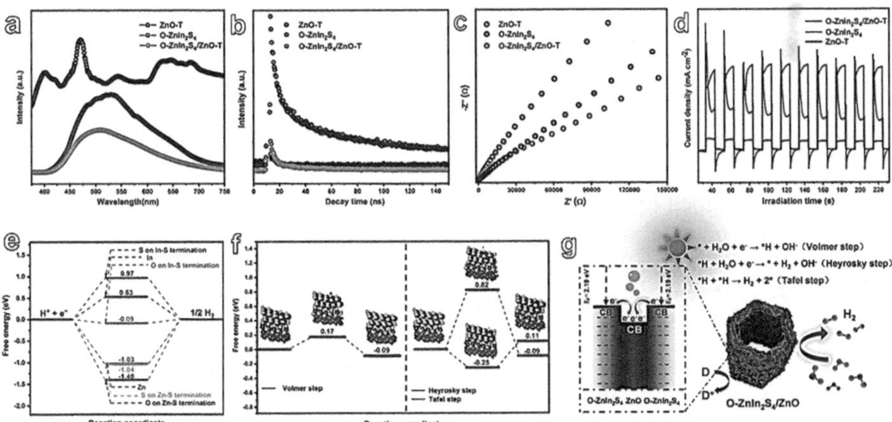

FIGURE 11.5 (a) Steady-state luminescence spectra of ZnO-T, O-$ZnIn_2S_4$ and O-$ZnIn_2S_4$/ ZnO-T samples, (b) time-resolved transient luminescence decay curves, (c) EIS changes, and (d) light current response. (e) DFT calculation of ΔGH* for the adsorption free energy of H at S, In, O, and Zn sites in O-$ZnIn_2S_4$/ZnO-T. (f) Energy barriers of the Volmer step, Tafel step and Heyrosky step of hydrogen evolution in the O-$ZnIn_2S_4$/ZnO-T system. (g) Schematic diagram of HER on O-$ZnIn_2S_4$/ZnO-T hypersurface. Ref. 19.

is primarily driven by their significant surface-to-volume ratio, which markedly enhances the number of active sites on the surface of nanoparticles (NPs) and their interactions with various molecules or organic moieties[21–23]. This has opened up numerous avenues for developing novel materials and streamlining chemical processes. Consequently, scientists worldwide are grappling with the integration of both homogeneous and heterogeneous catalysis to craft advanced nanomaterials, thereby bridging the gap between these two domains and merging their strengths[24]. Nanoengineering,[25] therefore, emerges as a branch of nanotechnology dedicated to manipulating nanostructures at sub-nano or atomic scale levels to optimize catalytic performance. In simpler terms, the pursuit involves crafting nanomaterials with meticulous control over size (quantum confinement), shape, composition (mono-, bi-, and trimetallic materials), and structure, which remains a primary research focus[26]. Engineered nanomaterials, typically ranging from 1 to 100 nm in size, are tailored or designed specifically to replicate their properties for targeted applications. Notably, nanomaterials are increasingly playing a pivotal role in pioneering new fields, in addition to their significant impact on enhancing hydrogen production and storage[27,28].

Hsu et al. conducted research where they utilized Ag_2S-coupled ZnO@ZnS core-shell nanorods to effectively produce hydrogen gas (H_2). They found that the highest rate of H_2 production occurred when the concentration of $AgNO_3$ was 2 millimolar (mM). Interestingly, increasing the concentration of $AgNO_3$ beyond this point led to a decrease in the rate of H_2 production. This suggests that there is an optimal concentration for increasing the efficiency of the process[16].

In another study, Yue et al. developed a new composite material consisting of Ag_2S, ZnS, and carbon nanofibers (CNF) with the aim of enhancing the rate of H_2 production even further compared to previously reported ZnS-based photocatalysts. They observed that the combined effect of Ag_2S and CNF played a crucial role in preventing the recombination of charge carriers, which is beneficial for maintaining high efficiency in H_2 generation[29].

Additionally, researchers found that nanosheets made of ZnS:Ag_2S exhibited a similar enhancement trend in H_2 production as the aforementioned core-shell nanorods. Yang et al. reported on the synthesis of porous ZnS:Ag_2S nanosheets, which were designed to have a large surface area. This increased surface area facilitated intimate contact between the material and the sacrificial solution, promoting the reaction between electrons and protons (H+) to produce H_2[30].

In another literature finding Song et al. observed that, in comparison with 2D Cu_2MoS_4 nanosheet with the exposed[31] facet, Cu_2MoS_4 nanotube with the exposed[32] facet exhibited effectively improved the performance for photocatalytic degradation and water splitting[33,34]. Shen et al. reported that the crystal facets of $ZnIn_2S_4$ with a 3D-hierarchical persimmon-like structure will influence the photocatalytic activity of $ZnIn_2S_4$. Extending the reaction time did not reveal any significant influences on the band gap or surface area of $ZnIn_2S_4$. Hence, the increase in the percentage of the [6] facet enhances hydrogen production[35]. Figure 11.6 depicts the generation of hydrogen on the [6] facets of the Pt loaded $ZnIn_2S_4$ photocatalysts.

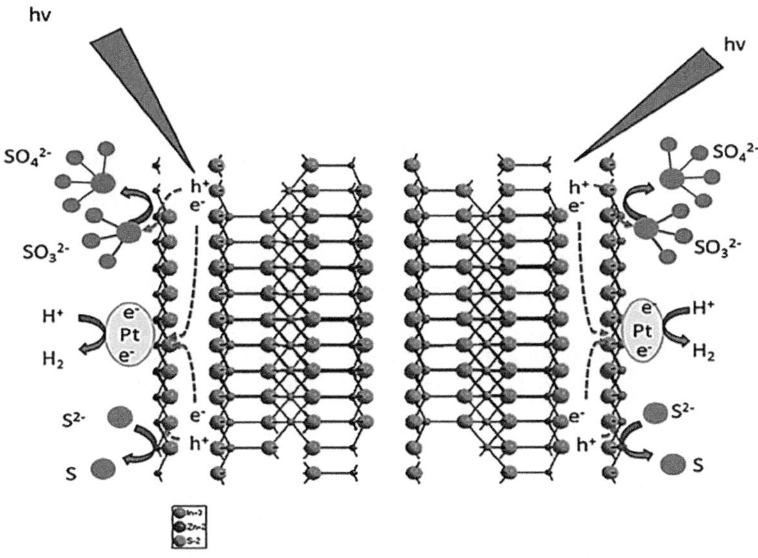

FIGURE 11.6 Hydrogen production reaction on the [6] facets of Pt loaded $ZnIn_2S_4$ photocatalysts. Copyright Elsevier (2012). Reproduced with permission from ref. 35.

11.4 GRAPHENE-BASED NANOCOMPOSITES FOR HYDROGEN PRODUCTION

The incorporation of single-atom catalysts into 2D materials and carbon-based materials, including graphene, carbon nanodots, fullerenes, and carbon nanotubes, has led to groundbreaking advancements across diverse applications[36–46]. This synergy enhances catalytic precision, impacting fields from energy storage to environmental remediation. In hydrogen generation, particularly in photocatalysis, the collaboration of carbon-based materials with single-atom catalysts shows promise, revolutionizing efficiency and offering a sustainable path for advanced hydrogen production[15]. These materials are effective across various optical spectra in enhancing hydrogen production. Research has demonstrated that employing carbon materials extends the absorption range of visible light and facilitates more efficient charge transfer. Graphene, specifically, offers advantages such as high charge carrier mobility, large surface area, excellent electrical and thermal conductivity, as well as robust physical and chemical stability, and it can be easily synthesized[47,48]. Li et al. synthesized S and N co-doped graphene quantum dots/TiO_2 (S,N-GQD/TiO_2) composite materials. These materials were found efficient for photocatalytic hydrogen production. Furthermore graphene-based nanocomposites exhibited superior photocatalytic activity compared to pure TiO_2, attributed to their enhanced visible light absorption and effective separation and migration of electrons and holes[49]. Additionally, Zhang et al. reported improved photocatalytic hydrogen evolution from water splitting using graphene/Ti nanocomposites[50]. Fan et al. developed TiO_2/RGO nanocomposites, while Hao et al.

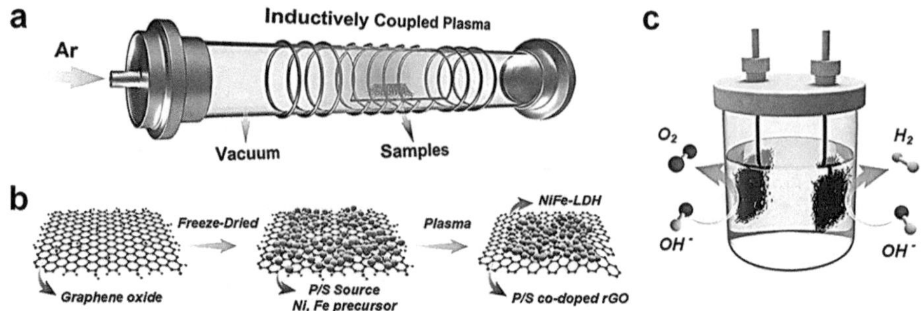

FIGURE 11.7 Schematic showing the process for fabricating the NiFe LDH/PX-rGO catalyst. Copyright Elsevier (2024). Reproduced with permission from Ref. 52.

anchored TiO_2 with graphene quantum dots, both resulting in enhanced photocatalytic hydrogen production[51]. This enhancement is attributed to the capabilities of graphene quantum dots to serve as efficient electron reservoirs and photosensitizers when combined with TiO_2. In Figure 11.7, a schematic details the stepwise process for fabricating the NiFe LDH/PX-rGO catalyst, beginning with the synthesis of precursor materials, such as NiFe Layered Double Hydroxides (LDH) and pristine reduced graphene oxide (PX-rGO), followed by a layered assembly of these components[52].

11.5 TWO-DIMENSIONAL (2D) COORDINATION NANOMATERIALS (2D-CNMs) IN ELECTROCATALYSIS APPLICATION

The use of two-dimensional (2D) coordination Nanomaterials (2D-CNMs) electrocatalysis has recently attracted significant interest[53–61]. Two-dimensional metal-organic frameworks (2D-MOFs) are regarded as ideal models, where metal atoms are evenly dispersed and organic ligands are attached in a relatively stable manner[62–66]. These materials are highly effective for CO_2 reduction reactions due to their extensive surface area, porous structure, and flexible topology[67]. These materials show high electrical conductivity as a result of long-range electron delocalization between conductive metal atoms and different ligands as well as conjugated benzene rings. This makes them excellent electrocatalysts for a variety of reactions such as evolution of hydrogen, reduction of oxygen[55], and carbon dioxide[53,67]. Recent studies have explored the utilization of 2D materials (e.g., nitrogen, phosphorus, and boron-doped carbon) as electrocatalysts for numerous valuable conversions. Ongoing efforts aim to enhance their catalytic properties further.[68] Among the various modifications attempted, introducing nitrogen atoms has proven particularly successful, leading to the formation of a covalently bonded carbon framework. However, existing techniques for nitrogen doping in graphene and graphene oxide often lack precise control over nitrogen content and oxidation state. Additionally, bottom-up strategies for constructing highly nitrogenized 2D materials with well-defined pore sizes are underexplored[55]. These materials are considered promising hosts for incorporating single metal atoms due to their desirable properties. Furthermore, they prove suitable for

CO2 reduction reactions (CO2RR), oxygen reduction reactions (ORR), and oxygen evolution reactions (OER), especially when coordinated with non-noble metals. This presents a greener and more cost-effective alternative to noble metals such as iridium, ruthenium, and platinum[69]. Non-precious metal-anchored materials provide a pivotal advantage, being not only more economical but also safer and highly sustainable as electrocatalysts[69].

11.6 CONCLUSION AND FUTURE PERSPECTIVE

Coordination nanomaterials, characterized by their unique structure of metal ligands bound to organic substrates, exhibit a wide range of applications across different sectors. Their distinctive qualities, such as increased surface area, porosity, and diverse chemical structures, make them highly valuable in catalytic water splitting processes for hydrogen generation. In particular, two-dimensional metal-organic frameworks and graphene-based nanomaterials demonstrate enhanced efficacy in the hydrogen manufacturing process.

To further advance the field, researchers should explore alternative methods and incorporate new substrates to optimize the catalytic qualities of coordination nanomaterials. This ongoing innovation holds the potential to broaden their applications in sustainable energy production and contribute to the development of eco-friendly technologies.

Looking ahead, the strategic use of coordination nanomaterials in the production of sustainable fuels, particularly hydrogen, is crucial for a more environmentally conscious energy landscape. Encouraging the exploration of new sources for hydrogen aligns with the goal of promoting synthetic fuels to establish a circular economy. The continued development and thoughtful application of these nanoparticles promise a brighter, more sustainable future in energy production.

REFERENCES

[1] Abbas, Y.; Ali, S.; Ali, S.; Zuhra, Z.; Haoliang, W.; Bououdina, M.; Zhenzhong, S., Amine-rich cyclotriphosphazene (P_3N_3) nano-cages for enhanced and selective Au(III) and Pd(II) recovery and hydrogen generation: Waste-to-resource tactics. *Chemical Engineering Journal* **2024,** *492*, 152127.

[2] Wang, C.; Wang, W.; Tan, J.; Zhang, X.; Yuan, D.; Zhou, H.-C., Coordination-based molecular nanomaterials for biomedically relevant applications. *Coordination Chemistry Reviews* **2021,** *438*, 213752.

[3] Shah, R.; Ali, S.; Raziq, F.; Ali, S.; Ismail, P. M.; Shah, S.; Iqbal, R.; Wu, X.; He, W.; Zu, X.; Zada, A.; Adnan; Mabood, F.; Vinu, A.; Jhung, S. H.; Yi, J.; Qiao, L., Exploration of metal organic frameworks and covalent organic frameworks for energy-related applications. *Coordination Chemistry Reviews* **2023,** *477*, 214968.

[4] Khan, S.; Iqbal, A., Organic polymers revolution: Applications and formation strategies, and future perspectives. *Journal of Polymer Science and Engineering* **2023,** *6* (1), 3125.

[5] Altundal, O. F.; Haslak, Z. P.; Keskin, S., Combined, G. C. M. C., MD, and DFT approach for unlocking the performances of COFs for methane purification. *Industrial & Engineering Chemistry Research* **2021,** *60* (35), 12999–13012.

[6] Robson, R., A net-based approach to coordination polymers. *Journal of the Chemical Society, Dalton Transactions* **2000,** (21), 3735–3744.

[7] Li, H.; Eddaoudi, M.; O'Keeffe, M.; Yaghi, O. M., Design and synthesis of an exceptionally stable and highly porous metal-organic framework. *Nature* **1999,** *402* (6759), 276–279.
[8] Tao, A. R.; Habas, S.; Yang, P., Shape control of colloidal metal nanocrystals. *Small* **2008,** *4* (3), 310–325.
[9] Lee, S. L.; Chang, C.-J., Recent progress on metal sulfide composite nanomaterials for photocatalytic hydrogen production. *Catalysts* **2019,** *9* (5), 457.
[10] Rueda-Navarro, C. M.; Ferrer, B.; Baldoví, H. G.; Navalón, S., Photocatalytic Hydrogen production from glycerol aqueous solutions as sustainable feedstocks using Zr-based UiO-66 materials under simulated sunlight irradiation. *Nanomaterials* **2022,** *12* (21), 3808.
[11] Parkinson, B.; Balcombe, P.; Speirs, J. F.; Hawkes, A. D.; Hellgardt, K., Levelized cost of CO2 mitigation from hydrogen production routes. *Energy & Environmental Science* **2019,** *12* (1), 19–40.
[12] Staffell, I.; Scamman, D.; Velazquez Abad, A.; Balcombe, P.; Dodds, P. E.; Ekins, P.; Shah, N.; Ward, K. R., The role of hydrogen and fuel cells in the global energy system. *Energy & Environmental Science* **2019,** *12* (2), 463–491.
[13] Quach, Q.; Biehler, E.; Abdel-Fattah, T. M., Synthesis of palladium nanoparticles supported over fused graphene-like material for hydrogen evolution reaction. *Catalysts* **2023,** *13* (7), 1117.
[14] Younas, M.; Shafique, S.; Hafeez, A.; Javed, F.; Rehman, F., An overview of hydrogen production: Current status, potential, and challenges. *Fuel* **2022,** *316*, 123317.
[15] Jawhari, A. H., Novel nanomaterials for hydrogen production and storage: Evaluating the futurity of graphene/graphene composites in hydrogen energy. *Energies* **2022,** *15* (23), 9085.
[16] Hsu, M.-H.; Chang, C.-J.; Weng, H.-T., Efficient H2 production using Ag2S-coupled ZnO@ZnS core–shell nanorods decorated metal wire mesh as an immobilized hierarchical photocatalyst. *ACS Sustainable Chemistry & Engineering* **2016,** *4* (3), 1381–1391.
[17] Khan, S.; Rahman, M.; Marwani, H. M.; Althomali, R. H.; Rahman, M. M., Bicomponent polymorphs of salicylic acid, their antibacterial potentials, intermolecular interactions, DFT and docking studies. *Zeitschrift für Physikalische Chemie* **2023,** *238* (01), 1–16.
[18] Cheng, W.; Zhang, H.; Luan, D.; Lou, X. W., Exposing unsaturated Cu1-O2 sites in nanoscale Cu-MOF for efficient electrocatalytic hydrogen evolution. **2021,** *7* (18), eabg2580.
[19] Zhu, Q.; Xu, S.; Wu, W.; Qi, Y.; Lin, Z.; Li, Y.; Qin, Y., Hierarchical hollow zinc oxide nanocomposites derived from morphology-tunable coordination polymers for enhanced solar hydrogen production. **2022,** *61* (29), e202205312.
[20] Parra, D.; Valverde, L.; Pino, F. J.; Patel, M. K., A review on the role, cost and value of hydrogen energy systems for deep decarbonisation. *Renewable and Sustainable Energy Reviews* **2019,** *101*, 279–294.
[21] Fiorio, J. L.; Gothe, M. L.; Kohlrausch, E. C.; Zardo, M. L.; Tanaka, A. A.; de Lima, R. B.; da Silva, A. G. M.; Garcia, M. A. S.; Vidinha, P.; Machado, G., Nanoengineering of catalysts for enhanced hydrogen production. *Hydrogen* **2022,** *3* (2), 218–254.
[22] Ali, S.; Zuhra, Z.; Ali, S.; Qi, H.; Ali, S.; Muhammad, A.; Zhongying, W. Ultra-deep removal of Pb by functionality tuned UiO-66 framework: A combined experimental, theoretical and HSAB approach. *Chemosphere* **2021,** *284*, 131305.
[23] Khan, S.; Zahoor, M.; Rahman, M. U.; Gul, Z., Cocrystals; basic concepts, properties and formation strategies. *Zeitschrift für Physikalische Chemie* **2023,** *237* (3), 273–332.
[24] Cui, X.; Li, W.; Ryabchuk, P.; Junge, K.; Beller, M., Bridging homogeneous and heterogeneous catalysis by heterogeneous single-metal-site catalysts. *Nature Catalysis* **2018,** *1* (6), 385–397.

[25] Khan, S., Phase engineering and impact of external stimuli for phase tuning in 2D materials. *Advanced Energy Conversion Materials* **2023,** *5* (1), 40–55.
[26] Camargo, P. H. C.; Rodrigues, T. S.; da Silva, A. G. M.; Wang, J., Controlled synthesis: Nucleation and growth in solution. In *Metallic Nanostructures: From Controlled Synthesis to Applications*, Xiong, Y.; Lu, X., Eds. Springer International Publishing: Cham, 2015; pp. 49–74.
[27] Epelle, E. I.; Desongu, K. S.; Obande, W.; Adeleke, A. A.; Ikubanni, P. P.; Okolie, J. A.; Gunes, B., A comprehensive review of hydrogen production and storage: A focus on the role of nanomaterials. *International Journal of Hydrogen Energy* **2022,** *47* (47), 20398–20431.
[28] Linping, B.; Yushuai, J.; Xiaohui, R.; Liu, X.; Dai, C.; Ali, S.; Bououdina, M.; Lu, Z.; Zeng, C., Cr dopants and S vacancies in ZnS to trigger efficient photocatalytic H_2 evolution and CO_2 reduction. *Journal of Materials Science & Technology* **2024,** *199*, 75–85.
[29] Yue, S.; Wei, B.; Guo, X.; Yang, S.; Wang, L.; He, J., Novel Ag2S/ZnS/carbon nanofiber ternary nanocomposite for highly efficient photocatalytic hydrogen production. *Catalysis Communications* **2016,** *76*, 37–41.
[30] Yang, X.; Xue, H.; Xu, J.; Huang, X.; Zhang, J.; Tang, Y.-B.; Ng, T.-W.; Kwong, H.-L.; Meng, X.-M.; Lee, C.-S., Synthesis of porous ZnS:Ag2S nanosheets by ion exchange for photocatalytic H2 generation. *ACS Applied Materials & Interfaces* **2014,** *6* (12), 9078–9084.
[31] Zhou, H.-C.; Long, R. J.; Yaghi, M. O., Introduction to metal–organic frameworks. *Chemical Reviews* **2012,** *112* (2), 673–674.
[32] Abate, S.; Arrigo, R.; Schuster, M. E.; Perathoner, S.; Centi, G.; Villa, A.; Su, D.; Schlögl, R., Pd nanoparticles supported on N-doped nanocarbon for the direct synthesis of H_2O_2 from H_2 and O_2. *Catalysis Today* **2010,** *157*, 280.
[33] Zhang, K.; Lin, Y.; Muhammad, Z.; Wu, C.; Yang, S.; He, Q.; Zheng, X.; Chen, S.; Ge, B.; Song, L., Active {010} facet-exposed Cu2MoS4 nanotube as high-efficiency photocatalyst. *Nano Research* **2017,** *10* (11), 3817–3825.
[34] Khan, S.; Ajmal, S.; Hussain, T.; Rahman, M. U., Clay-based materials for enhanced water treatment: Adsorption mechanisms, challenges, and future directions. *Journal of Umm Al-Qura University for Applied Sciences* **2023**, 1–16.
[35] Shen, J.; Zai, J.; Yuan, Y.; Qian, X., 3D hierarchical ZnIn2S4: The preparation and photocatalytic properties on water splitting. *International Journal of Hydrogen Energy* **2012,** *37* (22), 16986–16993.
[36] Ali, S.; Liu, T.; Lian, Z.; Li, B.; Su, D. S., The tunable effect of nitrogen and boron dopants on a single walled carbon nanotube support on the catalytic properties of a single gold atom catalyst: A first principles study of CO oxidation. *Journal of Materials Chemistry A* **2017,** *5* (32), 16653–16662.
[37] Ali, S.; Fu Liu, T.; Lian, Z.; Li, B.; Sheng Su, D., The effect of defects on the catalytic activity of single Au atom supported carbon nanotubes and reaction mechanism for CO oxidation. *Physical Chemistry Chemical Physics* **2017,** *19* (33), 22344–22354.
[38] Ali, S.; Liu, T.; Lian, Z.; Sheng Su, D.; Li, B., The stability and reactivity of transition metal atoms supported mono and di vacancies defected carbon based materials revealed from first principles study. *Applied Surface Science* **2019,** *473*, 777–784.
[39] Ali, S.; Olanrele, S.; Liu, T.; Lian, Z.; Si, C.; Yang, M.; Li, B., Single Au anion can catalyze acetylene hydrochlorination: Tunable catalytic performance from rational doping. *The Journal of Physical Chemistry C* **2019,** *123* (48), 29203–29208.
[40] Ali, S.; Haneef, M.; Akbar, J.; Ullah, I.; Ullah, S.; Samad, A., Single Au atom supported defect mediated boron nitride monolayer as an efficient catalyst for acetylene hydrochlorination: A first principles study. *Molecular Catalysis* **2021,** *511*, 111753.
[41] Ali, S.; Lian, Z.; Li, B., Density functional theory study of a graphdiyne-supported single Au atom catalyst for highly efficient acetylene hydrochlorination. *ACS Applied Nano Materials* **2021,** *4* (6), 6152–6159.

[42] Ali, S.; Xie, Z.; Xu, H., Stability and catalytic performance of single-atom supported on Ti2CO2 for low-temperature CO oxidation: A first-principles study. *ChemPhysChem* **2021,** *22* (22), 2352–2361.
[43] Bao, L.; Ren, X.; Liu, C.; Liu, X.; Dai, C.; Yang, Y.; Bououdina, M.; Ali, S.; Zeng, C., Modulating the doping state of transition metal ions in ZnS for enhanced photocatalytic activity. *Chemical Communications* **2023**, *59*, 11280–11283.
[44] Muhammad Ismail, P.; Ali, S.; Raziq, F.; Bououdina, M.; Abu-Farsakh, H.; Xia, P.; Wu, X.; Xiao, H.; Ali, S.; Qiao, L., Stable and robust single transition-metal atom catalyst for CO2 reduction supported on defective WS2. *Applied Surface Science* **2023**, 157073.
[45] Sun, M.; Ali, S.; Liu, C.; Dai, C.; Liu, X.; Zeng, C., Synergistic effect of Fe doping and oxygen vacancy in AgIO3 for effectively degrading organic pollutants under natural sunlight. *Environmental Pollution* **2024,** *344*, 123325.
[46] Chabira, F.; Ali, S.; Khan, A.; Humayun, M.; Bououdina, M., Comprehensive review on single-atom catalysts in electrochemical hydrogen-evolution reaction: Computational modelling and experimental investigation. *Philosophical Magazine Letters* **2024,** *104* (1), 2343665.
[47] Balandin, A. A.; Ghosh, S.; Bao, W.; Calizo, I.; Teweldebrhan, D.; Miao, F.; Lau, C. N., Superior thermal conductivity of single-layer graphene. *Nano Letters* **2008,** *8* (3), 902–907.
[48] Stoller, M. D.; Park, S.; Zhu, Y.; An, J.; Ruoff, R. S., Graphene-based ultracapacitors. *Nano Letters* **2008,** *8* (10), 3498–3502.
[49] Xie, H.; Hou, C.; Wang, H.; Zhang, Q.; Li, Y., S, N Co-doped graphene quantum dot/TiO2 composites for efficient photocatalytic hydrogen generation. *Nanoscale Research Letters* **2017,** *12* (1), 400.
[50] Fan, W.; Lai, Q.; Zhang, Q.; Wang, Y., Nanocomposites of TiO2 and reduced graphene oxide as efficient photocatalysts for hydrogen evolution. *The Journal of Physical Chemistry C* **2011,** *115* (21), 10694–10701.
[51] Hao, X.; Jin, Z.; Xu, J.; Min, S.; Lu, G., Functionalization of TiO2 with graphene quantum dots for efficient photocatalytic hydrogen evolution. *Superlattices and Microstructures* **2016,** *94*, 237–244.
[52] He, W.; Wu, S.; Zhang, Z.; Duan, P.; Yang, Q., Modulating P-S anions of graphene supported amorphous nickel-iron nanocomposites by plasma to optimize overall water splitting activity. *International Journal of Hydrogen Energy* **2024,** *51*, 44–54.
[53] Iqbal, R.; Akbar, M. B.; Ahmad, A.; Hussain, A.; Altaf, N.; Ibraheem, S.; Yasin, G.; Khan, M. A.; Tabish, M.; Kumar, A.; Majeed, M. K.; Saleem, A.; Ali, S., Exploring the synergistic effect of novel Ni-Fe in 2D bimetallic metal-organic frameworks for enhanced electrochemical reduction of CO2. *Advanced Materials Interfaces* **2022,** *9* (1), 2101505.
[54] Ali, S.; Yasin, G.; Iqbal, R.; Huang, X.; Su, J.; Ibraheem, S.; Zhang, Z.; Wu, X.; Wahid, F.; Ismail, P. M.; Qiao, L.; Xu, H., Porous aza-doped graphene-analogous 2D material a unique catalyst for CO_2 conversion to formic-acid by hydrogenation and electroreduction approaches. *Molecular Catalysis* **2022,** *524*, 112285.
[55] Iqbal, R.; Ali, S.; Yasin, G.; Ibraheem, S.; Tabish, M.; Hamza, M.; Chen, H.; Xu, H.; Zeng, J.; Zhao, W., A novel 2D $Co_3(HADQ)_2$ metal-organic framework as a highly active and stable electrocatalyst for acidic oxygen reduction. *Chemical Engineering Journal* **2022,** *430*, 132642.
[56] Iqbal, R.; Yasin, G.; Hamza, M.; Ibraheem, S.; Ullah, B.; Saleem, A.; Ali, S.; Hussain, S.; Anh Nguyen, T.; Slimani, Y.; Pathak, R., State of the art two-dimensional covalent organic frameworks: Prospects from rational design and reactions to applications for advanced energy storage technologies. *Coordination Chemistry Reviews* **2021,** *447*, 214152.
[57] Ali, S.; Iqbal, R.; Wahid, F.; Ismail, P. M.; Saleem, A.; Ali, S.; Raziq, F.; Ullah, S.; Ullah, I.; Tahir; Zahoor, M.; Wu, X.; Xiao, H.; Zu, X.; Qiao, L., Cobalt coordinated two-dimensional covalent organic framework a sustainable and robust electrocatalyst for selective CO_2 electrochemical conversion to formic acid. *Fuel Processing Technology* **2022,** *237*, 107451.

[58] Huang, X.; Gan, L.-Y.; Wang, J.; Ali, S.; He, C.-C.; Xu, H., Developing proton-conductive metal coordination polymer as highly efficient electrocatalyst toward oxygen reduction. *The Journal of Physical Chemistry Letters* **2021,** *12* (38), 9197–9204.
[59] Yasin, G.; Ibrahim, S.; Ibraheem, S.; Ali, S.; Iqbal, R.; Kumar, A.; Tabish, M.; Slimani, Y.; Nguyen, T. A.; Xu, H.; Zhao, W., Defective/graphitic synergy in a heteroatom-interlinked-triggered metal-free electrocatalyst for high-performance rechargeable zinc–air batteries. *Journal of Materials Chemistry A* **2021,** *9* (34), 18222–18230.
[60] Yasin, G.; Ibraheem, S.; Ali, S.; Arif, M.; Ibrahim, S.; Iqbal, R.; Kumar, A.; Tabish, M.; Mushtaq, M. A.; Saad, A.; Xu, H.; Zhao, W., Defects-engineered tailoring of tri-doped interlinked metal-free bifunctional catalyst with lower Gibbs free energy of OER/HER intermediates for overall water splitting. *Materials Today Chemistry* **2022,** *23*, 100634.
[61] Yasin, G.; Ali, S.; Ibraheem, S.; Kumar, A.; Tabish, M.; Mushtaq, M. A.; Ajmal, S.; Arif, M.; Khan, M. A.; Saad, A.; Qiao, L.; Zhao, W., Simultaneously engineering the synergistic-effects and coordination-environment of dual-single-atomic iron/cobalt-sites as a bifunctional oxygen electrocatalyst for rechargeable zinc-air batteries. *ACS Catalysis* **2023**, 2313–2325.
[62] Qadir, S.; Gu, Y.; Ali, S.; Li, D.; Zhao, S.; Wang, S.; Xu, H.; Wang, S., A thermally stable isoquinoline based ultra-microporous metal-organic framework for CH4 separation from coal mine methane. *Chemical Engineering Journal* **2022,** *428*, 131136.
[63] Ali, S.; Zuhra, Z.; Ali, S.; Han, Q.; Ahmad, M.; Wang, Z., Ultra-deep removal of Pb by functionality tuned UiO-66 framework: A combined experimental, theoretical and HSAB approach. *Chemosphere* **2021,** *284*, 131305.
[64] Zuhra, Z.; Ali, S.; Ali, S.; Xu, H.; Wu, R.; Tang, Y., Exceptionally amino-quantitated 3D MOF@CNT-sponge hybrid for efficient and selective recovery of Au(III) and Pd(II). *Chemical Engineering Journal* **2021**, 133367.
[65] Iqbal, R.; Ali, S.; Saleem, A.; Majeed, M. K.; Hussain, A.; Rauf, S.; Rehman Akbar, A.; Xu, H.; Qiao, L.; Zhao, W., Electrically conductive Pt-MOFs for acidic oxygen reduction: Optimized performance via altering conjugated ligands. *Chemical Engineering Journal* **2022**, 140799.
[66] Iqbal, R.; Ali, S.; Saleem, A.; Majeed, M. K.; Hussain, A.; Rauf, S.; Rehman Akbar, A.; Xu, H.; Qiao, L.; Zhao, W., Electrically conductive Pt-MOFs for acidic oxygen reduction: Optimized performance via altering conjugated ligands. *Chemical Engineering Journal* **2023,** *455*, 140799.
[67] Ali, S.; Ismail, P. M.; Wahid, F.; Kumar, A.; Haneef, M.; Raziq, F.; Ali, S.; Javed, M.; Khan, R. U.; Wu, X.; Xiao, H.; Yasin, G.; Qiao, L.; Xu, H., Benchmarking the two-dimensional conductive $Y_3(C_6X_6)_2$ (Y = Co, Cu, Pd, Pt; X = NH, NHS, S) metal-organic framework nanosheets for CO2 reduction reaction with tunable performance. *Fuel Processing Technology* **2022,** *236*, 107427.
[68] Gul, Z.; Salman, M.; Khan, S.; Shehzad, A.; Ullah, H.; Irshad, M.; Zeeshan, M.; Batool, S.; Ahmed, M.; Altaf, A. A., Single Organic ligands act as a bifunctional sensor for subsequent detection of metal and cyanide ions, a statistical approach toward coordination and sensitivity. *Critical Reviews in Analytical Chemistry* **2023**, 1–17.
[69] Balali, Y.; Stegen, S., Review of energy storage systems for vehicles based on technology, environmental impacts, and costs. *Renewable and Sustainable Energy Reviews* **2021,** *135*, 110185.

12 Coordination Nano-Architectures for Nitrogen Reduction Reactions

Abdul Haq, Muhammad Arif, Umair Azhar, Muhammad Sagir, Muhammad Asim Mushtaq, Unaiza Talib and Ghulam Yasin

12.1 INTRODUCTION

Coordination refers to the arrangement of metal ions or clusters with surrounding ligands, typically organic molecules or species capable of donating electron pairs[1]. These ligands coordinate with metal centers to form complexes, which are crucial for tailoring the properties of materials and playing a significant role in applications like catalysis, sensing, and molecular recognition[2]. In catalysis, coordination nanoarchitectures may enhance the catalytic activity of metal centers in reactions like nitrogen reduction to ammonia[3]. The specific coordination environment and ligand choice influence the catalyst's efficiency, selectivity, and stability. Coordination interactions between metal centers and ligands are essential for the material's architecture and functionality[4].

Coordination nanoarchitectures are a dynamic area where research is continuously focused on reforming these structures for energy-related technologies, sensing, and catalysis, among other applications[5]. The rational design of nanoscale structures in which metal ions are coordinated with ligands to improve catalytic activity in the conversion of nitrogen gas (N_2) into ammonia (NH_3) or other useful nitrogen-containing chemicals is known as coordination nanoarchitectures towards NRR processes. These structures are essential for overcoming the difficulties posed by conventional nitrogen fixation techniques, such as the pressure-and temperature-intensive Haber-Bosch process[6].

Because of their special advantage of high surface area, modified coordination settings, enhanced stability, and adaptability in catalyzing different electrochemical reactions, coordinated nanoarchitectures are useful electrocatalysts[7]. Their characteristics render them promising contenders for utilization in electrochemical devices such as fuel cells and electrolyzers. The latest developments often involve innovations in synthesis techniques, understanding structure-property relationships, and exploring novel applications in emerging fields[8]. In this chapter, coordinated

DOI: 10.1201/9781003345886-14

nanostructured materials are discussed while keeping in mind the DFT evaluations for selectivity of NRR route, designing and synthesis, characterization techniques, mechanism, significance of NRR, challenges, possible pathways, and the electrochemical measurements of coordinated nanoarchitectures. Finally, summary on the conclusion and perspectives of coordinated nanoarchitectures as electrocatalysts for electrochemical NRR is presented.

12.2 BACKGROUND

Energy is essential to human life as well as the survival of modern-day society. Concerns about depletion of fossil fuels, global warming, geopolitical instability, and growing fuel prices plague the use of fossil fuels[9,10]. Sustainable solutions to these problems include the use of renewable energy sources like solar, wind, biomass, wave, and tidal energy. These are plentiful, limitless, and eco-friendly resources[11]. The severe energy crisis and concerns about environmental contamination have made electrochemical technology an essential field of study for the development of better energy generation and conversion technologies[12]. Although ammonia is essential for fuels, fertilizers, and hydrogen carriers, its natural supply cannot keep up with the world's need. Because the traditional Haber-Bosch process uses a lot of energy and produces a lot of greenhouse gases, so there is a need for more ecologically friendly and energy-efficient ways to reduce nitrogen to NH_3[13]. Ammonia is an essential part of fertilizers and an important part of the world's nitrogen cycle[14].

12.3 SIGNIFICANCE OF NITROGEN REDUCTION REACTIONS (NRR)

An environmentally friendly substitute for the energy-intensive Haber-Bosch process, which produces ammonia, is the electrochemical nitrogen reduction reaction (NRR)[15]. The Haber-Bosch process currently dominates industrial ammonia synthesis, resulting in considerable energy consumption and environmental problems[16]. The majority of NH_3 production is supported by the conventional Haber-Bosch process, which involves severe reaction conditions. Green solvents known for their exceptional performance in the electrocatalytic NRR process and distinctive properties make ionic liquids (ILs) a viable substitute for the harsh Haber-Bosch process[17,18]. The production of ammonia using electrochemical synthesis has various advantages, such as reduced carbon dioxide emissions, the capacity to store energy from renewable sources in chemical bonds, and the possibility of producing ammonia. With these benefits, it is expected that the electrochemical synthesis of ammonia will be a competitive alternative to the Haber-Bosch process[19].

A vital component of fertilizer, ammonia is produced by the NRR along with alternative fuel and carbon-free hydrogen carrier. The conventional Haber-Bosch process produces a lot of energy and greenhouse gases, and the natural process is unable to supply the world's demand. Nitrogen reduction to NH_3 requires more ecological and energy-efficient techniques[13].

12.4 CHALLENGES IN NRR

Unfortunately, low faradic efficiency results from the competing hydrogen evolution reaction and high N_2 energy barrier activation which restricts NRR[20]. A barrier impeding the advancement of high-efficiency ammonia electrosynthesis is the inadequate ability of the electrocatalysts to cleave N-N bonds. Lately, there has been a lot of interest in the development of electrocatalysts for the N_2 reduction process (NRR)[21]. However, electrocatalytic reduction of N_2 to ammonia generation still has a number of useful applications. Nørskov and colleagues conducted an early density functional theory (DFT) investigation, which showed that on most transition metal surfaces, the HER is a strong competitive reaction for NRR[22].

However, the energy efficiency of the Haber-Bosch reaction is now far higher than that of electrocatalytic NRR. This low efficiency is mostly caused by poor Faradaic efficiencies, which result in energy loss as a result of the HER side reaction. Recent developments in this area include the sensible design of catalysts, reactor layout, electrolytes, etc. to inhibit the HER process[23].

The electrochemical NRR process is a technique that can be utilized to synthesize ammonia according to thermodynamics. Thus, every attempt has revealed that this technique has two main issues: a high overpotential and extremely low selectivity of NRR path[22]. The accurate and unbiased evaluation of the electrocatalyst's activity is key to the progress of NRR. The following parameters are used to quantify the activity and selectivity of the NH_3 manufacturing process: i) the yield rate, which indicates the reaction rate; ii) the Faradaic efficiency; and iii) the catalyst stability[22].

12.5 DESIGN PRINCIPLES FOR NANOARCHITECTURES IN NRR

For NRR process, coordination nanoarchitectures are designed by carefully creating metal centers and ligands at the nanoscale to maximize catalytic activity in nitrogen gas (N_2), which is converted into ammonia (NH_3) or other molecules containing nitrogen[24,25]. The process of creating coordination nanoarchitectures for NRR electrocatalysts entails choosing appropriate transition metals and adjusting ligands to provide the best possible electronic structure[26]. Reactant binding and catalytic performance are improved by giving priority to surface functionalization and active site accessibility[27]. Overall performance is improved by addressing pH sensitivity and streamlining charge transfer paths. Appropriate catalyst design is guided by an understanding of reaction processes, whilst in-situ characterization methods monitor structural changes[28]. While taking into account the sustainability and environmental impact, efficient development of catalysts for nitrogen reduction to ammonia is promised.

12.6 FUNDAMENTALS OF NRR

The nitrogen reduction reaction is the process that converts molecular nitrogen into ammonia or other nitrogenous compounds. It includes the difficult thermodynamic process of breaking the triple bond in N_2. The electrocatalysts facilitate this process by providing active sites for the activation and reduction of nitrogen[29]. Although the reaction usually happens in mild environments, there are certain difficulties, such as high ammonia selectivity. Thorough design considerations – such as metal selection, ligand engineering, and surface changes – are necessary for effective NRR electrocatalysts[30,31].

12.7 MECHANISM OF NRR FOR THE PRODUCTION OF AMMONIA

The mechanism of NRR towards ammonia production involves multiple electrochemical steps. The primary challenge lies in splitting the powerful triple bond of molecular nitrogen (N_2)[32,6a]. The overall reaction can be represented as:

$$N_2 + 6H^+ + 6e^- \rightarrow 2NH_3 \quad (12.1)$$

Here are the key steps in the NRR mechanism:

N_2 is adsorbed onto the catalytic surface, and the triple bond activation occurs. This step requires overcoming a substantial energy barrier due to the firm (N≡N) bond[33]. The activated nitrogen species undergoes protonation (gain of H^+), forming intermediates. Electron transfer: Electrons are transferred to the adsorbed nitrogen species, reducing it to form intermediate species[34]. The reduction continues, leading to the formation of ammonia (NH_3) and desorption from the catalyst surface. The reaction is influenced by factors such as catalyst composition, surface structure, and reaction conditions.

12.8 PATHWAYS INVOLVED IN MECHANISM OF NITROGEN REDUCTION REACTION

The overall reaction involves multiple pathways, influenced by catalyst composition, surface structure, and reaction conditions. Optimizing these factors is crucial for high NRR efficiency and selectivity. Understanding these pathways guides electrocatalyst design for sustainable ammonia synthesis, using advanced spectroscopic and computational techniques[35].

According to advanced techniques, the distal[6a], alternative, and enzymatic pathways are involved for ammonia synthesis through NRR as shown in Figure 12.1[19a].

12.9 ELECTROCATALYTIC APPLICATIONS OF NANOARCHITECTURE COMPLEXES FOR NRR

The applications of nanoarchitecture complexes for electrocatalytic NRR exhibit their potential to revolutionize nitrogen fixation processes, offering sustainable solutions for agriculture, energy, and environmental challenges. In this section, the experimental and computational studies, design and synthesis, and electrochemical performance of coordinated nanostructure materials for nitrogen reduction reaction towards ammonia synthesis are highlighted.

12.10 METAL-ORGANIC FRAMEWORKS (MOFs) FOR NRR

Metal ions are coupled with organic ligands to form porous structures called MOFs. They provide special characteristics and a large surface area for catalytic applications.

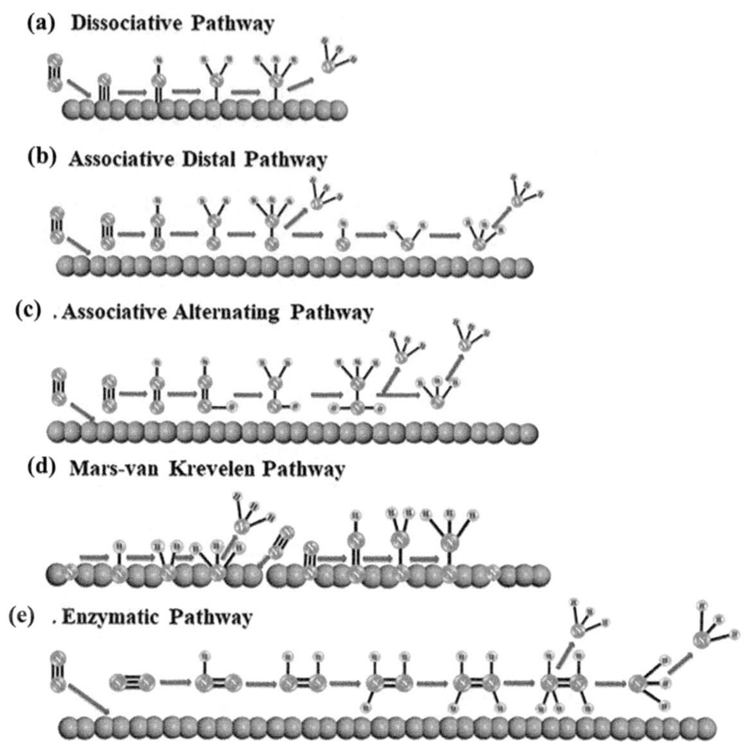

FIGURE 12.1 Various Pathways for NRR mechanism forming ammonia from nitrogen. (a) Dissociative route. (b) Associative distal route. (c) Associative alternating route. (d) Mars-van Krevelen route. (e) Enzymatic route.

Because of their easily modifiable pore size and shape ranging from microporous to mesoporous, high porosity, and designable crystalline structures, metal-organic frameworks (MOFs) are highly desired materials[36,37]. MOFs are known to be the best catalysts because of their organic linkage and metal nodes, which allow for proper manipulation for a given catalytic capacity. Moreover, improved catalytic efficacy can be attained because of the vast surface area and adjustable organic functional groups[38–40].

Because of their controllable and wide range of functional components, the utilization of metal-organic frameworks as electrocatalysts is a very promising one. Bimetal-MOFs (CoxFe-MOFs) in two dimensions and around 10 nm in thickness are synthesized in an alkaline medium for both NRR and OER. The modified Co_3Fe-MOF may outperform commercial RuO_2 by achieving a peak potential of 280 milli-volt at 10 mA/cm^2 current density at a reasonable tafel slope by adjusting the composition of CoxFe-MOF. On a glassy carbon electrode, the Co_3Fe-MOF also reaches actual NH_3 yields and faradaic efficiency (FE)[41].

Using density functional theory (DFT), a research team developed a conductive 'MOF' composed of molybdenum, the most prevalent element, that is employed as an electrocatalyst for the reaction involving the reduction of nitrogen to ammonia (Figure 12.2). The Mo-based MOF outperforms a range of previously developed 2D

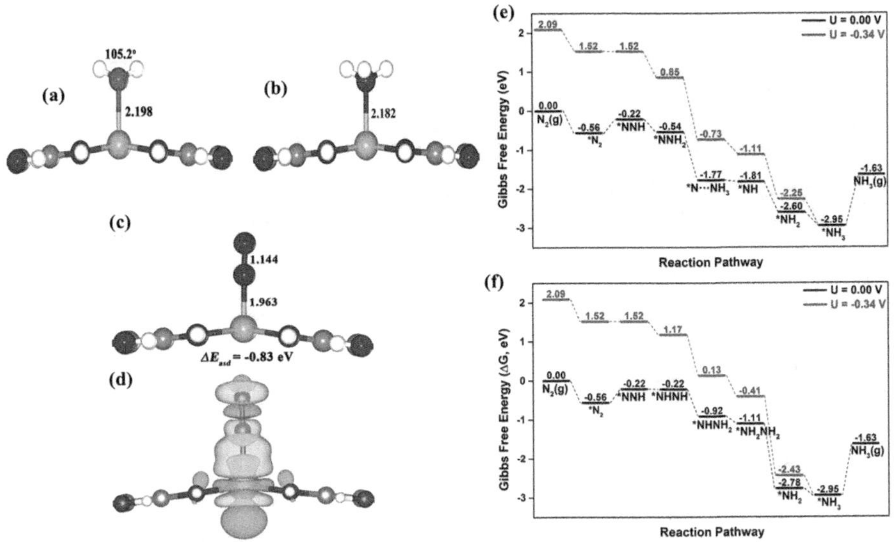

FIGURE 12.2 Top and side views for the geometric configurations of H_2O (a) and NH_3 (b) adsorbed on the catalyst (Mo-based MOF) with (Å) bond length. (c) Customized topology of N_2 adsorbed onto the catalyst. (d) The charge density difference with respect to (c). The Gibbs free energy diagram, the value of NRR for the catalyst is calculated by limiting potential through distal (e) and alternating (f) routes.

MOFs in its remarkable catalytic capacity to convert N_2 into NH_3 at ambient temperature with an incredibly low overpotential of 0.18 V. Its structure is extremely well organized, and it functions superbly in actual applications[42].

One such method is the pyrolysis of triphenylphosphine (TPP) with a "zeolitic imidazolate framework" (ZIF-8) in a nitrogen atmosphere, which yields a sequence of nitrogen and phosphorus co-doped porous carbon. The catalyst that was produced by heating ZIF-8 and TPP for three hours at a mass ratio of 1:15 was known as PN-C-ZIF-8. Using an acidic electrolyte at a voltage of negative 0.3 v, demonstrated an amazing ammonia synthesis capacity and Faraday efficacy of 16.67%. More notably, the PN-C-ZIF-8 catalyst showed exceptional selectivity and a 72-hour half-life, and no hydrazine by-products were found and outcomes are shown in Figure 12.3[43].

A research group reported a study on the preparation of a cobalt phosphide hollow nanocage (CoP HNC) from ZIF-67 as an NRR electrocatalyst. Initially, the ZIF-67 nanocrystal underwent solvothermal treatment in the presence of Co^{2+} to become a cobalt-layered double hydroxide hollow nanocage (CoP HNC), followed by thermally assisted phosphorization with NaH_2PO_2. A high FE of 7.36% at 0 V vs. RHE and a high NH_3 generation rate were achieved in the absence of hydrazine (by-product formation). Another group published a study on Co@N-doped carbon (Co@NC), an additional NRR catalyst produced by pyrolyzing ZIF-67[44]. Their results suggest that the Co-Nx structures in Co@NC may also be involved in the activation of N_2 molecules (since the nitrogen atom in Co-Nx is identical to the nitrogen of pyridine). Co@NC's remarkable catalytic efficiency was facilitated by its vast area of surface and

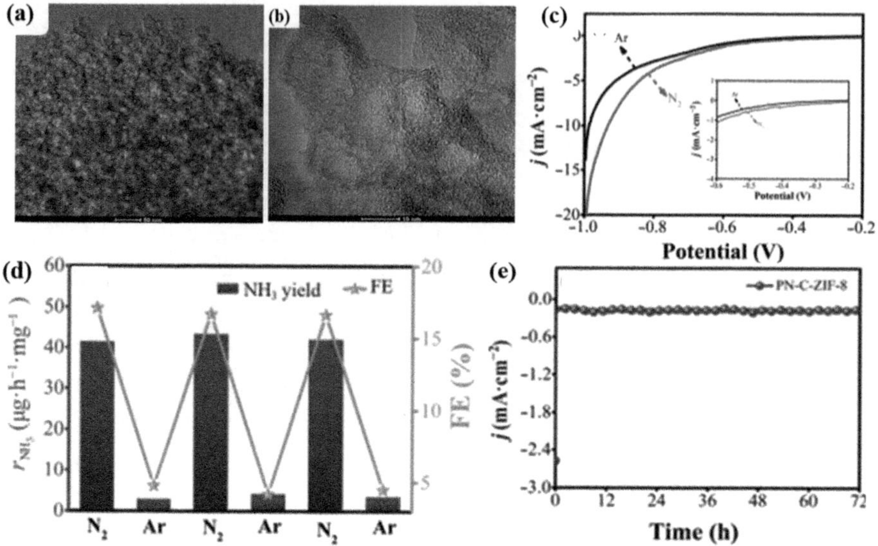

FIGURE 12.3 (a) SEM image. (b) TEM image of PN-C-ZIF-8. (c) LSV curves in 0.05 M H_2SO_4 that is Ar- and N_2-saturated. (d) NH_3 production and matching FEs for PN-C-ZIF-8 at 2-hour intervals. (e) Time-dependent current density profile at (-0.3 V) for catalyst.

availability of active positions. Figure 12.4 shows the synthesized catalyst's faradic efficiency and NH_3 production rate as well as cyclic voltammetry in addition to physicochemical analysis (SEM, TEM)[44].

Using a hydrothermal technique, numerous MOFs(M) were created, where *M* is equal to Fe, Co, and Cu. The synthesized MOFs exhibit high specific surface areas, many microscopic pores, and excellent crystalline forms. The produced MOFs were used to obtain NH_3 electrochemically at ambient conditions. MOF(Fe) had the highest catalytic activity of all the catalysts. Using fresh nitrogen and H_2O as raw substance (feed), the maximum production rate and columbic efficiency value for ammonia of the MOF (Fe) catalyst were promising at a potential of 1.2 V and 90°C temperature. Moreover, the main ingredients in the direct synthesis of ammonia were water and air. According to this initial investigation, the produced MOFs exhibited outstanding catalytic efficiency for low-temperature electrochemical ammonia production[45].

12.11 COFs-BASED COORDINATION NANOARCHITECTURES FOR NRR

A promising strategy for NRR activity is the use of covalent organic frameworks (COFs) in coordination with other nanoarchitectures. Sustainable nitrogen fixation could benefit from their controlled synthesis, which enables accurate mass transport, selectivity, and changing active sites. Nitrogen reduction processes involving electrocatalysis employ organic-based coordination structures, such as COFs, to generate sustainable ammonia[46,47]. These eco-friendly choices for nitrogen fixation offer a

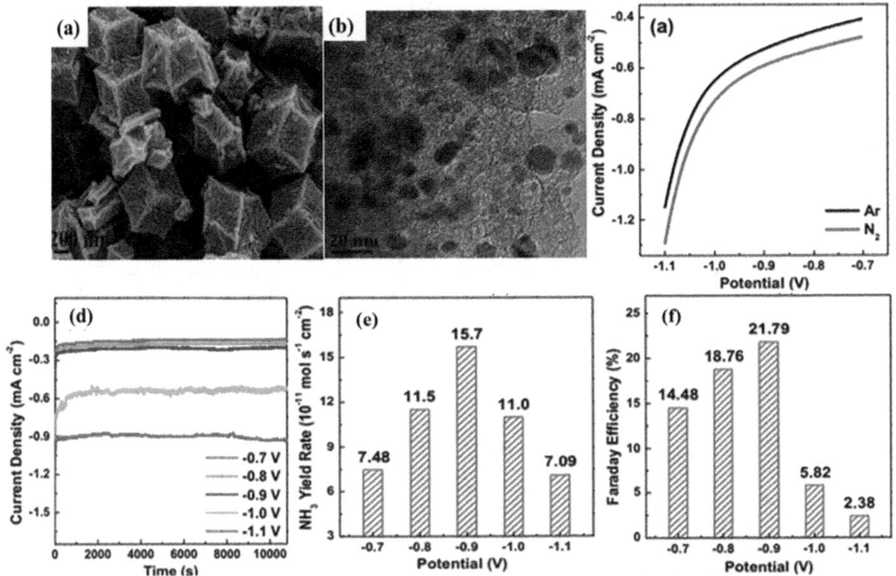

FIGURE 12.4 (a) SEM image of Co@NC. (b) TEM photograph. (c) CV graphs of Co@NC in an electrolyte saturated with argon and nitrogen. (d) Current density versus time curves of the catalyst (Co@NC) at different potentials. (e, f) Rates of ammonia synthesis and Faraday efficiencies at various voltage.

significant surface area and structural stability. However, high-performance NRR electrocatalysts require additional work[48]. Based on the experimental and computational work, here are some examples of efficient catalysts based on organic matter coordinated with transition metals towards ammonia synthesis via NRR[49,50].

In connection with the N_2 reduction reaction, a team of researchers examined the electrocatalytic activity of active and durable two-dimensional (2D) TM-linked covalent organic framework (TM-COF)[51]. Out of 20 combinations, the theoretical simulation indicates that the 2D active Mo-COF has the best power for electrocatalytic N_2 fixation, with an extraordinarily low excessive potential of 0.16 V. It can also effectively inhibit the process of competitive hydrogen evolution (HER). Because of its inherent advantages – strong spin moment and significant positive charge on the Mo atom, good conductive properties, and the appropriate bonding capacity for many NRR materials – the Mo-COF displays exceptional NRR activity and selectivity. Figure 12.5 depicts the theoretical study of molybdenum atoms in COFs and the chemical pathways utilized to assess NRR catalyst[51].

The pyrazine-connected metalphthalocyanine (MPc)-based 2D c-COFs have been shown by another research team to be effective electrocatalysts for raising NRR performance and selectivity in acidic electrolytes. They observed that the protonation of adsorbed N_2 to create *NNH is the rate-determining step of (MPc-pz), which is similar to its considerable contribution to the NRR. Furthermore, the intermediate $*NNH_2$ produced by the distal route on FePc-pz is less favorable than the alternative path that yields *NHNH intermediate. $*NHNH_2$ reduction to *N and NH_3 requires

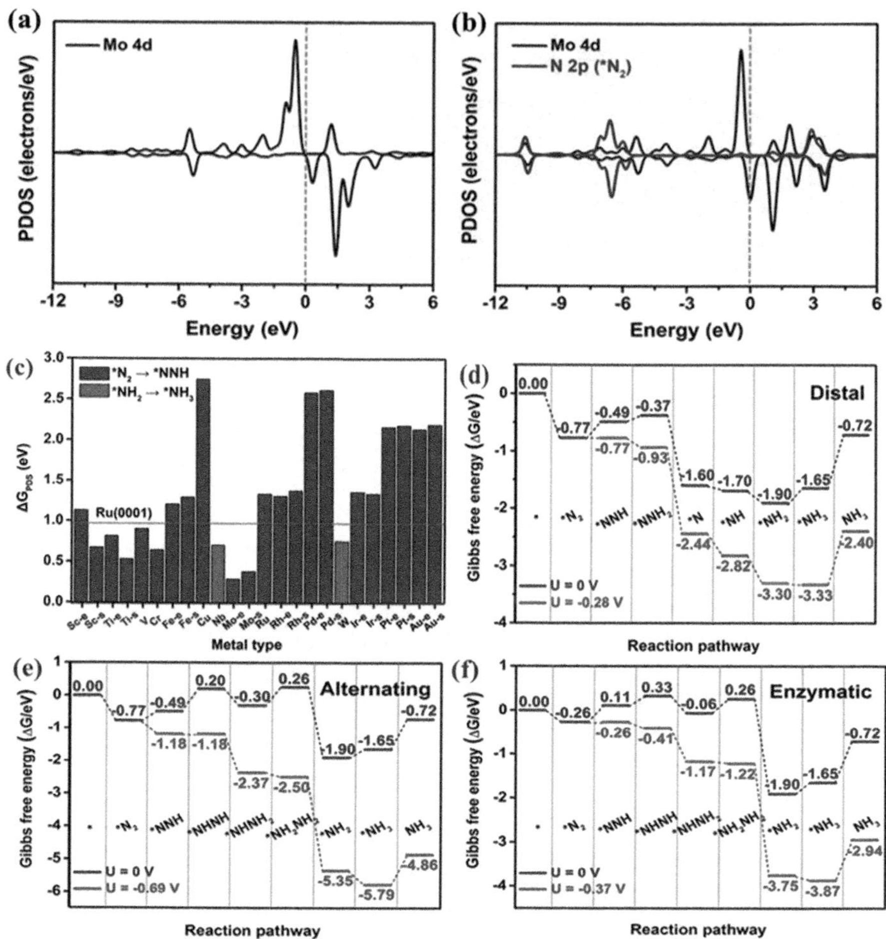

FIGURE 12.5 (a) The PDOS of Mo before and (Figure 12.5 (b)) for both $^*N_2$ and (Mo) after N_2 adsorption on the Mo-COF. (c) The difference of Gibbs free energy (ΔG) of PDS during the NRR. Figure 12.5 (d, e, and f) Gibbs free energy graph for N_2 reduction on the catalyst (Mo-COF) via three mechanisms at different voltages.

a large barrier energy, whereas *NHNH hydrogenation to $^*NHNH_2$ releases energy. Thus, NRR usually follows the alternative pathway instead of the "distal" route on FePc-pz. Pc-pz has the potential for NRR; however, low FE and the rate at which NH_3 is generated for CoPc-pz are caused by high HER competition, which severely reduces the NRR performance on Co-N_4-C sites. The free energy profiles indicate that the FePc-pz with Fe-N_4-C linkage is more efficient for electrocatalytic NRR as compared to MN_4-based COF catalysts. The catalyst (FePc-pz) exhibited remarkable catalytic activity at -0.1 V versus RHE, producing a high amount of ammonia at an FE of 31.9%. This was explained by its laminar packing of porous array, strong direct pi-conjugation with quick transfer of electrons, and availability of Fe-N_4-C sites. The

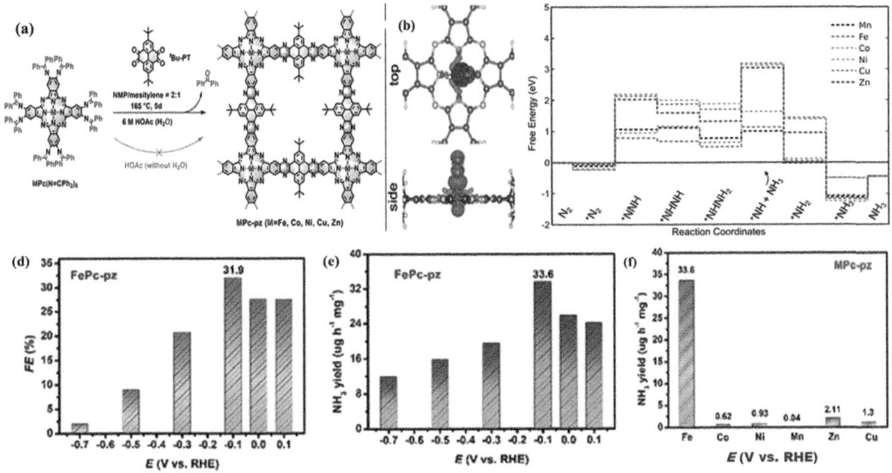

FIGURE 12.6 (a) The schematic synthesis process of MPc-pz. (b) Charge density differences of N_2 adsorbed on FePc-pz as compared to the isolated N_2 molecule and MPc-pz. (c) Free energy profiles of NRR along the alternating route on MPc-pz. FE and NH_3 yield rate of FePc-pz at various potentials in nitrogen-saturated 0.01 molar sulfuric acid electrolyte (d, e). NH_3 production capacity of (MPc-pz) (f).

unusual characteristics of the 'FeN_4' active positions for dinitrogen conversion and activation were discovered by theoretical processing, meticulously designed contrast electrochemical tests, in-situ XAS, and Raman analyses, and other techniques as shown in Figure 12.6[52].

The conjugated two-dimensional (2D) covalent organic frameworks (COFs) with controlled quasi-phthalocyanine nitrogen-coordinated transition metal centers have been shown to be one synthetic method of enhancing NRR performance. The Ti-COF catalyst is more active than Cu- and Co-COF because it can impede the HER and activate inert N_2 molecules. DFT calculations show that Ti-COF effectively activates and traps N_2 molecules while inhibiting HER[53].

The electrocatalytic NRR is a sustainable way to produce renewable ammonia in ambient conditions. Nevertheless, creating advanced NRR electrocatalysts that are both selective and efficient enough is still a difficult task. To achieve this, three-dimensional COFs based on metal-porphyrin were created, having metal-N_4 catalytic sites that had various three-dimensional spatial configurations ("scu" and "flu" geometries). The potential of 3D COFs with Fe-N_4 catalytic sites as NRR electrocatalysts is shown in Figure 12.7, wherein their higher NH_3 production and Faradaic efficiency against RHE are demonstrated in comparison to those with Cu-N_4 centres[54].

12.12 POROUS COORDINATION POLYMERS (PCPs)

PCPs are porous polymeric solids that can be one-, two-, or three-dimensional. They are made up of organic ligands acting as linkers via coordination bonds and metal ions or clusters acting as nodes[55]. PCPs are distinct from other inorganic porous

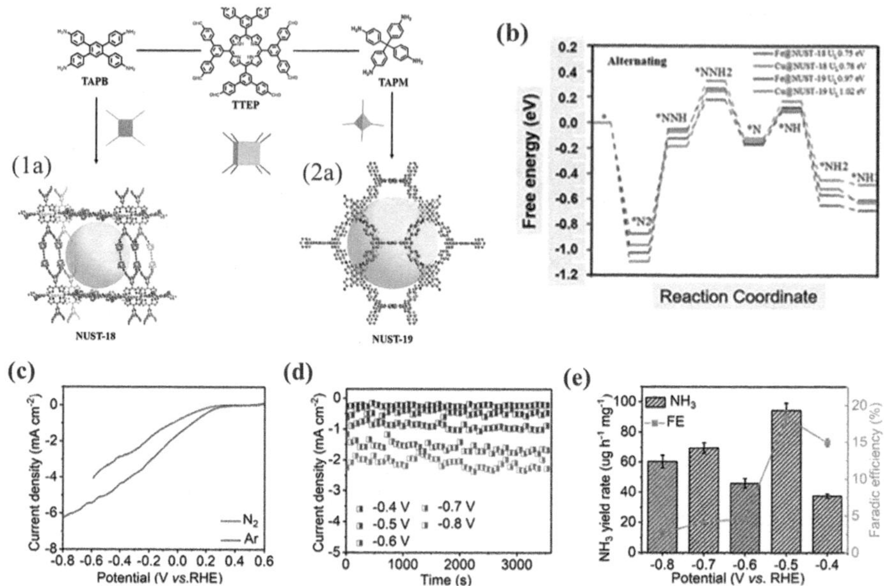

FIGURE 12.7 (a) Schematic synthetic of COFs. (1a and 2a) Structural characterization of NUST-18 and NUST-19 having [8 + 4] expanded network with "scu" and "flu" topology. (b) The NRR Gibbs free energy diagram of different catalysts. The NRR capabilities of Fe@ NUST-18. (c) LSV curves (d) Chronoamperometric curves. (e) NH_3 yield and FEs at various potentials in N_2-saturated electrolyte.

materials and conventional zeolites due to their excellent crystallinity, adjustable/ designable porosity, and structural flexibility[56]. Because of their special benefits, PCPs can be used in numerous applications, including medication delivery, ionic/ electronic conductivity, gas storage, separation, catalysis, and sensing[57,58]. Therefore, the fields of PCP chemistry and physics have been quite active since the 1990s and are currently expanding quickly. Diverse PCPs can be created by combining different metal ions/clusters with organic ligands[58,59]. The coordination orientation of the metal ions/clusters, the flexibility and shape of the bridge-forming ligands, and the size of the guest inclusions can all be used to regulate the frameworks[60].

To do this, a research group created NJUZ-1, the first coordination polymer based on zinc with a bridging N_2 anion ligand. This polymer remains highly effective even when exposed to impure air, which makes it the perfect catalyst for ambient nitrogen fixation. Both theoretical and experimental investigations have shown that the functional "Zn^{2+}-$(N{\equiv}N)^{-1}$-Zn^{2+}" sites can promote NH_3 synthesis and the resulting NH_3 can be separated to produce unsaturated "$Zn^{2+}\cdots Zn^{+}$" intermediates. It is rapidly produced by swift intermolecular electron transfer and exogenous N_2 capture. Through a mechanism, the "$Zn^{2+}\cdots Zn^{+}$" species anchored by the interconnected core shell-like "donor-acceptor-donor" architecture may permit prolonged catalytic cycling[61].

Carbon materials (CMs) derived from MOFs have garnered a significant attention in various fields of applications, including energy applications, restoration of the environment, and catalysis, due to their highly adjustable physicochemical properties and nanoporous structure that yields a large surface area[62,63]. High nitrogen content MOF-derived NPCs exhibit strong N_2 adsorption, tunable functionalization, and adjustable morphologies; nonetheless, typical synthetic approaches to NPC synthesis are expensive and time-consuming. Applications of MOF-obtained 'NPCs' for NRR electrocatalysis have been demonstrated[64].

A special family of porous coordination polymers, known as metal cluster-supported PCPs, are made by using organic linkers to connect poly-nuclear clusters, which serve as secondary building blocks. Because of their varied elements, stable architectures, predictable layouts, and adjustable functions, these PCPs provide a unique platform for discovering structure-property relationships and creating new functional materials with improved attributes[65].

Metal ions and polymeric ligands combine to generate coordination polymers. These complexes are made up of metal ions and synthetic polymers joined by coordination bonds[66]. Numerous artificial polymer-metal complexes have excellent heat resistance, semiconductivity, catalytic efficiency, and biological potential[67].

In summary, the purpose of these nanoscale coordination designs is to maximize the catalytic environment, boost selectivity, and increase the effectiveness of nitrogen reduction processes[68]. In order to promote sustainable nitrogen fixation reactions, researchers investigate different metal-ligand combinations and nanostructures.

Increasing their NRR activity and selectivity at the same time is still a formidable task, and the idea of precisely modifying the active sites has proven to be elusive. The first crystalline two-dimensional conjugated COFs (2D c-COFs) containing '$M–N_4–C$' centers have been created by a research team. These c-COFs greatly increase the activity and selectivity of NRR towards ammonia production. These catalysts are among the best NRR electrocatalysts because they exhibit higher ammonia generation and columbic efficiency against reversible hydrogen electrodes. They are made of organic (metal-phthalocyanine and pyrene) units bound by pyrazine couplings. The (Iron to N-C) edges have a different electronic structure from other M–N4–C centers, with confined states of electrons at the Fermi level. This allows for greater contact with N_2, which leads to accelerated N_2 stimulation and Nitrogen fixation (NRR) rates[69].

Nitrogen fixation (NRR) and other electrochemical processes benefit greatly from the use of single-atom catalysts (SACs) as they reduce the requirement for costly metal components by providing active and targeted areas for the transformation of nitrogen. SACs provide accurate control over catalytic sites, increasing productivity and developing environmentally friendly nitrogen fixation technology. As such, they are a good choice for optimizing the performance of the NRR process[70].

A study analyses the NRR catalytic function of graphene monolayers anchored by transition metals through first-principles calculations. Because of its great electron-donating capacity and concentrated electronic states at the Fermi level, the cobalt dimer-coordinated GDY monolayer (Co_2@GDY) shows the highest NRR catalytic operation. Additionally, the analysis shows a roughly linear relationship between the projected onset potential and the nitrogen adsorption energy, which suggests that this relationship could be a gauge of these catalysts' intrinsic NRR catalytic efficiency[71].

Recent studies have shown that controlling the coordination of single-atom catalysts (SACs) in specific electrochemical reactions via electronic structure tuning may be a useful strategy to further increase the catalytic efficacy of SACs. Using density functional theory (DFT) simulations, this study examined the relationship between a Fe–N–C catalyst's coordination structure and catalytic activity towards the nitrogen reduction process. The outcomes shown in (Figure 12.8) showed that coordinating with a boron (B) dopant can significantly increase the NRR performance on the fundamental Fe atom. Out of all the B-containing Fe–N–C catalysts, Fe–B_2N_2 has the lowest predicted cut-off potential (-0.65 V), indicating outstanding catalytic activity for NRR. It's noteworthy that by altering the way an individual Fe atom interfaces with the N_2H* entities, adding B coordination can dramatically increase the catalytic activity of NRR. Strong NRR selectivity as well as kinetic and thermodynamic inhibition of the hydrogen production path are exhibited by Fe–B_2N_2 as shown in Figure 12.8. Consequently, the N and B dual coordination singular Fe catalyst is a potential NRR electrocatalyst[72].

The study used density functional theory simulations to examine the catalytic effectiveness and stability of bimetals for N_2 fixation. Because of insufficient thermodynamic instability, carbon was determined to be the optimal coordinating element for bimetallic couples, while the other four coordination atoms were found to be insufficient. Bimetallic complexes MoTi-CG and TiV-CG, attached to C-coordinated graphene, were predicted to be effective NRR catalysts[73].

To increase its recyclability, the molecule "FePc" was uniformly grafted as an NRR electrocatalyst on an O-MWCNT surface. Following two hours of electrocatalytic

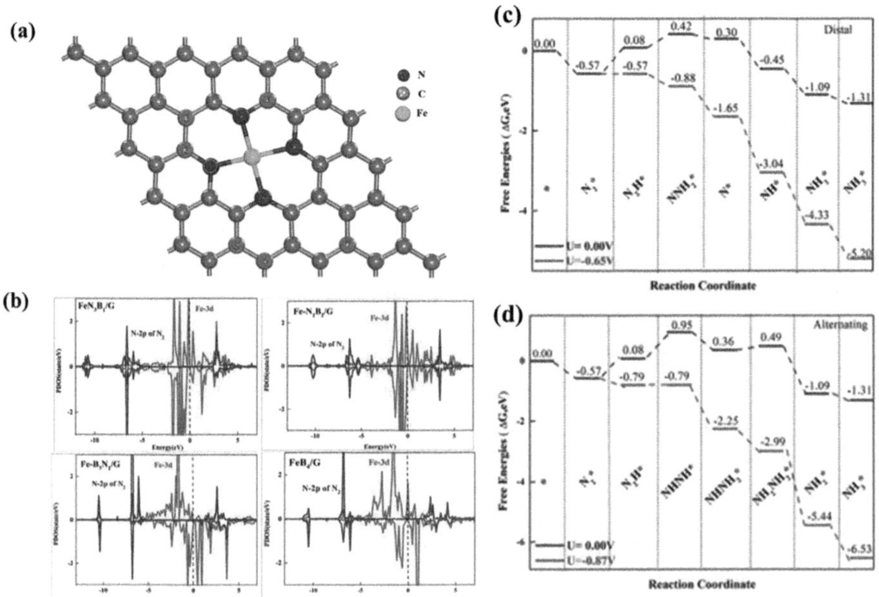

FIGURE 12.8 The calculated (a) charge density difference. Partial density of states of N_2 adsorption on Fe-BxN4-x/G (x = 1, 2, 3, and 4) (b). The free energy profiles of NRR (two paths followed; c, d).

activity in an acidic electrolyte, this catalyst showed good selection and remarkable electrocatalytic results.

In comparison to most of the materials described, it yielded a notably higher amount of NH_3. DFT simulations showed that the NRR preferred the alternate pathway and that the N_2 triggering to form N_2H^*, with an ΔG score of 1.79 eV, is the rate-limiting step. According to one source[74], iron phthalocyanine (FePc) with a clarified FeN_4 configuration can be used as a prototype catalyst to effectively carry out the electrochemical N_2 reduction reaction (NRR) employing Fe-NC materials. The most active location for NRR is found to be the Fe core in FeN_4, and the alternating pathway of N_2 on Fe is the preferred path. As a result, at minimum potential, the NH_3 yield rate is significant and is displayed in Figure 12.9.

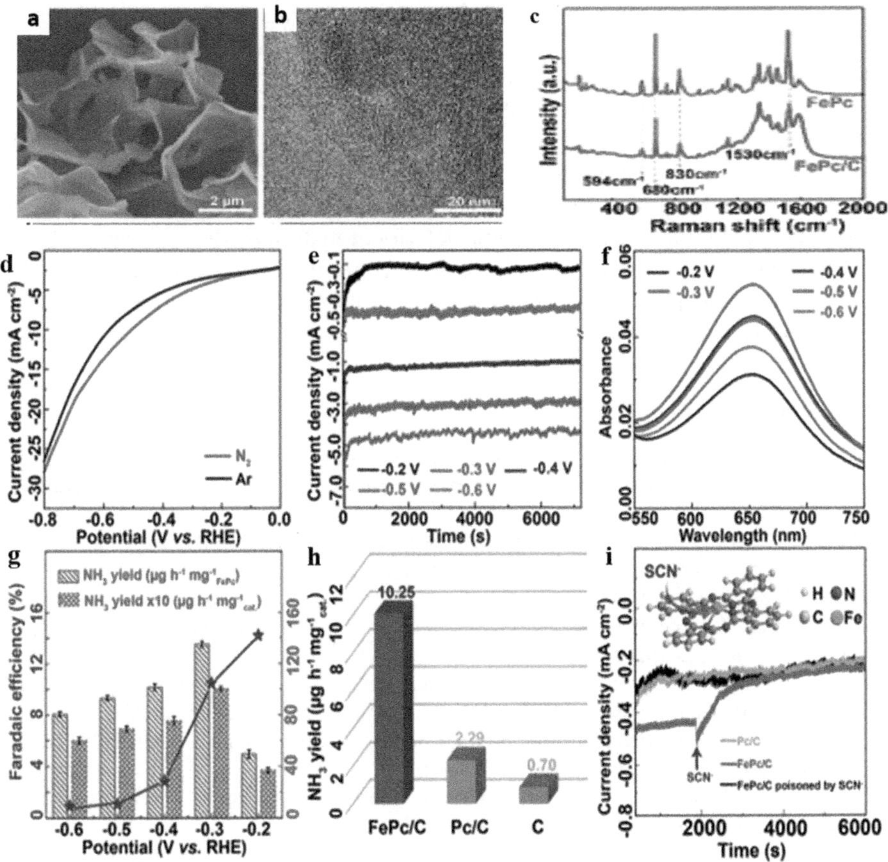

FIGURE 12.9 (a) SEM and (b)TEM images of FePc/C (d) LSV curves in both the nitrogen and argon saturated electrolyte at a scan rate of 5 mV s–1, (e) chronoamperometry curves. (f) UV–vis absorption spectra of the electrolytes. (g) The rate for NH_3 synthesis and F.E. for FePc/C. (h) Comparative graph for NH_3 production rate of catalysts (FePc/C, Pc/C, and porous carbon (C)). (i) Chrono-amperometric curves at –0.3 V in the nitrogen-saturated solution (electrolyte).

12.13 CONCLUSION

Coordinated nanostructure materials are composed of several components and provide active sites for nitrogen fixation. Precise synchronization enables them to maximize electron transport channels, diminishing NRR energy obstacles and augmenting the efficiency of electrocatalysts in the nitrogen-to-ammonia conversion processes. Therefore, these coordinated nanostructured materials are crucial for improving the efficiency and selectivity of NRR. A brief summary of recent work is provided on organic framework-based NRR electrocatalysts, such as transition metal-coordinated MOFs, COFs, and porous coordination polymers (PCPs). The selectivity and faradic efficiency of NRR catalysts towards the electrochemical production of ammonia are highlighted, along with density functional theories and experimental studies of these electrocatalysts.

Prospects for producing value-added fuel and sustainable nitrogenous fertilizers are promising, according to recent research on coordinated nanoarchitectures for NRR. The issues that need to be solved are catalyst stability, scalability, and cost-effectiveness. Furthermore, work needs to be done on tuning the already-existing nanostructure designs, investigating new catalyst compositions, and delving into complex mechanisms in order to fully realize the promise of the NRR process. Peculiar active locations are made possible by the precise coordination of several components at the nanoscale, which improves the catalytic activity of materials. Researchers will probably investigate and improve the properties of a variety of these structures, including their chemical composition, electrical features, and surface characteristics, in order to boost NRR efficacy.

REFERENCES

[1] Robin, A. Y.; Fromm, K. M., Coordination polymer networks with O- and N-donors: What they are, why and how they are made. *Coordination Chemistry Reviews* **2006,** *250* (15), 2127–2157.

[2] Gao, C.; Wang, J.; Xu, H.; Xiong, Y., Coordination chemistry in the design of heterogeneous photocatalysts. *Chemical Society Reviews* **2017,** *46* (10), 2799–2823.

[3] Iqbal, M.; Kaneti, Y. V.; Kim, J.; Yuliarto, B.; Kang, Y.-M.; Bando, Y.; Sugahara, Y.; Yamauchi, Y., Chemical design of palladium-based nanoarchitectures for catalytic applications. *Small* **2019,** *15* (6), 1804378.

[4] Castillo-Blas, C.; Montoro, C.; Platero-Prats, A. E.; Ares, P.; Amo-Ochoa, P.; Conesa, J.; Zamora, F., Chapter three – The role of defects in the properties of functional coordination polymers. In *Advances in Inorganic Chemistry*, Ruiz-Molina, D.; van Eldik, R., Eds. Academic Press: 2020; Vol. 76, pp. 73–119.

[5] Ariga, K., Materials Nanoarchitectonics at dynamic interfaces: Structure formation and functional manipulation. *Materials* **2024,** *17* (1), 271.

[6] (a) Arif, M.; Yasin, G.; Luo, L.; Ye, W.; Mushtaq, M. A.; Fang, X.; Xiang, X.; Ji, S.; Yan, D., Hierarchical hollow nanotubes of NiFeV-layered double hydroxides@CoVP heterostructures towards efficient, pH-universal electrocatalytical nitrogen reduction reaction to ammonia. *Applied Catalysis B: Environmental* **2020,** *265*, 118559; (b) Mushtaq, M. A.; Kumar, A.; Yasin, G.; Arif, M.; Tabish, M.; Ibraheem, S.; Cai, X.; Ye, W.; Fang, X.; Saad, A.; Zhao, J.; Ji, S.; Yan, D., 3D interconnected porous Mo-doped WO3@CdS hierarchical hollow heterostructures for efficient photoelectrochemical nitrogen reduction to ammonia. *Applied Catalysis B: Environmental* **2022,** *317*, 121711.

[7] Zhu, C.; Du, D.; Eychmüller, A.; Lin, Y., Engineering ordered and nonordered porous noble metal nanostructures: Synthesis, assembly, and their applications in electrochemistry. *Chemical Reviews* **2015,** *115* (16), 8896–8943.

[8] Unnikrishnan, V.; Zabihi, O.; Ahmadi, M.; Li, Q.; Blanchard, P.; Kiziltas, A.; Naebe, M., Metal–organic framework structure–property relationships for high-performance multifunctional polymer nanocomposite applications. *Journal of Materials Chemistry A* **2021,** *9* (8), 4348–4378.

[9] (a) Pathania, A.; Haldhar, R.; Kim, S.-C., Chapter 18 – Current application of biomolecules in biomolecular engineering. In *Handbook of Biomolecules*, Verma, C.; Verma, D. K., Eds. Elsevier: 2023; pp. 371–383; (b) Dalby, S., Anthropocene formations: Environmental security, geopolitics and disaster. *Theory, Culture & Society* **2017,** *34* (2–3), 233–252.

[10] Ye, W.; Yang, Y.; Arif, M.; Yang, S.; Fang, X.; Mushtaq, M. A.; Chen, X.; Yan, D., Fe, Mo–N/C hollow porous nitrogen-doped carbon nanorods as an effective electrocatalyst for N2 reduction reaction. *ACS Sustainable Chemistry & Engineering* **2020,** *8* (42), 15946–15952.

[11] Asif, M.; Muneer, T., Energy supply, its demand and security issues for developed and emerging economies. *Renewable and Sustainable Energy Reviews* **2007,** *11* (7), 1388–1413.

[12] (a) Yang, C.; Huang, H.; He, H.; Yang, L.; Jiang, Q.; Li, W., Recent advances in MXene-based nanoarchitectures as electrode materials for future energy generation and conversion applications. *Coordination Chemistry Reviews* **2021,** *435*, 213806; (b) Shakeel, M.; Zhang, X.; Yasin, G.; Arif, M.; Abbas, Z.; Zaman, U.; Li, B., Fabrication of amorphous BiOCl/TiO2-C3N4 heterostructure for efficient water oxidation. *ChemistrySelect* **2019,** *4* (28), 8277–8282.

[13] Qiao, Z.; Johnson, D.; Djire, A., Challenges and opportunities for nitrogen reduction to ammonia on transitional metal nitrides via Mars-van Krevelen mechanism. *Cell Reports Physical Science* **2021,** *2* (5), 100438.

[14] Zhang, X.; Ward, B. B.; Sigman, D. M., Global nitrogen cycle: Critical enzymes, organisms, and processes for nitrogen budgets and dynamics. *Chemical Reviews* **2020,** *120* (12), 5308–5351.

[15] Mushtaq, M. A.; Arif, M.; Yasin, G.; Tabish, M.; Kumar, A.; Ibraheem, S.; Ye, W.; Ajmal, S.; Zhao, J.; Li, P.; Liu, J.; Saad, A.; Fang, X.; Cai, X.; Ji, S.; Yan, D., Recent developments in heterogeneous electrocatalysts for ambient nitrogen reduction to ammonia: Activity, challenges, and future perspectives. *Renewable and Sustainable Energy Reviews* **2023,** *176*, 113197.

[16] Zhao, R.; Xie, H.; Chang, L.; Zhang, X.; Zhu, X.; Tong, X.; Wang, T.; Luo, Y.; Wei, P.; Wang, Z.; Sun, X., Recent progress in the electrochemical ammonia synthesis under ambient conditions. *EnergyChem* **2019,** *1* (2), 100011.

[17] Chebrolu, V. T.; Jang, D.; Rani, G. M.; Lim, C.; Yong, K.; Kim, W. B., Overview of emerging catalytic materials for electrochemical green ammonia synthesis and process. *Carbon Energy* **2023**, e361.

[18] Tian, Y.; Liu, Y.; Wang, H.; Liu, L.; Hu, W., Electrocatalytic reduction of nitrogen to ammonia in ionic liquids. *ACS Sustainable Chemistry & Engineering* **2022,** *10* (14), 4345–4358.

[19] (a) Arif, M.; Babar, M.; Azhar, U.; Sagir, M.; Bilal Tahir, M.; Asim Mushtaq, M.; Yasin, G.; Mubashir, M.; Wei Roy Chong, J.; Shiong Khoo, K.; Loke Show, P., Rational design and modulation strategies of Mo-based electrocatalysts and photo/electrocatalysts towards nitrogen reduction to ammonia (NH3). *Chemical Engineering Journal* **2023,** *451*, 138320; (b) Wu, T.; Fan, W.; Zhang, Y.; Zhang, F., Electrochemical synthesis of ammonia: Progress and challenges. *Materials Today Physics* **2021,** *16*, 100310.

[20] Agour, A. M.; Elkersh, E.; Khedr, G. E.; El-Aqapa, H. G.; Allam, N. K., Fe-single-atom catalysts on nitrogen-doped carbon nanosheets for electrochemical conversion of nitrogen to ammonia. *ACS Applied Nano Materials* **2023,** *6* (17), 15980–15989.

[21] Arif, M.; Kumar, A.; Asim Mushtaq, M.; Azhar, U.; Sagir, M.; Bilal Tahir, M.; Talib, U.; Ajmal, S.; Alotaibi, K. M.; Yasin, G., Edge-hosted CoFeB active sites with graphene nanosheets for highly selective nitrogen reduction reaction towards ambient ammonia synthesis. *Chemical Engineering Journal* **2023,** *473*, 145368.

[22] Chen, G.-F.; Ren, S.; Zhang, L.; Cheng, H.; Luo, Y.; Zhu, K.; Ding, L.-X.; Wang, H., Advances in electrocatalytic N2 reduction – Strategies to tackle the selectivity challenge. *Small Methods* **2019,** *3* (6), 1800337.

[23] Hou, J.; Yang, M.; Zhang, J., Recent advances in catalysts, electrolytes and electrode engineering for the nitrogen reduction reaction under ambient conditions. *Nanoscale* **2020,** *12* (13), 6900–6920.

[24] Ye, W.; Arif, M.; Fang, X.; Mushtaq, M. A.; Chen, X.; Yan, D., Efficient photoelectrochemical route for the ambient reduction of N2 to NH3 based on nanojunctions assembled from MoS2 nanosheets and TiO2. *ACS Applied Materials & Interfaces* **2019,** *11* (32), 28809–28817.

[25] Kumar, S.; Chauhan, C.; Kumar, R.; Kalra, N.; Saini, A.; Sharma, S.; Singh, A., Concomitant role of metal clusters and ligands in the synthesis and control of porosity in Metal-Organic Frameworks: A literature review. *Results in Chemistry* **2023,** *6*, 101206.

[26] Feng, D.; Zhou, L.; White, T. J.; Cheetham, A. K.; Ma, T.; Wei, F., Nanoengineering metal–organic frameworks and derivatives for electrosynthesis of ammonia. *Nano-Micro Letters* **2023,** *15* (1), 203.

[27] Cui, W.-G.; Hu, T.-L., Incorporation of active metal species in crystalline porous materials for highly efficient synergetic catalysis. *Small* **2021,** *17* (22), 2003971.

[28] Chee, S. W.; Lunkenbein, T.; Schlögl, R.; Roldán Cuenya, B., Operando electron microscopy of catalysts: The missing cornerstone in heterogeneous catalysis research? *Chemical Reviews* **2023,** *123* (23), 13374–13418.

[29] Iriawan, H.; Andersen, S. Z.; Zhang, X.; Comer, B. M.; Barrio, J.; Chen, P.; Medford, A. J.; Stephens, I. E.; Chorkendorff, I.; Shao-Horn, Y., Methods for nitrogen activation by reduction and oxidation. *Nature Reviews Methods Primers* **2021,** *1* (1), 56.

[30] Guo, X.; Du, H.; Qu, F.; Li, J., Recent progress in electrocatalytic nitrogen reduction. *Journal of Materials Chemistry A* **2019,** *7* (8), 3531–3543.

[31] Cui, X.; Tang, C.; Zhang, Q., A review of electrocatalytic reduction of dinitrogen to ammonia under ambient conditions. *Advanced Energy Materials* **2018,** *8* (22), 1800369.

[32] Yang, B.; Ding, W.; Zhang, H.; Zhang, S., Recent progress in electrochemical synthesis of ammonia from nitrogen: Strategies to improve the catalytic activity and selectivity. *Energy & Environmental Science* **2021,** *14* (2), 672–687.

[33] Du, C.; Qiu, C.; Fang, Z.; Li, P.; Gao, Y.; Wang, J.; Chen, W., Interface hydrophobic tunnel engineering: A general strategy to boost electrochemical conversion of N2 to NH3. *Nano Energy* **2022,** *92*, 106784.

[34] Liu, Y.; Wang, L.; Chen, L.; Wang, H.; Jadhav, A. R.; Yang, T.; Wang, Y.; Zhang, J.; Kumar, A.; Lee, J., Unveiling the protonation kinetics-dependent selectivity in nitrogen electroreduction: Achieving 75.05% selectivity. *Angewandte Chemie International Edition* **2022,** *61* (50), e202209555.

[35] Shen, H.; Choi, C.; Masa, J.; Li, X.; Qiu, J.; Jung, Y.; Sun, Z., Electrochemical ammonia synthesis: Mechanistic understanding and catalyst design. *Chem* **2021,** *7* (7), 1708–1754.

[36] Venkateswarlu, S.; Vallem, S.; Umer, M.; Jyothi, N.; Babu, A. G.; Govindaraju, S.; Son, Y.; Kim, M. J.; Yoon, M., Recent progress on MOF/MXene nanoarchitectures: A new era in coordination chemistry for energy storage and conversion. *Journal of Energy Chemistry* **2023**, 86 (2023) 409–436.

[37] Khan, N. A.; Hasan, Z.; Ahmed, I.; Jhung, S. H., Chapter 2.2 – Metal-organic frameworks for nanoarchitectures: Nanoparticle, composite, core-shell, hierarchical, and hollow structures. In *Advanced Supramolecular Nanoarchitectonics*, Ariga, K.; Aono, M., Eds. William Andrew Publishing: 2019; pp. 151–194.
[38] He, H.; Li, H.-K.; Zhu, Q.-Q.; Li, C.-P.; Zhang, Z.; Du, M., Hydrophobicity modulation on a ferriporphyrin-based metal–organic framework for enhanced ambient electrocatalytic nitrogen fixation. *Applied Catalysis B: Environmental* **2022,** *316*, 121673.
[39] Zhong, H.; Wang, M.; Chen, G.; Dong, R.; Feng, X., Two-dimensional conjugated metal–organic frameworks for electrocatalysis: Opportunities and challenges. *ACS Nano* **2022,** *16* (2), 1759–1780.
[40] Wang, X.; Wu, F., Chapter 4 – MOF-derived carbonaceous materials. In *Metal Organic Frameworks and Their Derivatives for Energy Conversion and Storage*, Guan, C., Ed. Elsevier: 2024; pp. 63–84.
[41] Li, W.; Fang, W.; Wu, C.; Dinh, K. N.; Ren, H.; Zhao, L.; Liu, C.; Yan, Q., Bimetal–MOF nanosheets as efficient bifunctional electrocatalysts for oxygen evolution and nitrogen reduction reaction. *Journal of Materials Chemistry A* **2020,** *8* (7), 3658–3666.
[42] Cui, Q.; Qin, G.; Wang, W.; K. R, G.; Du, A.; Sun, Q., Mo-based 2D MOF as a highly efficient electrocatalyst for reduction of N2 to NH3: A density functional theory study. *Journal of Materials Chemistry A* **2019,** *7* (24), 14510–14518.
[43] Zhao, S.; Wang, J.; Wang, P.; Wang, S.; Li, J., One-step synthesis of N, P co-doped porous carbon electrocatalyst for highly efficient nitrogen fixation. *Nano Research* **2022,** *15* (3), 1779–1785.
[44] Yin, F.; Lin, X.; He, X.; Chen, B.; Li, G.; Yin, H., High Faraday efficiency for electrochemical nitrogen reduction reaction on Co@N-doped carbon derived from a metal-organic framework under ambient conditions. *Materials Letters* **2019,** *248*, 109–113.
[45] Zhao, X.; Yin, F.; Liu, N.; Li, G.; Fan, T.; Chen, B., Highly efficient metal–organic-framework catalysts for electrochemical synthesis of ammonia from N2 (air) and water at low temperature and ambient pressure. *Journal of Materials Science* **2017,** *52* (17), 10175–10185.
[46] Wang, L.; Liu, L.; Li, Y.; Xu, Y.; Nie, W.; Cheng, Z.; Zhou, Q.; Wang, L.; Fan, Z., Molecular-level regulation strategies toward efficient charge separation in donor– acceptor type conjugated polymers for boosted energy-related photocatalysis. *Advanced Energy Materials* **2023**, 2303346.
[47] Cai, W.; Liu, X.; Wang, L.; Wang, B., Design and synthesis of noble metal–based electrocatalysts using metal–organic frameworks and derivatives. *Materials Today Nano* **2022,** *17*, 100144.
[48] Hu, H.; Miao, R.; Yang, F.; Duan, F.; Zhu, H.; Hu, Y.; Du, M.; Lu, S., Intrinsic activity of metalized porphyrin-based covalent organic frameworks for electrocatalytic nitrate reduction. *Advanced Energy Materials n/a* (n/a), 2302608.
[49] Cui, Y.; Sun, C.; Qu, Y.; Dai, T.; Zhou, H.; Wang, Z.; Jiang, Q., The development of catalysts for electrochemical nitrogen reduction toward ammonia: Theoretical and experimental advances. *Chemical Communications* **2022,** *58* (74), 10290–10302.
[50] Guo, W.; Zhang, K.; Liang, Z.; Zou, R.; Xu, Q., Electrochemical nitrogen fixation and utilization: Theories, advanced catalyst materials and system design. *Chemical Society Reviews* **2019,** *48* (24), 5658–5716.
[51] Wang, C.; Zhao, Y.-N.; Zhu, C.-Y.; Zhang, M.; Geng, Y.; Li, Y.-G.; Su, Z.-M., A two-dimensional conductive Mo-based covalent organic framework as an efficient electrocatalyst for nitrogen fixation. *Journal of Materials Chemistry A* **2020,** *8* (44), 23599–23606.

[52] Zhong, H., Boosting the electrocatalytic conversion of nitrogen to ammonia on metal-phthalocyanine-based two-dimensional conjugated covalent organic frameworks. **2021,** *143* (47), 19992–20000.
[53] Jiang, M.; Han, L.; Peng, P.; Hu, Y.; Xiong, Y.; Mi, C.; Tie, Z.; Xiang, Z.; Jin, Z., Quasi-phthalocyanine conjugated covalent organic frameworks with nitrogen-coordinated transition metal centers for high-efficiency electrocatalytic ammonia synthesis. *Nano Letters* **2022,** *22* (1), 372–379.
[54] Shan, Z.; Sun, Y.; Wu, M.; Zhou, Y.; Wang, J.; Chen, S.; Wang, R.; Zhang, G., Metal-porphyrin-based three-dimensional covalent organic frameworks for electrocatalytic nitrogen reduction. *Applied Catalysis B: Environmental* **2024,** *342*, 123418.
[55] Pettinari, C.; Tăbăcaru, A.; Galli, S., Coordination polymers and metal–organic frameworks based on poly(pyrazole)-containing ligands. *Coordination Chemistry Reviews* **2016,** *307*, 1–31.
[56] Kim, C. R.; Uemura, T.; Kitagawa, S., Inorganic nanoparticles in porous coordination polymers. *Chemical Society Reviews* **2016,** *45* (14), 3828–3845.
[57] Foo, M. L.; Matsuda, R.; Kitagawa, S., Functional hybrid porous coordination polymers. *Chemistry of materials* **2014,** *26* (1), 310–322.
[58] Zhang, W.-X.; Liao, P.-Q.; Lin, R.-B.; Wei, Y.-S.; Zeng, M.-H.; Chen, X.-M., Metal cluster-based functional porous coordination polymers. *Coordination Chemistry Reviews* **2015,** *293*, 263–278.
[59] Horike, S.; Kitagawa, S., Design of porous coordination polymers/metal–organic frameworks: Past, present and future. *Metal-Organic Frameworks: Applications from Catalysis to Gas Storage* **2011**, 1–21.
[60] Cai, H.; Huang, Y.-L.; Li, D., Biological metal–organic frameworks: Structures, host–guest chemistry and bio-applications. *Coordination Chemistry Reviews* **2019,** *378*, 207–221.
[61] Xiong, Y.; Li, B.; Gu, Y.; Yan, T.; Ni, Z.; Li, S.; Zuo, J.-L.; Ma, J.; Jin, Z., Photocatalytic nitrogen fixation under an ambient atmosphere using a porous coordination polymer with bridging dinitrogen anions, Nature Chemistry, 15 (2023) 286–293.
[62] Wang, C.; Yao, Y.; Li, J.; Yamauchi, Y., Metal–organic frameworks: A robust platform for creating nanoarchitectured carbon materials. *Accounts of Materials Research* **2022,** *3* (4), 426–438.
[63] Wang, C.; Kim, J.; Tang, J.; Kim, M.; Lim, H.; Malgras, V.; You, J.; Xu, Q.; Li, J.; Yamauchi, Y., New strategies for novel MOF-derived carbon materials based on nanoarchitectures. *Chem* **2020,** *6* (1), 19–40.
[64] Wang, Y.; Li, Q.; Shi, W.; Cheng, P., The application of metal-organic frameworks in electrocatalytic nitrogen reduction. *Chinese Chemical Letters* **2020,** *31* (7), 1768–1772.
[65] Jiao, L.; Seow, J. Y. R.; Skinner, W. S.; Wang, Z. U.; Jiang, H.-L., Metal–organic frameworks: Structures and functional applications. *Materials Today* **2019,** *27*, 43–68.
[66] Du, M.; Li, C.-P.; Liu, C.-S.; Fang, S.-M., Design and construction of coordination polymers with mixed-ligand synthetic strategy. *Coordination Chemistry Reviews* **2013,** *257* (7–8), 1282–1305.
[67] Kaliyappan, T.; Kannan, P., Co-ordination polymers. *Progress in Polymer Science* **2000,** *25* (3), 343–370.
[68] Wu, J.; Wu, D.; Li, H.; Song, Y.; Lv, W.; Yu, X.; Ma, D., Tailoring the coordination environment of double-atom catalysts to boost electrocatalytic nitrogen reduction: A first-principles study. *Nanoscale* **2023,** *15* (39), 16056–16067.
[69] Zhong, H.; Wang, M.; Ghorbani-Asl, M.; Zhang, J.; Ly, K. H.; Liao, Z.; Chen, G.; Wei, Y.; Biswal, B. P.; Zschech, E.; Weidinger, I. M.; Krasheninnikov, A. V.; Dong, R.; Feng, X., Boosting the electrocatalytic conversion of nitrogen to ammonia on metal-phthalocyanine-based two-dimensional conjugated covalent organic frameworks. *Journal of the American Chemical Society* **2021,** *143* (47), 19992–20000.

[70] Feng, Z.; Tang, Y.; Chen, W.; Wei, D.; Ma, Y.; Dai, X., O-doped graphdiyne as metal-free catalysts for nitrogen reduction reaction. *Molecular Catalysis* **2020,** *483*, 110705.

[71] Ma, D.; Zeng, Z.; Liu, L.; Huang, X.; Jia, Y., Computational evaluation of electrocatalytic nitrogen reduction on TM single-, double-, and triple-atom catalysts (TM = Mn, Fe, Co, Ni) based on graphdiyne monolayers. *The Journal of Physical Chemistry C* **2019,** *123* (31), 19066–19076.

[72] Jiao, D.; Liu, Y.; Cai, Q.; Zhao, J., Coordination tunes the activity and selectivity of the nitrogen reduction reaction on single-atom iron catalysts: A computational study. *Journal of Materials Chemistry A* **2021,** *9* (2), 1240–1251.

[73] Hu, R.; Li, Y.; Zeng, Q.; Wang, F.; Shang, J., Bimetallic pairs supported on graphene as efficient electrocatalysts for nitrogen fixation: Search for the optimal coordination atoms. *ChemSusChem* **2020,** *13* (14), 3636–3644.

[74] He, C.; Wu, Z.-Y.; Zhao, L.; Ming, M.; Zhang, Y.; Yi, Y.; Hu, J.-S., Identification of FeN4 as an efficient active site for electrochemical N2 reduction. *ACS Catalysis* **2019**, *9* (8), 7311–7317.

13 Coordination Nanomaterial-Based Nanohybrids for Solar Cells

Muhammad Zahoor, Shahab Khan,
Muhammad Ikram, Sajjad Ali and Ghulam Yasin

13.1 INTRODUCTION

Nanomaterials are materials that have at least one dimension in the nanoscale range, typically between 1 and 100 nanometers. This size range gives these materials unique properties and behaviors compared to bulk materials[1]. At the nanoscale, materials may exhibit enhanced mechanical, electrical, magnetic, optical, or other properties that make them valuable for various applications in science and technology. Nanomaterials can be classified into different categories, including nanoparticles, nanotubes, nanowires, and nanosheets, based on their shapes and structures. Due to their special characteristics, nanomaterials find applications in a wide range of fields such as electronics, medicine, energy, catalysis, and materials science. However, the unique properties of nanomaterials also raise concerns about potential risks and environmental impacts, leading to ongoing research into their safe use and disposal[2].

In recent years, the field of molecular nanomaterials has bridged the gap between atoms/molecules and traditional nanomaterials. In Figure 13.1 a simple route is represented, which shows how basic material constituents proceed towards various nanomaterials. Unlike ordinary nanomaterials, molecular nanomaterials are constructed with precise molecular weights, well-defined chemical structures, and excellent synthetic repeatability. This precision allows for further enhancement and optimization through precise modifications. Coordination-based molecular nanostructures, known for their diverse characteristics, have garnered attention in both basic research and potential applications. Early advancements in nanomaterials were marked by the success of numerous inorganic and organic nanoparticles, driven by rapid progress in synthetic technologies and characterization techniques. These materials have found applications in basic research[3], industry[4], agriculture[5], biomedical sciences[6], and catalysis.[7–10] Despite their contributions to biomedicine, concerns about potential negative health effects persist. Recent decades have witnessed significant progress in biologically active membrane, DNA, and peptide-based nanomaterials. Nano-drugs,

DOI: 10.1201/9781003345886-15

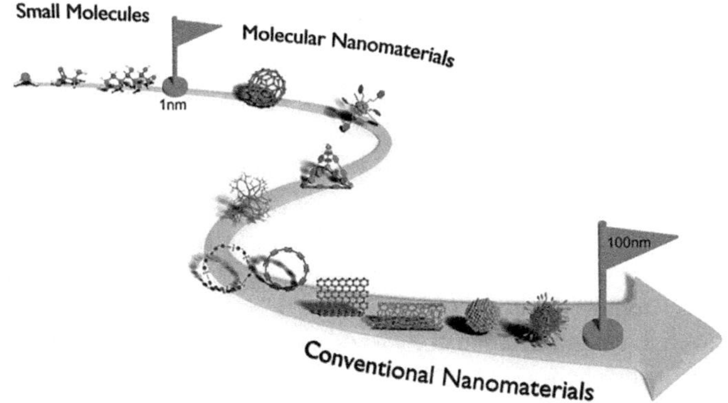

FIGURE 13.1 Molecular nanomaterials bridge the transition between small molecules and conventional nanomaterials. Specifically, coordination-based molecular nanomaterials demonstrate promising applications in the field of biomedicine. Copyright Elsevier (2021). Reproduced with permission from ref. 18.

with advantages over small-molecule medications, offer effective solutions in diagnosis and therapy. Factors such as increased permeability and retention (EPR) in tumor tissues and strong imaging signals contribute to their success. However, traditional nanomaterials still face challenges like hazy surfaces and poor synthetic repeatability.

In contrast, coordination nanomaterials emerge as a solution to these issues, offering improved precision and repeatability. Innovative techniques such as mass spectrometry imaging enable the tracking of these materials in complex biological environments.[11] Solar energy is a promising renewable energy source that has gained significant attention in recent years due to the increasing demand for clean and sustainable energy.[12–16] Solar cells, also known as photovoltaic cells, are at the forefront of solar energy conversion. Traditional solar cells have limitations in terms of efficiency and cost-effectiveness, which has led to the exploration of new materials and technologies.[17] In this chapter, we focus on the application of coordination nanomaterial-based nanohybrids for solar cells. We will discuss the properties, synthesis methods, and performance of these nanohybrids in improving solar cell efficiency. Furthermore, we highlight the challenges and future prospects of coordination nanomaterial-based nanohybrids in solar cell technology. The ever-increasing demand for clean and sustainable energy sources has spurred intensive research into renewable energy technologies. Among these, solar energy stands out as a compelling solution to address the challenges of climate change, energy security, and greenhouse gas emissions. Solar photovoltaic (PV) technology, in particular, has witnessed remarkable growth in recent years. Traditional silicon-based solar cells have played a pivotal role in harnessing solar energy, driving significant advancements in the solar industry. However, the limitations of silicon, including material scarcity, manufacturing costs, and efficiency constraints, have prompted researchers to explore alternative materials and innovative approaches.[17]

13.1.1 Nanohybrids and Coordination Materials

Nanohybrids are materials that combine nanoscale components to produce composite structures with unique properties and functionalities.[19,20] They are at the forefront of interdisciplinary research, bridging the gap between different fields of science and engineering.[21] Nanohybrids offer a synergistic combination of properties from their constituent materials, resulting in enhanced or entirely new properties. They offer tunability in terms of composition, size, shape, and structure, allowing researchers to precisely control the ratio of each component[22] and tailor the hybrid's properties to meet specific requirements[23]. Nanohybrids can serve multiple functions simultaneously — such as magnetic and catalytic properties – and can enhance the stability and durability of the hybrid. They find application across various fields, including nanoelectronics, drug delivery, diagnostics, imaging, catalysis, energy-related applications, and sensors and detectors.[24] As interdisciplinary collaboration continues to drive progress in nanoscience and nanotechnology, the potential applications of nanohybrids are expected to expand, contributing to advancements in various industries and scientific disciplines[25–30]. Coordination materials, also known as metal-organic frameworks (MOFs) or coordination polymers, are crystalline materials with a highly ordered, porous, and tunable structure. These materials are characterized by their high surface area and porosity, which are crucial for applications like gas adsorption and storage. Their tunable pore size allows for precise control over properties, making them adaptable to various applications. Coordination materials also have remarkable chemical stability, maintaining structural integrity under various conditions. Their applications include gas adsorption and storage, catalysis, chemical sensing, drug delivery, and separation and filtration. Their high surface area, tunable pore size, and chemical stability make them ideal for gas separation, storage, and purification. They can also be used for catalysis, chemical sensing, drug delivery, and separation and filtration. As researchers continue to explore and engineer coordination materials, their potential impact on technological advancements is expected to grow significantly.[9,26]

13.2 SOLAR ENERGY AND ITS IMPORTANCE

Solar energy is a renewable and sustainable source of power that is derived from the sun[14,29,31,32]. It is an important form of clean energy that does not produce greenhouse gas emissions or contribute to air pollution. Solar energy is harnessed using solar panels or photovoltaic cells, which convert sunlight into electricity

There are several reasons why solar energy is important:

1. Renewable and sustainable: Solar energy is a renewable resource, meaning it will never run out as long as the sun continues to shine. It is also a sustainable source of power, as it does not deplete any natural resources.[33]
2. Environmentally friendly: Solar energy is a clean form of energy that does not produce any harmful emissions or pollutants. It helps to reduce greenhouse gas emissions and combat climate change.
3. Energy independence: Solar energy provides individuals and communities with the opportunity to generate their own electricity, reducing dependence

on fossil fuels and the electricity grid. This can lead to greater energy independence and resilience.[33]

4. Cost-effective: While the initial investment in solar panels can be high, the cost of solar energy has been decreasing over the years. Once installed, solar panels have low operating and maintenance costs, making them a cost-effective energy solution in the long run.
5. Job creation: The solar energy industry has been growing rapidly, creating numerous job opportunities. The installation, maintenance, and manufacturing of solar panels require a skilled workforce, contributing to economic growth and employment.
6. Remote power generation: Solar energy can be harnessed in remote areas where connecting to the electricity grid may be difficult or expensive. It provides a reliable and clean source of power for off-grid communities, improving their quality of life and enabling development.
7. Diverse applications: Solar energy can be used for a variety of purposes, including heating water, powering homes and businesses, and generating electricity for large-scale projects.[34] It is a versatile energy source that can be integrated into different sectors of the economy.
8. Solar energy plays a crucial role in the transition towards a sustainable and low-carbon future. Its renewable nature, environmental benefits, and diverse applications make it an important form of clean energy. As technology continues to advance, the potential for solar energy to meet a significant portion of global energy needs is becoming increasingly promising.[33]

13.3 AN OVERVIEW OF SOLAR CELLS AND ROLE OF NANOMATERIALS IN SOLAR CELL TECHNOLOGY

Solar cells, also known as photovoltaic cells, convert sunlight into electricity, serving as the foundation of solar panels. They work through the photovoltaic effect, where sunlight excites electrons in semiconductor materials, creating an electric current that powers various devices. Silicon is the most common material due to its efficiency and availability. Solar cells have a p-n junction, combining p-type and n-type semiconductors, which generates electricity. The absorbed photons transfer energy to the electrons, creating an electric field that separates positive and negative charges. Metal contacts collect the generated electricity, which can be used for various applications. Solar cells have revolutionized the renewable energy industry, providing a clean and sustainable source of electricity.[33] Nanomaterials have revolutionized the field of solar cell technology by offering unprecedented opportunities for enhancing efficiency, reducing costs, and expanding the scope of solar energy applications. This chapter explores the critical role of nanomaterials in advancing solar cell technology. It delves into the unique properties of nanomaterials, their synthesis methods, and their applications in various types of solar cells. By harnessing the advantages of nanomaterials, we can address key challenges in solar energy conversion and pave the way for a sustainable energy future. Solar energy, as a clean and renewable resource, holds immense potential in mitigating climate change and meeting global energy demands. Traditional solar cells based on bulk materials, such as crystalline silicon, have made significant progress. However, they face limitations in efficiency, cost, and scalability. Nanomaterials, with their unique properties stemming from nanoscale dimensions, offer innovative solutions to these challenges. In this chapter, we explore the pivotal role of nanomaterials in transforming solar cell technology.[35]

13.3.1 Nanomaterial Integration into Solar Cells

Nanomaterials are integrated into solar cells through:

1. Nanostructured electrodes: Nanowire or nanotube-based transparent conductive electrodes improve light transmission.
2. Photoactive layers: Nanoparticles or nanocomposites enhance light absorption and charge separation.
3. Electron and hole transport layers: Nanomaterials facilitate efficient charge transport.
4. Plasmonic enhancements: Metal nanoparticles enhance light trapping and absorption in thin-film solar cells.

13.3.2 Coordination Nanomaterials for Solar Cells

Coordination nanomaterials, also known as metal-organic frameworks (MOFs) and coordination polymers, represent a fascinating class of materials at the intersection of chemistry, materials science, and nanotechnology. These nanomaterials are characterized by their intricate and highly tunable structures, consisting of metal ions or clusters coordinated with organic ligands at the nanoscale. Coordination nanomaterials have emerged as a thriving field of research due to their remarkable properties, which make them versatile candidates for a wide array of applications. The origins of coordination nanomaterials can be traced back to the early 20th century when inorganic coordination polymers were first synthesized. However, it wasn't until the late 1990s that the field gained significant momentum with the discovery of a new class of materials MOFs. The synthesis of MOFs marked a pivotal moment in the development of coordination nanomaterials, opening up avenues for innovative research and applications. Coordination nanomaterials owe their unique characteristics to their precisely engineered structures. At their core, they consist of metal ions or clusters as nodes, connected by organic ligands forming the linkers.[36,37] The result is a highly ordered, porous, and crystalline structure. The remarkable properties of coordination nanomaterials include:

- High surface area: Their porous nature leads to an exceptionally high surface area, providing ample sites for adsorption, catalysis, and guest molecule interaction.[38]
- Tunable pore size: Researchers can tailor the size and geometry of the pores by selecting specific metal ions and organic ligands, allowing for applications in gas separation and storage.
- Chemical versatility: Coordination nanomaterials exhibit a broad chemical diversity, with a wide range of metal ions and ligands available for synthesis.

This versatility enables the design of materials with customized properties for various applications.

- Stability: many coordination nanomaterials are chemically stable, even in harsh environments. This stability makes them valuable in applications requiring robust materials.

13.4 APPLICATIONS OF METAL-ORGANIC FRAMEWORK NANOMATERIALS

In the present era of cutting-edge technology, which demands significant power consumption while also considering environmental concerns, recent research has been primarily directed towards the advancement of metal-organic frameworks (MOFs) for clean energy applications. These applications encompass a wide range of areas such as lithium-based batteries (including lithium-ion, lithium-sulfur, and lithium-oxide), fuel cells, supercapacitors, solar cells, and hydrogen production and storage. The field of electrochemical energy storage devices and the synthesis of novel materials with enhanced properties and improved efficiency pose considerable challenges.

Higher charge/discharge rates, bigger theoretical capacity, enhanced electrical stability, and the properties of anode/cathode materials all play a role in determining the best performance and efficiency of energy storage devices. As a result, scientists have invested a lot of time and energy into creating next-generation electrochemical energy storage systems and making revolutionary strides in solar cell device power conversion efficiency[39,40]. Due to the synergistic interactions between metal clusters and multifunctional organic ligands, which allow MOFs to exhibit a range of properties suited for different devices, MOFs are viewed as promising materials for numerous technical applications. Beyond uses in energy conversion and storage, MOFs are also promising in areas like sensing, proton conduction, and medication delivery. The numerous uses of MOF materials in technological breakthroughs to fulfill the current energy demands of the modern world are summarized in this section. Figure 13.2 provides a graphic depiction of the potential uses for MOFs.

13.4.1 MOFs Applications in Electrochemical Energy Conversion Devices

Environmentally benign electrochemical processes are essential to the conversion of energy into sustainable forms. In particular, solar cells, fuel cells, and white light-emitting diodes have demonstrated the potential of metal-organic framework (MOF) materials and their derivatives in a variety of electrochemical energy conversion devices. A viable substitute for non-renewable energy sources like coal and fossil fuels is solar energy. Since their invention in 1954, crystalline silicon-based solar cells have produced power conversion efficiencies greater than 26%. The emergence of second and third generation solar cells, such as organic and perovskite solar cells, can be attributed to their reduced material content and affordability. Metrics such as power conversion efficiency, fill factor, open-circuit voltage, and short-circuit current density can all be improved by MOF materials. MOF thin films are suitable for absorber layers or photoanode applications due to their large surface area and exceptional light-harvesting capabilities[42,43]. Figure 13.3 illustrates the utilization of

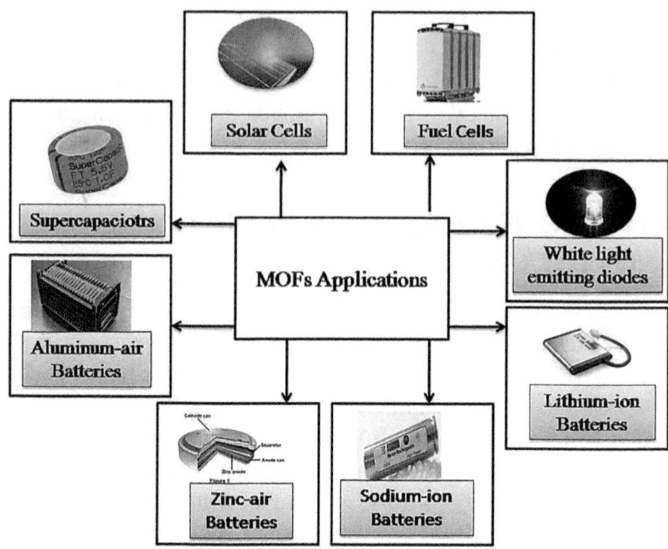

FIGURE 13.2 Visual representation of the potential applications of MOF materials and their derivatives. Copyright Elsevier (2021). Reproduced with permission from ref. 41

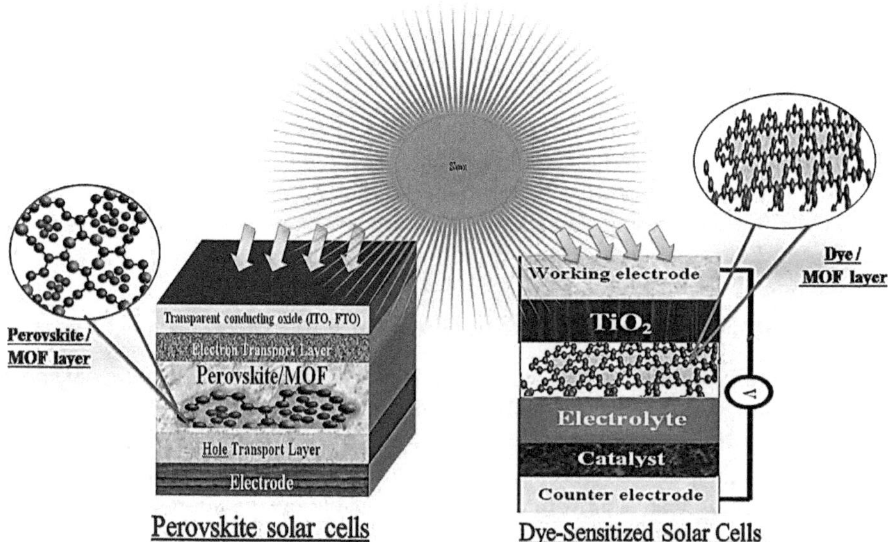

FIGURE 13.3 The role of MOF thin films in perovskite and dye-sensitized solar cells. Copyright Elsevier (2021). Reproduced with permission from ref. 41.

MOF thin films in perovskite and dye-sensitized solar cell architectures for light absorption or photoanode applications, depending on the technology's requirements.

Layer-by-layer (LBL) technology can be used to create Co-MOF thin films, which can then be used as MOF absorber layers in solar cells. These films were initially

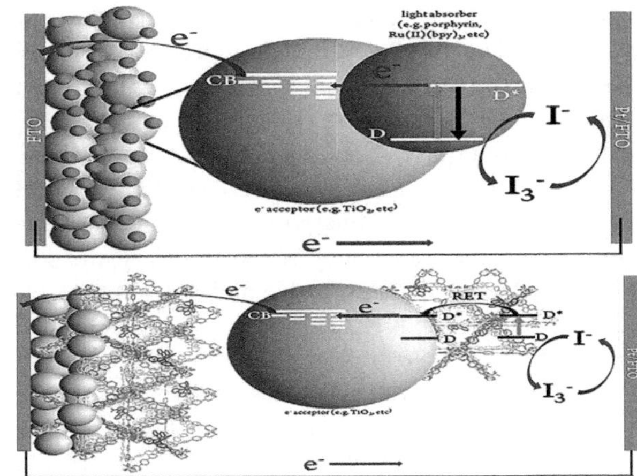

FIGURE 13.4 An example of a conventional DSSC (top) and a Ru-MOF-sensitized solar cell device (bottom). Copyright Elsevier (2021). Reproduced with permission from ref. 41.

insulators, but after being doped with a hole, their conductivity changed to p-type. These created layers were effectively used in TiO2-based solar cells as light-harvesting absorber films, improving power conversion efficiency. To characterize the MOF thin films, chemicals such as cobalt (II) 2,6-naphthalenedicarboxylic acid (Co-NDC) and cobalt (II) benzendicarboxylate (Co-BDC) can be used as reference materials. Recently, hierarchical porous anatase TiO_2 photoanodes were developed to improve the power conversion efficiency of DSSCs[44]. Both conventional DSSCs and Ru-MOF sensitized solar cells (MOFSCs) are shown in Figure 13.4[45].

13.5 COORDINATION MONOHYBRIDS IN DYE-SENSITIZED SOLAR CELLS

Global warming is causing environmental concerns, prompting the search for alternative energy sources. Sunlight is a cost-effective and eco-friendly green energy option, which can be harnessed through photovoltaic (PV) devices. Solar cells, composed of semiconductors, capture and convert solar radiation into electrical energy. Despite advancements in solar cells, the widespread adoption of dye-sensitized solar cells (DSSCs) has been hindered by high material costs and limited power conversion efficiency. Lanthanide-based upconversion nanoparticles (UCNPs) have shown potential in DSSCs due to their adjustable light absorption properties. The morphological characteristics, crystal structure, and interactions with plasmonic nanoparticles, graphene materials, and mesoporous titanium dioxide (TiO_2) shells influence the overall power conversion efficiency. One-dimensional nanostructures and high-specific-surface-area mesoporous structures enhance charge collection efficiency and photovoltaic current output. The use of UCNPs and tailored surface functionalities can enhance photovoltaic current output in DSSC devices. This analysis also highlights the potential for optimizing electron transport properties to increase light-harvesting efficiency of UCNP-based DSSCs. Current research focuses the employment of MOFs and COF materials in solar technology for ultimate energy resources. The incorporation of inorganic structures with organic pigments like dyes[46] makes a special material know as light harvesting nanohybrids (LHNs), which can be used in solar cells (as the active layer). Soumik Sarkar et al.[47] synthesized hematoporphyrin-ZnO-nanohybrids as efficient light harvesting nanohybrids, which were further used as active layer in dye-sanitized solar cells. Sometimes referred to as the third generation of solar cells, dye-sensitized solar cells (DSSCs) are an extremely advantageous kind of organic solar cell. A redox electrolyte or ionic liquid solution, organic dye molecules (photosensitizers), a glass substrate coated in indium-tin-oxide (ITO) or fluorine-tin-oxide (FTO) as a conducting glass, and a porous nanocrystalline film of titanium dioxide (TiO_2) serve as the photo-electrode in dye-sensitive solar cells (DSSCs). By converting NIR solar photons into high-energy photons, upconversion materials can improve incident light processing and boost photocurrent in DSSCs.[48,49]. Figure 13.5, as illustrated by Sengupta et al., provides a visual representation of the functioning principle of DSSC devices[50]. In this setup, the electrolyte located at the counter electrode is regenerated when an external load completes the circuit. The desired processes for photo-excitation, transportation, and the renewal of electrons within the DSSC are depicted sequentially in Figure 13.5 through paths 1 to 6.

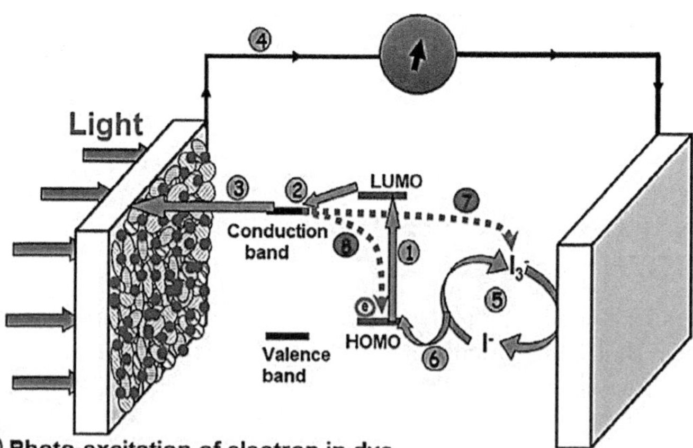

① Photo-excitation of electron in dye
② Injection of electron from dye to the conduction band of TiO_2
③ Transfer of electron from metal oxide to transparent conducting glass substrate
④ Conduction of electron from photo-anode to cathode
⑤ Reduction of tri-iodide to iodide
⑥ Oxidation of iodid to tri-iodide and regeneration of dye
⑦ Recombination of photo-excited electron with the oxidized dye
⑧ Recombination of photo-excited electron with the oxidized tri-iodide

FIGURE 13.5 Diagrammatic illustration of the dye-sensitized solar cell's operating principle. Copyright Elsevier (2016). Reproduced with permission from ref. 50.

13.5.1 Constructional Components and Operational Mechanism of the DSSCs

Dye-sensitized solar cells (DSSCs) are cost-effective, environmentally friendly, and easy-to-fabricate alternatives to silicon solar cells. They consist of a sandwich-type structure with photosensitizers, an electrolyte, a semiconductor absorber, and a counter electrode. The structure is composed of a mesoporous semiconductor TiO_2 layer on a transparent conducting glass electrode, coated with a monolayer of photosensitizers. The redox couple is a liquid electrolyte or ionic liquid containing iodide/iodine. Platinum serves as the counter electrode. The TiO_2 layer absorbs sunlight, increasing photocurrent.[51] Efficiency can be improved by controlling factors like nano-structural morphology, photosensitizers, and redox solution composition. Techniques like sunlight-harvesting molecules can enhance conversion efficiency, contributing to improved photocurrent and photovoltage efficiency.[49]

13.5.2 Conducting Glass Substrate (Film/Electrode)

To manufacture dye-sensitized solar cells (DSSCs), two sheets of conductive transparent glass substrates (films/electrodes) are essential for the subsequent deposition of semiconductor metallic oxides such as TiO_2, ZrO_2, CeO_2, ZnO, and photosensitizers/photo-catalysts, which can be organic molecules or dyes. The most commonly utilized films/electrodes for this purpose are made of materials like Indium Tin Oxide (ITO) or Fluorine Tin Oxide (FTO). These materials involve the doping of indium and fluorine in tin oxide and are applied as a very thin, highly transparent layer on soda-lime glass substrates, as depicted in Figures 13.6–8[52]. ITO or FTO sheets are the preferred choices for transparent conductive glass electrodes in the fabrication of DSSCs. These substrate electrodes exhibit impressive characteristics, including 75–90% transparency and a sheet transmittance of 8–25 X/cm^2 in the visible region of the electromagnetic spectrum. This high level of transparency is advantageous as it allows optimal sunlight to reach the active part of the DSSC device. Additionally, these substrate electrodes offer low resistivity, signifying their

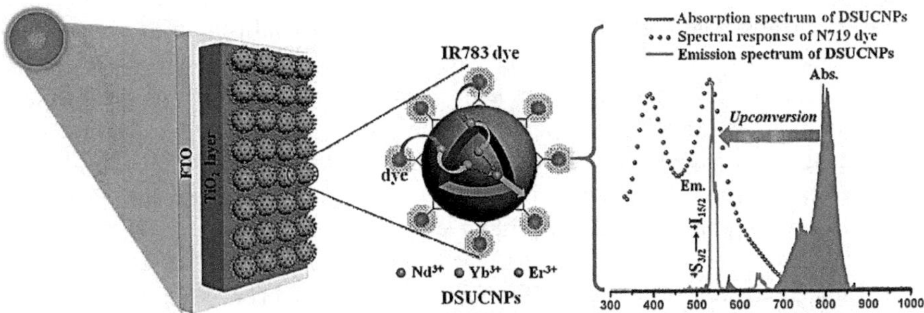

FIGURE 13.6 For the purpose of improving a DSSC device, broadband near-infrared sunlight harvesting and subsequent spectral conversion into visible range to activate N719 dye. Copyright Elsevier (2021). Reproduced with permission from ref. 49.

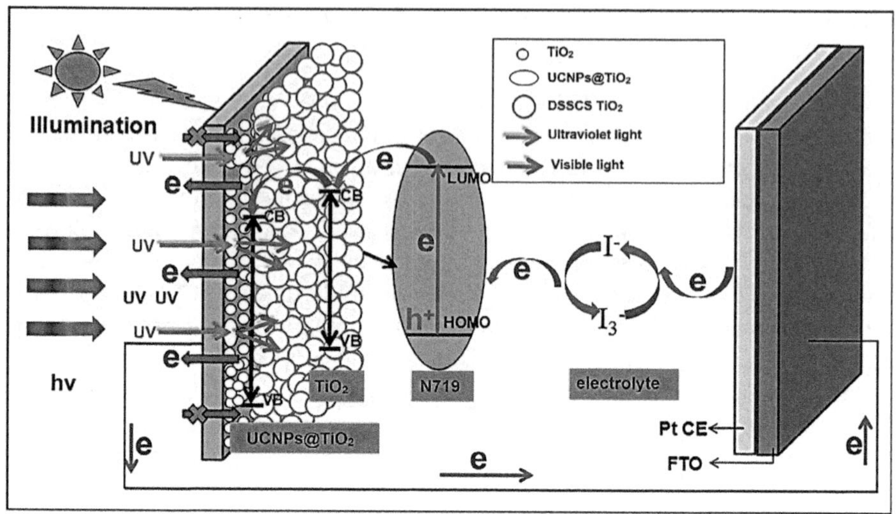

FIGURE 13.7 Diagrammatic design of a DSSC device based on UCNP. Copyright Elsevier (2016). Reproduced with permission from ref. 49.

capacity for efficient electron and photon transfer within the DSSCs, leading to minimal energy loss during the conversion process[50]. Moreover, their low resistivity enhances their ability to transport electrons effectively.

13.6 PERFORMANCE OF COORDINATION NANOMATERIAL-BASED NANOHYBRIDS IN SOLAR CELLS

13.6.1 Efficiency Enhancement through Nanohybrids

Nanohybrids in solar cells can enhance efficiency and performance. Perovskite solar cells (PSCs) (Figure 13.9) are a promising technology with high-performance due to their absorption coefficient, defect tolerance, and long charge carrier lifetime. However, stability is a challenge due to moisture, light, oxygen, and temperature degradation. The hole-transport layer (HTL) plays a crucial role in extracting photo-generated holes from the perovskite layer and transporting them to metal electrodes. Recently, tri-iodide ions have been introduced to improve the performance of organic-inorganic lead halide perovskite devices. R. Rejeev et al. used interface engineering to improve the effectiveness and moisture stability of PSCs, resulting in a 41% efficiency improvement and increased device stability.[53] Here are some of the key ways in which nanohybrids can enhance the efficiency of solar cells:

1. Enhanced light absorption: Nanohybrids often incorporate nanoparticles with precisely engineered sizes and shapes that can efficiently scatter and trap incoming sunlight. This results in a longer optical path for photons within the solar cell, increasing the probability of absorption and leading to improved efficiency.

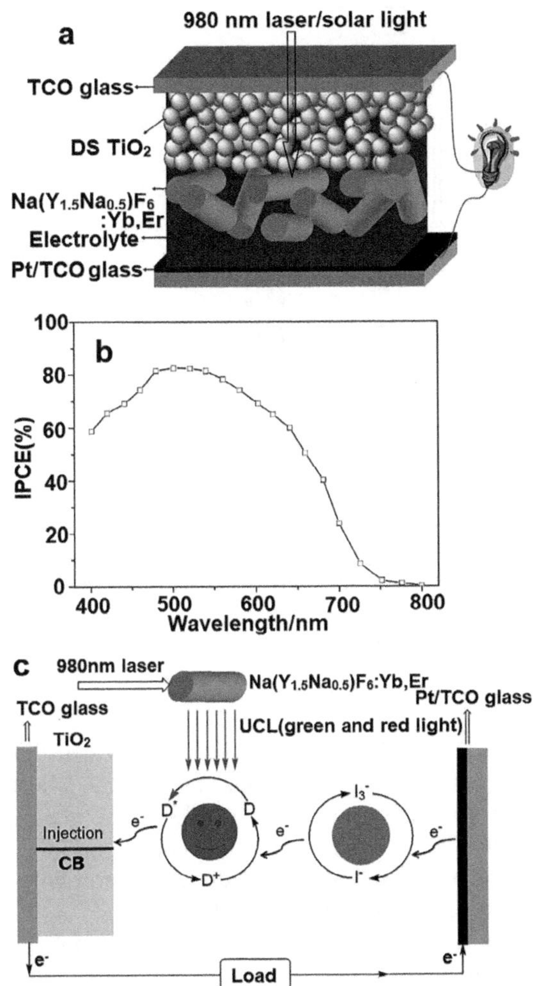

FIGURE 13.8 a) The photovoltaic cell with the $Na(Y_{1.5}Na_{0.5})$ F_6:Yb,Er film is shown schematically. b) The IPCE of the resultant $Na(Y_{1.5}Na_{0.5})F_6$:Yb,Er nanorod layer-containing photovoltaic cells. The N_3-sensitized TiO_2 film was excited by monochromatic light in the 400–800 nm range, which was produced by a series of filters and a 1,000 W xenon lamp used as the light source. c) The operation of 980LD-PVCs using a light source that is a 980 nm laser. Copyright Elsevier (2016). Reproduced with permission from ref. 49.

2. Expanded spectral response: By integrating semiconductor nanoparticles like quantum dots into solar cells, nanohybrids can extend the spectral response of the cell. This enables the capture of a broader range of wavelengths, including those in the infrared and ultraviolet regions, which traditional solar cells may not utilize effectively.
3. Reduced reflectance: Nanoscale materials in nanohybrids can minimize reflectance at the surface of solar cells, ensuring that more incident sunlight is

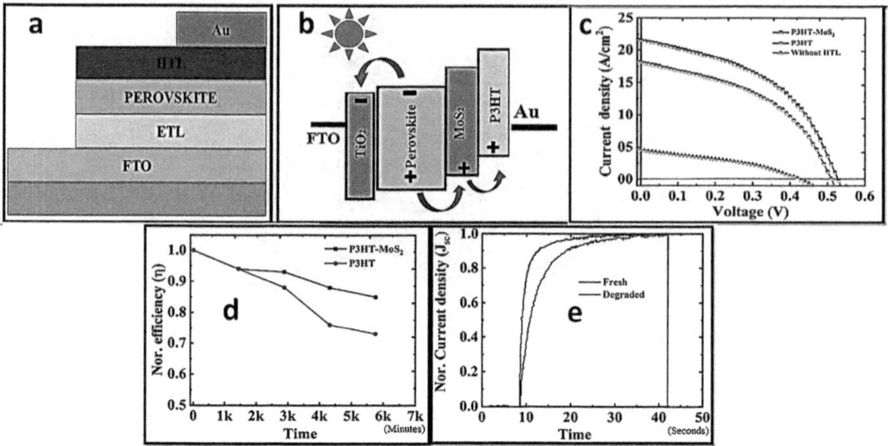

FIGURE 13.9 (a) Diagram showing how the solar cells were made. (b) A band diagram showing the various layers of the manufactured P3HT-MoS_2 solar cells. (c) J-V characteristics for solar cells that were manufactured with and without HTL (using P3HT-MoS_2 and P3HT as HTM). (d) The artificial PSCs' relative self-stability (using pure P_3HT and P_3HT-MoS_2 as HTL). (e) The fresh and degraded PSCs' transient current density and JSC (normalized at maximum). Copyright Elsevier (2019). Reproduced with permission from ref. 53.

absorbed rather than being bounced back. This is particularly beneficial for thin-film solar cells, which are inherently more susceptible to reflectance losses.

4. Improved charge separation and transport: The high surface area and unique properties of nanoscale components in nanohybrids facilitate efficient charge separation and transport. This reduces electron-hole recombination, a major source of energy loss in solar cells.
5. Quantum effects: In some cases, nanoparticles in nanohybrids can exhibit quantum confinement effects. This phenomenon can enhance charge carrier generation by confining charge carriers to discrete energy levels, increasing the efficiency of photovoltaic conversion.
6. Plasmonic enhancements: Incorporating metal nanoparticles, such as gold or silver nanoparticles, into solar cell structures can lead to plasmonic enhancements. These enhancements can enhance light trapping and absorption, particularly in thin-film and organic solar cells.
7. Tunable bandgaps: Nanohybrids allow for the precise tuning of semiconductor bandgaps. This enables the customization of the solar cell to match the solar spectrum, optimizing absorption and efficiency.
8. Efficient charge collection: Nanohybrids with improved charge transport properties ensure that generated charge carriers (electrons and holes) are efficiently collected and directed to the electrodes, minimizing resistive losses and improving overall efficiency.
9. Reduced material thickness: In thin-film and flexible solar cell technologies, nanohybrids can reduce the thickness of active layers while maintaining or

even improving efficiency. This can lead to cost savings and lightweight, flexible solar panels.

10. Tandem cell applications: Nanohybrids play a critical role in tandem solar cells, which stack multiple solar cell materials on top of each other. Each layer is designed to capture specific portions of the solar spectrum, increasing overall efficiency.
11. Improved environmental stability: Certain nanohybrids can enhance the stability and longevity of solar cells by providing protection against environmental factors, such as moisture and UV radiation.
12. Cost-effective materials: While nanohybrids may introduce added complexity, their potential to boost efficiency can offset the initial investment by enabling the use of cost-effective, thinner, or less expensive materials in the solar cell design.
13. Integration into existing technology: Nanohybrids can be integrated into existing solar cell manufacturing processes, making it feasible to enhance the performance of current solar technologies without a complete overhaul.
14. Versatility: Nanohybrids can be tailored for various types of solar cells, including silicon-based, organic, perovskite, and thin-film cells, among others, making them versatile tools for efficiency enhancement.

Efficiency enhancement through nanohybrids represents a dynamic and promising field in solar cell research and development. As researchers continue to innovate and refine nanohybrid technologies, we can anticipate further improvements in solar cell efficiency, contributing to the widespread adoption of solar energy as a clean and sustainable power source.

13.6.2 Light Absorption and Charge Separation in Nanohybrids

Coordination nanohybrids, also known as coordination material-based nanohybrids, have shown great potential in improving light absorption and charge separation processes in solar cells. These unique materials, which combine coordination materials (e.g., metal-organic frameworks or MOFs) with nanoscale components, offer several advantages that can enhance the efficiency of solar energy conversion. In this discussion, we'll explore how coordination nanohybrids impact light absorption and charge separation, both of which are crucial aspects of solar cell performance.[54]

13.6.2.1 Light Absorption in Coordination Nanohybrids

1. Enhanced light harvesting: Coordination nanohybrids often incorporate nanoscale components, such as nanoparticles or quantum dots, which can be engineered to efficiently absorb light across a broad spectrum of wavelengths. These nanoscale materials act as light harvesters, capturing photons from the visible and even near-infrared regions.[54]
2. Precise tuning of bandgaps: The choice of nanoscale materials in coordination nanohybrids allows for precise tuning of the material's bandgap. This tuning ensures that the nanoscale components can absorb photons in a complementary manner to the coordination material, maximizing overall light absorption.

3. Multiple exciton generation: In some cases, nanoscale components like quantum dots exhibit a phenomenon called multiple exciton generation (MEG). MEG allows a single photon to generate multiple electron-hole pairs, increasing the efficiency of light absorption.
4. Reduced reflectance: The integration of nanoscale materials can help reduce reflectance at the surface of the solar cell. This minimizes the loss of incident photons that would otherwise be reflected away, increasing the number of photons available for absorption.
5. Scattering and trapping: Nanoscale components within coordination nanohybrids can scatter and trap incoming light effectively. This results in a longer optical path for photons within the solar cell, increasing the probability of absorption and reducing the likelihood of light escaping before it can be converted into electrical energy.[54]

13.6.2.2 Charge Separation in Coordination Nanohybrids

1. High surface area: Coordination nanohybrids often possess a high surface area due to the presence of nanoscale components. This large surface area provides ample space for charge separation to occur. Photogenerated electron-hole pairs are more likely to be separated efficiently on the surfaces of nanoscale materials.[55]
2. Efficient charge transport: The unique properties of nanoscale materials, such as high charge carrier mobility, facilitate the efficient transport of charge carriers (electrons and holes) to the respective electrodes. Efficient charge transport minimizes recombination losses.
3. Tunable energy levels: Nanoscale components can have energy levels that align well with the energy levels of the coordination material. This alignment ensures that photogenerated charges are separated and transported to the electrodes with minimal energy loss.
4. Reduced electron-hole recombination: Nanoscale materials can act as electron or hole acceptors, reducing the likelihood of electron-hole recombination. This means that more photogenerated charge carriers reach the electrodes, contributing to higher efficiency.
5. Quantum effects: In some cases, nanoscale components exhibit quantum confinement effects. These effects can lead to efficient charge separation by confining charge carriers to discrete energy levels, preventing them from recombining before reaching the electrodes.
6. Integration with coordination materials: Coordination nanohybrids are designed to ensure strong interactions between the nanoscale components and the coordination materials. This integration promotes effective charge separation, allowing for efficient charge carrier generation and transport.
7. Coordination nanohybrids hold significant promise for improving light absorption and charge separation in solar cells. Their ability to combine the unique properties of nanoscale materials with coordination materials opens up new possibilities for enhancing the efficiency of solar energy conversion. As research in this field continues to advance, coordination nanohybrids may play a crucial role in the development of more efficient and sustainable solar cell technologies.[55]

13.6.3 Role of Coordination Nanomaterials in Electron Transport

Coordination nanohybrids play a crucial role in enhancing electron transport in solar cells and other electronic devices. Their unique properties and structural features make them highly effective in facilitating the efficient movement of electrons within these systems. Here, we explore the key roles of coordination nanohybrids in electron transport:[49]

1. High surface area: Many coordination nanohybrids incorporate nanoscale components, such as nanoparticles or nanowires, which significantly increase the surface area of the material. This enlarged surface area provides more sites for electron adsorption, facilitating the initial step of electron transport.
2. Charge carrier mobility: Nanoscale components in coordination nanohybrids often exhibit excellent charge carrier mobility. Electrons can move through these nanoscale materials with minimal resistance, reducing the likelihood of electron trapping or recombination. This property is particularly valuable in thin-film solar cells and other electronic devices.
3. Energy level alignment: Coordination nanohybrids are designed to have well-aligned energy levels between the nanoscale components and the coordination materials. This alignment ensures that electrons can transfer between these materials efficiently without significant energy barriers.
4. Effective charge separation: In solar cells, coordination nanohybrids assist in separating photogenerated electron-hole pairs. The nanoscale components can act as electron acceptors, effectively capturing electrons and preventing them from recombining with holes. This separation process ensures that electrons are available for transport to the electrode.
5. Reduced electron-hole recombination: By promoting efficient charge separation, coordination nanohybrids reduce the chances of electron-hole recombination. Electrons are more likely to reach the electron-collecting electrode without prematurely recombining with holes, increasing the overall efficiency of the device.[49]
6. Improved conductivity: Some coordination nanohybrids incorporate highly conductive nanoscale materials like carbon nanotubes or graphene. These materials can serve as efficient pathways for electron transport, reducing resistive losses and improving overall conductivity.
7. Quantum effects: Nanoscale components in coordination nanohybrids can exhibit quantum confinement effects. These effects lead to discrete energy levels for charge carriers, enhancing electron transport efficiency by preventing carriers from scattering or recombining before reaching the electrode.
8. Integration with electrodes: Coordination nanohybrids are often designed for strong integration with electrode materials. This ensures a seamless pathway for electrons to travel from the nanohybrid material to the electrode, further reducing resistive losses.
9. Tandem cell applications: In tandem solar cells, where multiple layers of materials with different bandgaps are stacked, coordination nanohybrids can optimize electron transport between these layers. This enhances the overall efficiency of the tandem cell by ensuring efficient charge transfer.

10. Doping and functionalization: Coordination nanohybrids can be doped or functionalized to further enhance their electron transport properties. For example, introducing specific dopants or functional groups can modify the electronic structure to facilitate charge transport.
11. Environmental stability: Some coordination nanohybrids are designed to be environmentally stable, ensuring that their electron transport properties remain consistent over time and under various environmental conditions.
12. Coordination nanohybrids are valuable materials for improving electron transport in solar cells and electronic devices. Their ability to combine high surface area, charge carrier mobility, energy level alignment, and efficient charge separation makes them instrumental in achieving high-performance solar cells and other electronic systems. As research in this field continues to progress, coordination nanohybrids are expected to play an increasingly significant role in advancing electron transport technologies.

13.6.4 Stability and Longevity of Nanohybrid-Based Solar Cells

The stability and durability of nanohybrid-based solar cells are critical factors in determining their practicality and long-term performance. While nanohybrids offer significant advantages in enhancing solar cell efficiency, their ability to withstand various environmental and operational conditions is a crucial concern. To ensure the stability and longevity of nanohybrid-based solar cells, several considerations and strategies must be taken into account. These include addressing environmental stability by enhancing moisture resistance, providing corrosion protection, ensuring UV stability, managing thermal stress, ensuring chemical compatibility, and implementing anti-oxidation measures. Additionally, longevity considerations involve conducting accelerated testing, selecting durable materials, employing encapsulation and passivation layers, establishing monitoring and maintenance protocols, considering sealing and hermetic packaging, adhering to quality control standards, assessing environmental impact, and continuously investing in research and development efforts. By addressing these factors comprehensively, nanohybrid-based solar cells can maximize their return on investment and become reliable and sustainable sources of clean energy.[56,57]

13.7 CHALLENGES AND FUTURE PROSPECTS

Achieving long-term stability and durability remains a significant challenge for coordination nanohybrids. Environmental factors, including moisture, temperature fluctuations, and UV radiation, can impact the performance of nanohybrid-based solar cells over time. Scaling up the production of nanohybrids for commercial use is a complex challenge. Developing cost-effective and large-scale synthesis methods while maintaining the quality and consistency of nanohybrids is essential. Efficiently integrating nanohybrids into existing solar cell technologies, such as silicon-based or thin-film solar cells, without compromising performance can be challenging. Ensuring proper material compatibility and interfaces is crucial. The cost-effectiveness of nanohybrids, including the materials used and the synthesis process, must be addressed to make them competitive with traditional solar cell technologies.

Evaluating and mitigating the environmental impact of nanohybrid production and disposal is essential for sustainable development. This includes considering the toxicity of materials used and recycling methods. Developing industry standards for the synthesis, characterization, and performance evaluation of coordination nanohybrids is necessary for ensuring consistency and reliability across different research and manufacturing efforts. Enhancing the energy efficiency of coordination nanohybrids during the synthesis process is crucial. Energy-intensive synthesis methods can offset the benefits gained from improved solar cell efficiency.

Continued research and development efforts are likely to lead to nanohybrid-based solar cells with even higher conversion efficiencies. Precise control over material properties and interfaces will play a crucial role in achieving this. Future nanohybrids may be engineered to serve multiple functions within a solar cell, such as simultaneously improving light absorption, charge separation, and electron transport. Integration of nanohybrids into tandem solar cell designs, which stack multiple layers of materials to capture a broader spectrum of light, holds great promise for achieving high efficiencies. Development of flexible and adaptable nanohybrids will enable their use in a wide range of solar cell applications, including lightweight and flexible solar panels for various surfaces and environments. Advances in materials science and surface passivation techniques are expected to enhance the stability and longevity of nanohybrid-based solar cells, making them more practical for real-world applications. Researchers will continue to explore environmentally friendly synthesis methods and materials, reducing the ecological footprint of nanohybrid production. With ongoing research, nanohybrids are likely to move closer to commercialization, offering a viable alternative to traditional solar cell technologies. Combining nanohybrids with other renewable energy technologies, such as energy storage systems and advanced electronics, could create integrated hybrid systems with improved energy harvesting and utilization capabilities. Wider adoption of nanohybrid-based solar cells could significantly contribute to the global transition to clean and sustainable energy sources, reducing dependence on fossil fuels. In short words the coordination nanohybrids hold great promise for advancing solar cell technology. While facing challenges related to stability, scalability, and cost, ongoing research and innovation are expected to lead to more efficient, durable, and sustainable nanohybrid-based solar cells in the future. These advancements will play a vital role in addressing the growing energy demands of society while reducing the environmental impact of energy production.

13.8 POTENTIAL FOR COMMERCIALIZATION AND FUTURE GROWTH

The potential for the commercialization and future growth of coordination nanohybrids in the field of solar cells is substantial. While research in this area is ongoing, several factors suggest that coordination nanohybrids have a promising future in commercial photovoltaic applications:

1. High efficiency: Coordination nanohybrids have demonstrated the potential to significantly enhance the efficiency of solar cells. Improved light absorption, charge separation, and electron transport properties make them

attractive for commercial solar cell manufacturers looking to produce more efficient panels.

2. Versatility: Coordination nanohybrids are versatile and can be tailored for specific solar cell designs and materials. This adaptability allows them to complement various types of solar cells, including silicon-based, thin-film, perovskite, and organic solar cells.
3. Emerging solar cell technologies: As emerging solar cell technologies, such as perovskite solar cells and tandem solar cells, gain traction, coordination nanohybrids can play a pivotal role in enhancing their performance. This opens up new markets and opportunities for commercialization.
4. Environmental benefits: The potential for higher efficiency in coordination nanohybrid-based solar cells can lead to a reduced carbon footprint. This environmental benefit aligns with the growing demand for clean and sustainable energy sources.
5. Research and development: Ongoing research and development efforts are focused on addressing stability and scalability challenges associated with coordination nanohybrids. As these challenges are overcome, the commercial viability of nanohybrid-based solar cells will increase.
6. Investment and funding: The renewable energy sector continues to attract substantial investment and funding, creating opportunities for the development and commercialization of innovative technologies like coordination nanohybrids.
7. Government incentives: Many governments worldwide offer incentives and subsidies for renewable energy projects and technologies. These incentives can encourage the adoption of advanced solar cell technologies, including coordination nanohybrids.
8. Global energy transition: The global transition towards cleaner and more sustainable energy sources is driving the demand for high-efficiency solar cells. Coordination nanohybrids have the potential to meet this demand and play a crucial role in the energy transition.
9. Collaborations and partnerships: Collaborations between research institutions, universities, and commercial entities are fostering innovation and accelerating the development of coordination nanohybrid-based solar cells.
10. Consumer demand: Increasing awareness of climate change and environmental concerns has led to greater consumer demand for eco-friendly energy solutions. High-efficiency solar cells, enabled by coordination nanohybrids, can meet this demand.

13.9 CONCLUSION

In conclusion, coordination nanohybrids represent a cutting-edge approach to revolutionizing solar cell technology. These innovative materials, formed by combining nanoscale components with coordination materials, offer the potential to significantly enhance the efficiency, stability, and versatility of solar cells. Throughout this chapter, we have explored various aspects of coordination nanohybrids for solar cells, from their synthesis and characterization to their applications and future prospects.

Coordination materials, such as metal-organic frameworks (MOFs) and coordination polymers, serve as robust scaffolds that can be precisely engineered to accommodate nanoscale components, including nanoparticles, quantum dots, and nanowires. This marriage of nanoscale and coordination materials opens up new possibilities for improving key aspects of solar cell performance, such as light absorption, charge separation, and electron transport. The benefits of coordination nanohybrids in solar cells are substantial. They offer the potential for higher efficiency through enhanced light harvesting, reduced electron-hole recombination, and efficient charge transport. These materials can be tailored to absorb specific wavelengths of light, making them versatile for various solar cell designs. Coordination nanohybrids have also found applications in perovskite solar cells, dye-sensitized solar cells (DSSCs), and other emerging photovoltaic technologies. Examples of coordination material-based nanohybrids include MOF-based nanohybrids for photovoltaics, perovskite solar cells stabilized with MOFs, MOF-quantum dot nanohybrids, polymer-MOF nanohybrids, and multi-layered MOF-perovskite structures. Each of these examples showcases the potential of coordination nanohybrids to enhance solar cell performance through unique material combinations and structural designs. While coordination nanohybrids hold great promise, they also face challenges, including stability and longevity concerns, scalability issues, and cost considerations. Addressing these challenges will be crucial for realizing the full potential of nanohybrid-based solar cells. Looking ahead, the future prospects of coordination nanohybrids in solar cell technology are bright. Researchers and engineers are actively working to overcome existing challenges and push the boundaries of efficiency and sustainability. This includes improving materials, optimizing synthesis techniques, and exploring new applications, such as flexible and tandem solar cells. In an era marked by a growing need for clean and sustainable energy sources, coordination nanohybrids have the potential to play a pivotal role in transforming the solar energy landscape. As these materials continue to evolve and mature, they may contribute significantly to meeting the world's energy demands while reducing our carbon footprint. Coordination nanohybrids are a testament to the power of interdisciplinary research and innovation in shaping a more sustainable future.

In this chapter, we investigated the world of coordination nanohybrids for solar cells, exploring their synthesis, benefits, applications, and challenges. Coordination nanohybrids, formed by combining nanoscale components with coordination materials like metal-organic frameworks (MOFs) and coordination polymers, have emerged as a promising avenue for improving the efficiency and functionality of solar cells. Coordination materials serve as versatile scaffolds for incorporating nanoscale components. MOFs, in particular, offer a well-defined structure that can be tailored to enhance solar cell performance. Coordination nanohybrids offer several advantages in solar cells, including improved light absorption, efficient charge separation, and enhanced electron transport. These benefits can lead to higher solar cell efficiency. Coordination nanohybrids find applications in various types of solar cells, including perovskite solar cells, dye-sensitized solar cells (DSSCs), and tandem solar cells. Their versatility allows them to be tailored for specific solar cell designs and materials. Several examples of coordination material-based nanohybrids were discussed, including MOF-based nanohybrids for photovoltaics, perovskite solar cells stabilized with MOFs, MOF-quantum dot nanohybrids, polymer-MOF nanohybrids, and multi-layered MOF-perovskite structures. Coordination nanohybrids face challenges related to stability, scalability, and cost. Long-term stability, especially under environmental conditions, remains a key concern. Developing scalable synthesis methods while maintaining cost-effectiveness is another challenge. The future of coordination nanohybrids in solar cell technology is promising. Ongoing research aims to address stability issues, optimize synthesis techniques, and explore new applications, including flexible and tandem solar cells. Considerations for the environmental impact of nanohybrid production and disposal are essential for sustainable development. Research into environmentally friendly synthesis methods and recycling is ongoing. Coordination nanohybrids represent an innovative approach to clean energy generation. They have the potential to significantly contribute to meeting global energy demands while reducing the carbon footprint. Coordination nanohybrids offer a compelling avenue for advancing solar cell technology. As researchers and engineers continue to address challenges and explore new possibilities, coordination nanohybrids may play a pivotal role in the transition to sustainable and efficient solar energy production.

REFERENCES

[1] Rao, C.; Müller, A.; Cheetham, A. K. *The Chemistry of Nanomaterials. Synthesis, Properties and Applications* (208–275), Wiley VCH, 2004.

[2] Wallyn, J.; Anton, N.; Akram, S.; Vandamme, T. F. Biomedical imaging: Principles, technologies, clinical aspects, contrast agents, limitations and future trends in nanomedicines. *Pharmaceutical Research* **2019,** *36*, 1.

[3] Ismail, P. M.; Ali, S.; Ali, S.; Li, J.; Liu, M.; Yan, D.; Raziq, F.; Wahid, F.; Li, G.; Yuan, S.et al. Photoelectron "Bridge" in Van Der Waals heterojunction for enhanced photocatalytic CO2 conversion under visible light. *Advanced Materials* **2023,** *35* (38), 2303047.

[4] Chen, J. S.; Lou, X. W. SnO2-based nanomaterials: Synthesis and application in lithium-ion batteries. *small* **2013,** *9* (11), 1877.

[5] Khot, L. R.; Sankaran, S.; Maja, J. M.; Ehsani, R.; Schuster, E. W. Applications of nanomaterials in agricultural production and crop protection: A review. *Crop Protection* **2012,** *35*, 64.

[6] Elieh-Ali-Komi, D.; Hamblin, M. R. Chitin and chitosan: Production and application of versatile biomedical nanomaterials. *International Journal of Advanced Research* **2016,** *4* (3), 411.

[7] Ali, S.; Iqbal, R.; Wahid, F.; Ismail, P. M.; Saleem, A.; Ali, S.; Raziq, F.; Ullah, S.; Ullah, I.; Tahir et al. Cobalt coordinated two-dimensional covalent organic framework a sustainable and robust electrocatalyst for selective CO2 electrochemical conversion to formic acid. *Fuel Processing Technology* **2022,** *237*, 107451.

[8] Ali, S.; Ismail, P. M.; Wahid, F.; Kumar, A.; Haneef, M.; Raziq, F.; Ali, S.; Javed, M.; Khan, R. U.; Wu, X.et al. Benchmarking the two-dimensional conductive $Y_3(C_6X_6)_2$ (Y = Co, Cu, Pd, Pt; X = NH, NHS, S) metal-organic framework nanosheets for CO2 reduction reaction with tunable performance. *Fuel Processing Technology* **2022,** *236*, 107427.

[9] Qadir, S.; Gu, Y.; Ali, S.; Li, D.; Zhao, S.; Wang, S.; Xu, H.; Wang, S. A thermally stable isoquinoline based ultra-microporous metal-organic framework for CH4 separation from coal mine methane. *Chemical Engineering Journal* **2022,** *428*, 131136.

[10] Ali, S.; Yasin, G.; Iqbal, R.; Huang, X.; Su, J.; Ibraheem, S.; Zhang, Z.; Wu, X.; Wahid, F.; Ismail, P. M. et al. Porous aza-doped graphene-analogous 2D material a unique catalyst for CO_2 conversion to formic-acid by hydrogenation and electroreduction approaches. *Molecular Catalysis* **2022,** *524*, 112285.

[11] Li, J.; Pu, K. Development of organic semiconducting materials for deep-tissue optical imaging, phototherapy and photoactivation. *Chemical Society Reviews* **2019,** *48* (1), 38.
[12] Sun, M.; Ali, S.; Liu, C.; Dai, C.; Liu, X.; Zeng, C. Synergistic effect of Fe doping and oxygen vacancy in AgIO3 for effectively degrading organic pollutants under natural sunlight. *Environmental Pollution* **2024,** *344*, 123325.
[13] Muhammad Ismail, P.; Ali, S.; Raziq, F.; Bououdina, M.; Abu-Farsakh, H.; Xia, P.; Wu, X.; Xiao, H.; Ali, S.; Qiao, L. Stable and robust single transition-metal atom catalyst for CO2 reduction supported on defective WS2. *Applied Surface Science* **2023**, 157073.
[14] Bao, L.; Ren, X.; Liu, C.; Liu, X.; Dai, C.; Yang, Y.; Bououdina, M.; Ali, S.; Zeng, C. Modulating the doping state of transition metal ions in ZnS for enhanced photocatalytic activity. *Chemical Communications* **2023**, *59*, 11280–11283.
[15] Wahid, F.; Ali, S.; Ismail, P. M.; Raziq, F.; Ali, S.; Yi, J.; Qiao, L. Metal single atom doped 2D materials for photocatalysis: Current status and future perspectives. *Progress in Energy* **2022**, *5*, 012001.
[16] Ali, S.; Iqbal, R.; Khan, A.; Rehman, S. U.; Haneef, M.; Yin, L. Stability and catalytic performance of single-atom catalysts supported on doped and defective graphene for CO_2 hydrogenation to formic acid: A first-principles study. *ACS Applied Nano Materials* **2021,** *4* (7), 6893.
[17] Kannan, N.; Vakeesan, D. Solar energy for future world:-A review. *Renewable and Sustainable Energy Reviews* **2016,** *62*, 1092.
[18] Wang, C.; Wang, W.; Tan, J.; Zhang, X.; Yuan, D.; Zhou, H.-C. Coordination-based molecular nanomaterials for biomedically relevant applications. *Coordination Chemistry Reviews* **2021,** *438*, 213752.
[19] Zhuhra, Z.; Ali, S.; Ali, S.; Xu, H.; Wu, R.; Tang, Y. Exceptionally amino-quantitated 3D MOF@CNT-sponge hybrid for efficient and selective recovery of Au(III) and Pd(II). *Chemical Engineering Journal* **2024**, *492*, 152127.
[20] Abbas, Y.; Ali, S.; Ali, S.; Zareen, Z.; Haoliang, W.; Bououdina, M.; Zhenzhong, S. Amine-rich cyclotriphosphazene (P_3N_3) nano-cages for enhanced and selective Au(III) and Pd(II) recovery and hydrogen generation: Waste-to-resource tactics. *Chemical Engineering Journal* **2022,** *431* (4), 133367.
[21] Khan, S. Phase engineering and impact of external stimuli for phase tuning in 2D materials. *Advanced Energy Conversion Materials* **2023,** *5* (1), 40.
[22] Khan, S.; Rahman, M.; Marwani, H. M.; Althomali, R. H.; Rahman, M. M. Bicomponent polymorphs of salicylic acid, their antibacterial potentials, intermolecular interactions, DFT and docking studies. *Zeitschrift für Physikalische Chemie* **2023,** *238* (01), 1.
[23] Lizundia, E.; Puglia, D.; Nguyen, T.-D.; Armentano, I. Cellulose nanocrystal based multifunctional nanohybrids. *Progress in Materials Science* **2020,** *112*, 100668.
[24] Gul, Z.; Salman, M.; Khan, S.; Shehzad, A.; Ullah, H.; Irshad, M.; Zeeshan, M.; Batool, S.; Ahmed, M.; Altaf, A. A. Single organic ligands act as a bifunctional sensor for subsequent detection of metal and cyanide ions, a statistical approach toward coordination and sensitivity. *Critical Reviews in Analytical Chemistry* **2023**, 1.
[25] Shah, R.; Ali, S.; Ali, S.; Xia, P.; Raziq, F.; Adnan; Mabood, F.; Shah, S.; Zada, A.; Ismail, P. M.et al. Amino functionalized metal-organic framework/rGO composite electrode for flexible Li-ion batteries. *Journal of Alloys and Compounds* **2023,** *936*, 168183.
[26] Shah, R.; Ali, S.; Raziq, F.; Ali, S.; Ismail, P. M.; Shah, S.; Iqbal, R.; Wu, X.; He, W.; Zu, X.et al. Exploration of metal organic frameworks and covalent organic frameworks for energy-related applications. *Coordination Chemistry Reviews* **2023,** *477*, 214968.
[27] Wu, L.; Li, Y.; Zhou, B.; Liu, J.; Cheng, D.; Guo, S.; Xu, K.; Yuan, C.; Wang, M.; Hong Melvin, G. J.et al. Vertical graphene on rice-husk-derived SiC/C composite for highly selective photocatalytic CO2 reduction into CO. *Carbon* **2023,** *207*, 36.

[28] Abid, A.; Haneef, M.; Ali, S.; Dahshan, A. A study of 2H and 1T phases of Janus monolayers and their van der Waals heterostructure with black phosphorene for optoelectronic and thermoelectric applications. *Journal of Solid State Chemistry* **2022,** *311*, 123159.

[29] Ali, S.; Ali, S.; Ismail, P. M.; Shen, H.; Zada, A.; Ali, A.; Ahmad, I.; Shah, R.; Khan, I.; Chen, J.et al. Synthesis and bader analyzed cobalt-phthalocyanine modified solar UV-blind β-Ga_2O_3 quadrilateral nanorods photocatalysts for wide-visible-light driven H_2 evolution. *Applied Catalysis B: Environmental* **2022,** *307*, 121149.

[30] Ali, S.; Ismail, P. M.; Humayun, M.; Bououdina, M.; Qiao, L. Tailoring 2D metal-organic frameworks for enhanced CO_2 reduction efficiency through modulating conjugated ligands. *Fuel Processing Technology* **2024,** *225*, 108049.

[31] Raziq, F.; Khan, K.; Ali, S.; Ali, S.; Xu, H.; Ali, I.; Zada, A.; Muhammad Ismail, P.; Ali, A.; Khan, H.et al. Accelerating CO_2 reduction on novel double perovskite oxide with sulfur, carbon incorporation: Synergistic electronic and chemical engineering. *Chemical Engineering Journal* **2022,** *446*, 137161.

[32] Raziq, F.; Aligayev, A.; Shen, H.; Ali, S.; Shah, R.; Ali, S.; Bakhtiar, S. H.; Ali, A.; Zarshad, N.; Zada, A. et al. Exceptional photocatalytic activities of rGO modified (B, N) co-doped WO3, coupled with CdSe QDs for one photon Z-scheme system: A joint experimental and DFT study. *Advanced Science* **2022,** *9* (2), 2102530.

[33] Chel, A.; Kaushik, G. Renewable energy for sustainable agriculture. *Agronomy for Sustainable Development* **2011,** *31*, 91.

[34] Khan, S.; Ajmal, S.; Hussain, T.; Rahman, M. U. Clay-based materials for enhanced water treatment: Adsorption mechanisms, challenges, and future directions. *Journal of Umm Al-Qura University for Applied Sciences* **2023**, 1.

[35] Brennan, L. J.; Byrne, M. T.; Bari, M.; Gun'ko, Y. K. Carbon nanomaterials for dye-sensitized solar cell applications: A bright future. *Advanced Energy Materials* **2011,** *1* (4), 472.

[36] Ali, S.; Zareen, Z.; Ali, S.; Qi, H.; Ali, S.; Muhammad, A.; Zhongying, W. Ultra-deep removal of Pb by functionality tuned UiO-66 framework: A combined experimental, theoretical and HSAB approach. *Chemosphere* **2021,** *284*, 131305.

[37] Khan, S.; Iqbal, A. Organic polymers revolution: Applications and formation strategies, and future perspectives. *Journal of Polymer Science and Engineering* **2023,** *6* (1), 3125.

[38] Khan, S.; Zahoor, M.; Rahman, M. U.; Gul, Z. Cocrystals; basic concepts, properties and formation strategies. *Zeitschrift für Physikalische Chemie* **2023,** *237* (3), 273.

[39] Yang, A.-N.; Lin, J. T.; Li, C.-T. Electroactive and sustainable Cu-MoF/PEDOT composite electrocatalysts for multiple redox mediators and for high-performance dye-sensitized solar cells. *ACS Applied Materials & Interfaces* **2021,** *13* (7), 8435.

[40] Zheng, S.; Li, Q.; Xue, H.; Pang, H.; Xu, Q. A highly alkaline-stable metal oxide@ metal–organic framework composite for high-performance electrochemical energy storage. *National Science Review* **2020,** *7* (2), 305.

[41] Chuhadiya, S.; Suthar, D.; Patel, S.; Dhaka, M. Metal organic frameworks as hybrid porous materials for energy storage and conversion devices: A review. *Coordination Chemistry Reviews* **2021,** *446*, 214115.

[42] Chueh, C.-C.; Chen, C.-I.; Su, Y.-A.; Konnerth, H.; Gu, Y.-J.; Kung, C.-W.; Wu, K. C.-W. Harnessing MOF materials in photovoltaic devices: Recent advances, challenges, and perspectives. *Journal of Materials Chemistry A* **2019,** *7* (29), 17079.

[43] Heo, D. Y.; Do, H. H.; Ahn, S. H.; Kim, S. Y. Metal-organic framework materials for perovskite solar cells. *Polymers* **2020,** *12* (9), 2061.

[44] Dou, J.; Li, Y.; Xie, F.; Ding, X.; Wei, M. Metal–organic framework derived hierarchical porous anatase TiO2 as a photoanode for dye-sensitized solar cell. *Crystal Growth & Design* **2016,** *16* (1), 121.

[45] Maza, W. A.; Haring, A. J.; Ahrenholtz, S. R.; Epley, C. C.; Lin, S.; Morris, A. J. Ruthenium (II)-polypyridyl zirconium (IV) metal–organic frameworks as a new class of sensitized solar cells. *Chemical Science* **2016,** *7* (1), 719.

[46] Rahman, F. U.; Khan, S.; Rahman, M. U.; Zaib, R.; Rahman, M. U.; Ullah, R.; Zahoor, M.; Kamran, A. W. Effect of ionic strength on DNA–dye interactions of Victoria blue B and methylene green using UV–visible spectroscopy. *Zeitschrift für Physikalische Chemie* **2023**, *238* (1), 173–186.

[47] Sarkar, S.; Makhal, A.; Bora, T.; Lakhsman, K.; Singha, A.; Dutta, J.; Pal, S. K. Hematoporphyrin–ZnO nanohybrids: Twin applications in efficient visible-light photocatalysis and dye-sensitized solar cells. *ACS Applied Materials & Interfaces* **2012,** *4* (12), 7027.

[48] Gong, J.; Liang, J.; Sumathy, K. Review on dye-sensitized solar cells (DSSCs): Fundamental concepts and novel materials. *Renewable and Sustainable Energy Reviews* **2012,** *16* (8), 5848.

[49] Ansari, A. A.; Nazeeruddin, M.; Tavakoli, M. M. Organic-inorganic upconversion nanoparticles hybrid in dye-sensitized solar cells. *Coordination Chemistry Reviews* **2021,** *436*, 213805.

[50] Sengupta, D.; Das, P.; Mondal, B.; Mukherjee, K. Effects of doping, morphology and film-thickness of photo-anode materials for dye sensitized solar cell application–A review. *Renewable and Sustainable Energy Reviews* **2016,** *60*, 356.

[51] Ali, S.; Ismail, P. M.; Khan, M.; Dang, A.; Ali, S.; Zada, A.; Raziq, F.; Khan, I.; Khan, M. S.; Ateeq, M.et al. Charge transfer in TiO2-based photocatalysis: Fundamental mechanisms to material strategies. *Nanoscale* **2024**, *16*, 4352–4377.

[52] Li, Q.; Lin, J.; Wu, J.; Lan, Z.; Wang, Y.; Peng, F.; Huang, M. Enhancing photovoltaic performance of dye-sensitized solar cell by rare-earth doped oxide of Lu2O3:(Tm3+, Yb3+). *Electrochimica Acta* **2011,** *56* (14), 4980.

[53] Ray, R.; Sarkar, A. S.; Pal, S. K. Improving performance and moisture stability of perovskite solar cells through interface engineering with polymer-2D MoS2 nanohybrid. *Solar Energy* **2019,** *193*, 95.

[54] Zhou, Y.; Zhang, T.; Zhu, W.; Qin, L.; Kang, S.-Z.; Li, X. Enhanced light absorption and electron transfer in dimensionally matched carbon nitride/porphyrin nanohybrids for photocatalytic hydrogen production. *Fuel* **2023,** *338*, 127394.

[55] Guldi, D. M.; Rahman, G. A.; Qin, S.; Tchoul, M.; Ford, W. T.; Marcaccio, M.; Paolucci, D.; Paolucci, F.; Campidelli, S.; Prato, M. Versatile coordination chemistry towards multifunctional carbon nanotube nanohybrids. *Chemistry–A European Journal* **2006,** *12* (8), 2152.

[56] Wang, D.; Baral, J. K.; Zhao, H.; Gonfa, B. A.; Truong, V. V.; El Khakani, M. A.; Izquierdo, R.; Ma, D. Controlled fabrication of PbS quantum-dot/carbon-nanotube nanoarchitecture and its significant contribution to near-infrared photon-to-current conversion. *Advanced Functional Materials* **2011,** *21* (21), 4010.

[57] Narayanasamy, M.; Hu, L.; Kirubasankar, B.; Liu, Z.; Angaiah, S.; Yan, C. Nanohybrid engineering of the vertically confined marigold structure of rGO-VSe2 as an advanced cathode material for aqueous zinc-ion battery. *Journal of Alloys and Compounds* **2021,** *882*, 160704.

14 Coordination Nanomaterials for Nanogenerators Technology

Harimohan Sharma, Monika Singh, Dipak Kumar Das, Jasvinder Kaur, Anuj Kumar and Ghulam Yasin

14.1 INTRODUCTION

14.1.1 NANOGENERATORS: AN EMERGING TECHNOLOGY

Due to the rapid growth of the Internet of Things (IoT) over the past two decades, enormous small electronic devices such as sensors, actuators, and wireless transmitters have become ubiquitous across the globe. [1] These devices have a wide range of applications, including health monitoring, biochemical detection, environmental protection, remote control, wireless transmission, and security. A portable, eco-friendly, and easily accessible power source is necessary due to the modest power requirements of these devices, which range from microwatts (μW) to milliwatts (mW). [2] Historically, batteries have been commonly used to provide the necessary energy for operating these devices. [3] Although conventional lithium-ion batteries are functional, they have drawbacks in terms of environmental impact and require regular battery replacement. It is imperative to include existing renewable energy sources found in nature and the human body, such as wind, wave, solar, acoustic, kinetic, and thermal energy. [4] To ensure a sustainable future, it is of utmost importance to identify and implement practical substitutes for traditional batteries. The critical nature of this situation has prompted the worldwide scientific community to emphasize not only the development of alternatives that effectively harness green energy but also the search for highly efficient energy storage technologies with broader applications in self-powered electronic devices and systems.

The milestone work has been carried out by Wang et al. [5] addressing the significant and ongoing need for energy and technology for small-scale power applications. This research has led to the development of compact electronics that are equipped with self-sustaining, long-lasting, and low-maintenance power sources. The energy requirements of a nation, city, or airline, which are quantified in megawatts (MW) or gigawatts (GW), are significantly distinct from the concept of nano-energy

 DOI: 10.1201/9781003345886-16

(Figure 14.1(a)). To efficiently transform these many forms of small-scale energy into electrical power, Wang et al. have pioneered the invention of the first NG and have subsequently created multiple variations of NGs over the past ten years. These NGs are based on the principles of piezoelectricity, triboelectricity, pyroelectricity, thermoelectricity, and ion streams, among others. In addition, NGs can incorporate energy storage units like batteries and super-capacitors into a self-powered system, which can then be used to provide power to functional devices [6]. In this system, the energy collected by the NGs can be stored temporarily in energy storage units. Subsequently, the energy can be utilized for various purposes such as environmental sensing, transmission of data processing, and controlling (Figure 14.1(b)). Wang et al. utilized a micro/nanobelt made of Sb-doped ZnO to fabricate a thermoelectric nanogenerator [7]. Recently, there has been greater attention to nanogenerators that exploit piezoelectric and triboelectric characteristics, owing to their practicality in self-sufficient electronic systems. Specifically, there has been a rise in innovative power solutions for the IoT and wearable devices. Additionally, the advancement of medical monitors and human-machine interface (HMI) systems has been greatly expedited by the introduction of self-powered sensors and sensing systems [8–10]. The working mechanism of the PENG is based on the fundamental idea that piezoelectric materials generate electric charges in response to mechanical stress [11]. This technology demonstrates the capability to effectively convert small-scale energy from various environmental sources into electrical power. NGs possess several advantages compared to traditional generators due to their unique properties, including cost-effectiveness, high dependability, simple structure, diverse shapes, and exceptional flexibility. They have a revolutionary impact in different fields by enabling the development of self-powered electronic devices and systems, offering an efficient and sustainable means of harvesting energy from the environment. Notably, NGs possess a diverse range of applications, including the harvesting of mechanical energy, powering wearable electronic devices, self-powered sensors, IoT devices,

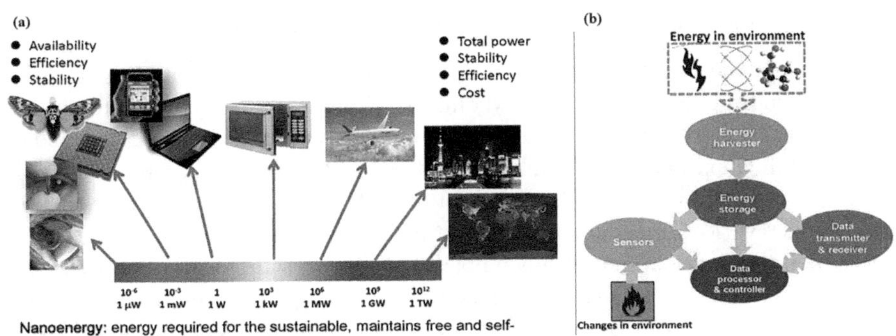

FIGURE 14.1 (a) The power levels and corresponding applications. Taken from [21], copyright 2013 The American Chemical Society. (b) The energy conversion process including energy harvesting by nanogenerators, energy storage, and applications. Taken from [6], copyright 2011 The American Chemical Society.

environmental monitoring, biomedical devices, flexible and transparent electronics, and contributing to the establishment of sustainable energy solutions by harnessing power from the surroundings rather than relying on conventional power grids. Recent developments in nanotechnology, developing efficient CNs, have facilitated the development of NGs, which transform mechanical energy into electricity mainly based on two effects: piezoelectricity and triboelectricity.

Within this framework, Yang et al. reported to have prepared ZnO nanowires to construct a pyroelectric nanogenerator capable of harvesting thermoelectric energy [12]. In this line, Wang et al. [13] developed the first piezoelectric ZnO nanowire (NW)-based power generator (PENG), which demonstrated a power efficiency of 17–30%. Computational studies also play a key role in researching materials and their mechanisms [14–16]. Reed et al. used density-functional theory to find that MoS_2, $MoSe_2$, $MoTe_2$, WS_2, and WSe_2 had interesting piezoelectric properties on a monolayer [17]. Wang et al. [18] conducted the initial experimental study on the piezoelectric properties of monolayer MoS_2 and published their findings. During the bending of the PET flexible substrate, the monolayer MoS_2 flake exhibited an output voltage of 15 mV, a current of 20 pA, and a power density of 2 $mW{\cdot}m^{-2}$. The bending resulted in a deformation of 0.53%. The combination of piezoelectric materials with flexible materials allows for the creation of flexible devices, which offer numerous possibilities, such as flexible electronics [19, 20].

The previously mentioned investigations have demonstrated than CNs can significantly influence the functioning of TENGs and PENGs. However, there is a paucity of literature concerning the characteristics and features of CNs that are relevant to their application in NGs. This chapter provides a comprehensive summary of the most recent advancements in the domain of NGs utilizing CNs. Additionally, relevant characteristics of CNs for their potential application in PENGs and TENGs are discussed in detail. Furthermore, the present study delineates the current state of development, potential applications, and challenges associated with materials in the context of sensing technology and NGs.

14.1.2 Role of Coordination Nanomaterials (CNs) in NGs

The development of NGs has been accompanied by several material-related challenges, including the integration of NG materials into existing systems and devices, the maintenance of long-term reliability and efficiency by ensuring mechanical durability, and the flexibility and stretchability of NG materials for wearable and flexible electronics. Additionally, the development of materials that can sustain their properties under various mechanical deformations, the improvement of low-efficiency levels, and the development of biocompatible NG materials for direct contact with biological systems are all crucial factors for practical applications. To enable large-scale production, scalable and cost-effective synthesis methods for NG materials are required, and these materials must remain stable under various environmental conditions, including temperature, humidity, and gas exposure. The tailoring of materials to exhibit desired electrical, mechanical, and piezoelectric properties is also crucial for optimal NG performance, and finding cost-effective alternatives and sustainable materials is important for widespread adoption. The lack of standardized testing methods can lead to inconsistencies in reported performance metrics, and the understanding and mitigation of the effects of wear and fatigue are essential for the durability of NG devices. Overcoming these challenges requires collaborative efforts among materials scientists, physicists, engineers, and other experts. Advances in material science, fabrication techniques, and a deeper understanding of the fundamental properties of NG materials will contribute to addressing these challenges and realizing the full potential of NG technologies. These problems and challenges have prompted researchers to investigate some novel materials for addressing these challenges related to materials used in NG. CNs have provided a promising solution to address these problems.

CNs can offer solutions to some of the challenges related to materials used in NG. By tailoring the coordination environment and choosing appropriate ligands, researchers can design CNs with properties optimized for NG applications. Some CNs, such as metal-organic frameworks (MOFs) and coordination polymers, can be synthesized in flexible and stretchable forms. This flexibility makes them suitable for integration into flexible NG devices, especially in applications where mechanical deformation is prevalent. The tunable nature of CNs allows for the selection of biocompatible components. Some CNs can be synthesized using scalable methods, addressing challenges related to large-scale production. CNs often exhibit robust and stable structures, addressing concerns related to material degradation. The unique electronic properties of certain CNs can contribute to high energy conversion efficiency in NG. The ability to tailor electronic structures allows for the optimization of charge transport and energy conversion processes. CNs can be integrated into NG devices using a variety of fabrication techniques. CNs can be designed with mechanical stability in mind. Addressing concerns related to the environmental impact of NG materials, sustainable synthesis methods for CNs can be explored. Additionally, ongoing research in the field of CNs continues to explore new materials and synthesis strategies to overcome existing challenges and further enhance the performance of NG.

14.2 COORDINATION NANOMATERIALS (CNs)

CNs like transition metals oxides (TMOs), sulfides (TMSs), phosphides (TMPs), MXenes, and metal-organic coordination polymers (MOCPs) and metal-organic framework (MOF) have played a crucial role in the advancement of NGs technologies due to their unique electronic, structural, and mechanical properties. Here are some examples of CNs used in nanogenerators. Certain coordination polymers exhibit piezoelectric or ferroelectric properties that can be utilized in nanogenerators. Layered coordination compounds, such as certain transition metal dichalcogenides, have been explored for their unique electronic and mechanical properties. Nanoparticles with a coordination structure, such as certain metal oxides or chalcogenides, can be synthesized to exhibit piezoelectric or ferroelectric properties. One-dimensional coordination polymer nanowires can possess enhanced piezoelectric properties.

CNs possess several characteristics that make them beneficial for applications in NG. These materials, formed by the coordination of metal ions with organic ligands, exhibit unique properties at the nanoscale that can be harnessed for efficient energy conversion. Here are some key characteristics of CNs that are advantageous in NG.

CNs often allow precise tuning of their electronic structure by choosing specific metal ions and ligands. This tunability can be advantageous for optimizing charge transport and energy conversion processes in NG. Some CNs exhibit piezoelectric or ferroelectric properties, enabling them to convert mechanical energy into electrical energy efficiently. This characteristic is crucial for the operation of NG, especially those relying on piezoelectric effects for energy harvesting. CNs come in a wide variety of structures, including MOFs, coordination polymers, and complex ions. This structural diversity provides options for selecting materials tailored to specific NG applications. The flexibility in ligand design allows for the tailoring of mechanical properties, making coordination nanomaterials suitable for applications where flexibility and mechanical durability are important, such as in flexible NG devices. MOFs, a type of CNs, often exhibit high surface area due to their porous structures. This characteristic can be advantageous in applications where surface interactions are critical, such as in the adsorption of mechanical stress for piezoelectric NG. CNs often involve strong coordination bonds, contributing to their stability and durability under various environmental conditions. This characteristic is essential for the longevity and reliability of NG. The choice of ligands and metals in CNs can be tailored to enhance biocompatibility, making them suitable for integration into biomedical devices or for use in wearable technologies that come into contact with the human body. The synthesis of CNs can be scalable, facilitating their production in larger quantities. This characteristic is important for practical applications where NG may be deployed on a larger scale.

14.3 PROPERTIES OF COORDINATION NANOMATERIALS

14.3.1 Electrical Properties

CNs, including MOFs and coordination polymers, are increasingly being investigated for their potential use in nanogenerator applications. These devices are designed to convert mechanical energy into electrical energy, and CNs possess a range of electrical properties that make them suitable for this purpose. Piezoelectric properties, for example, allow certain CNs to generate an electric charge in response to mechanical deformation or pressure. This property is crucial for nanogenerators, as it enables the conversion of mechanical energy into electrical energy. Coordination of nanomaterials with intrinsic electrical conductivity can also aid in the efficient transfer and collection of electrical charges, improving overall performance. Semiconducting CNs are particularly useful in nanogenerators due to their tunable band gaps, which allow for controlled electrical conductivity and better regulation of the energy conversion process. The dielectric constant of is also important for capacitive nanogenerators, as high dielectric constants can increase energy storage capacity and efficiency. Ferroelectric behavior in coordination nanomaterials, where they possess a spontaneous electric polarization that can be reversed by an external electric field, can also enhance the efficiency of nanogenerators. Furthermore, the coordination of nanomaterials with photoconductive properties can generate electrical charges in response to light exposure, making them useful for applications where both mechanical and light energy can be harnessed for energy conversion. Thermoelectric properties are also exhibited by certain coordination polymers, allowing them to convert temperature differences into electrical energy. High carrier mobility is crucial for efficient charge transport within the material, and coordination of nanomaterials with good carrier mobility can contribute to the overall effectiveness of nanogenerators in converting mechanical energy to electrical energy. Additionally, some CNs can be used as electrode materials in electrochemical nanogenerators, with their electrochemical properties contributing to efficient energy conversion. Overall, the various electrical properties of CNs make them promising candidates for the development of high-performance nanogenerators.

14.3.2 Mechanical Properties

Piezoelectric materials possess the ability to produce electrical charges in response to mechanical deformation. In recent years, there has been a growing interest in using CNs as the active component in piezoelectric nanogenerators due to their unique properties. Some CNs exhibit piezoelectric properties, which make them suitable for energy harvesting applications. The generation of electrical energy in these materials occurs due to charge separation induced by mechanical stress, such as bending or compressing the material. To be effective as nanogenerators, CNs must possess flexible and mechanically robust structures. Flexibility enables the material to withstand mechanical deformation, while high bending strength ensures the structural integrity of the material during bending or flexing movements. Elastic materials can deform reversibly under mechanical stress and return to their original shape, which is advantageous for nanogenerator applications that involve cyclic mechanical deformations. Elastic CNs demonstrate the ability to withstand repeated strains without permanent deformation. Tensile strength is another essential property for CNs used in nanogenerator devices, as it ensures the durability and reliability of the material during mechanical tension. Some CNs exhibit specific mechanical resonance frequencies that can optimize the energy conversion process. Utilizing materials with resonance matching the frequency of ambient mechanical vibrations can enhance the efficiency of nanogenerators. Compliance is yet another crucial property for CNs used in nanogenerators. Materials with good compliance can efficiently capture mechanical energy from irregular or varying mechanical motions. Low frictional properties can reduce energy losses in mechanical components of nanogenerators, making materials with suitable frictional properties preferred for energy harvesting applications that involve sliding or rubbing motions. Good adhesion properties are also important for ensuring that the CNs stay in place and maintain their structural integrity during mechanical deformation. Proper adhesion is crucial for the long-term stability and performance of nanogenerators. By understanding and tailoring these mechanical properties, CNs can be optimized for efficient energy conversion in nanogenerator devices. The ability to design and manipulate these properties makes CNs promising candidates for advancements in the field of nanogenerators and energy harvesting at the nanoscale.

14.3.3 Surface/Geometrical Features

The performance of CNs in nanogenerators, which involve the conversion of mechanical energy into electrical energy, is heavily influenced by their geometrical and structural properties. In this context, several properties are relevant to nanogenerator applications, including the overall crystal structure of CNs, their porosity, unit cell dimensions, flexibility, elasticity, specific nanostructure morphology, presence of defects and dislocations, and crystal symmetry. Proper tuning of these properties can optimize the mechanical response of the material to external forces, improve its performance in nanogenerators, and achieve the right balance of mechanical flexibility, stability, and electrical conductivity for efficient energy conversion. CNs with unique crystal structures, optimized porosity, and tailored morphological features can enhance their surface area and accessibility for mechanical interactions and provide enhanced piezoelectric effects or influence charge transport mechanisms. The flexibility and elasticity of CNs are essential for nanogenerators that undergo mechanical deformation. Furthermore, certain crystal symmetries may exhibit preferential directions for mechanical deformation or charge transport, which can be exploited through careful design and synthesis. In summary, the geometrical and structural properties of CNs play a crucial role in determining their suitability for nanogenerator applications.

14.4 COORDINATION NANOMATERIALS FOR NANOGENERATORS

14.4.1 TMO-Based NGs

Transition metal oxides (TMOs) have proven to be beneficial in nanogenerators due to their unique properties and potential applications. TMOs, such as transition metal oxide-based nanocomposites (TMONCs), offer enhanced surface area, increased surface activities, and reduced electron-hole recombination, which contribute to improved practical application efficiencies. Additionally, TMOs exhibit resistive switching properties, multiple metal-insulator transitions (MITs), and hysteresis, providing extra mechanisms for memory applications and offering additional functionalities in permissive hardware. The morphology of TMOs can be controlled by adjusting reaction parameters, allowing for tailored properties and applications. For instance, ZnO, a semiconductor, possesses a lattice constant of 0.325 and a radius of 0.52 nm. It exhibits a hexagonal wurtzite structure and belongs to the C6v-P63mc space group. At room temperature, it exhibits a significant electron-hole binding energy of 60 meV and a large, direct band gap of 3.37 eV [22]. The predominant bonding in its structure is ionic, with Zn^{2+} having a radius of 0.074 nm and O^{2-} having a radius of 0.140 nm. The presence of polar Zn-O bonds induces a significant piezoelectric effect on the z-axis, resulting in opposing charges on the oxygen and zinc planes. The extensive use of ZnO nanoparticles with one-dimensional structures, such as nanowires (NWs), nanoribbons, nanorods, and nanobelts, together with their remarkable physical and chemical properties, has stimulated significant research in this field [23, 24].

Several methods have been explored for the production of one-dimensional ZnO nanostructures, including physical vapour deposition, electrospinning, molecular beam epitaxy, wet chemical processes, and molecular beam epitaxy [25]. Figure 14.2 (a) ((i)-(ii)) displays images of ZnO nanostructures that were produced using thermal vapour deposition and photographed using typical TEM and SEM equipment. The nanobelt's width (Figure 14.2 (a)), remains consistent over its entirety, depicted in

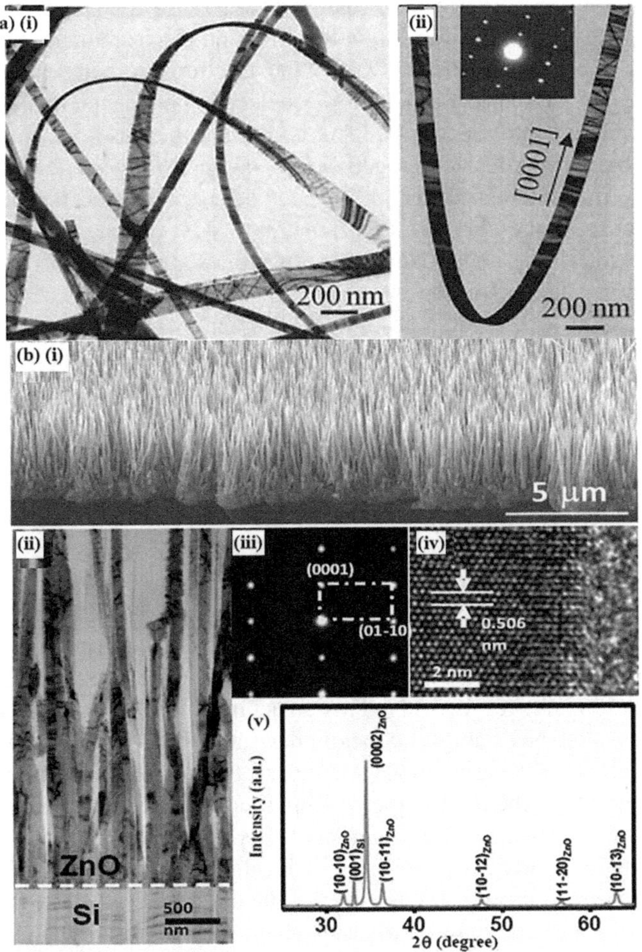

FIGURE 14.2 Characterization of one-dimensional ZnO nanostructures. Panel A: (i) TEM images and (ii) selected-area electron diffraction pattern of ZnO nanobelts showing their geometrical shape. Reproduced with permission. Copyright 2001, American Association for the Advancement of Science. Panel B: (i) SEM image of aligned phosphorus doped ZnO NW arrays grown on Si substrate; (ii) cross-section TEM image of phosphorus-doped NW arrays; (iii) selected-area electron diffraction pattern from a single phosphorus-doped ZnO NW; (iv) HRTEM image from the side of a phosphorus-doped NW, and (v) x-ray powder diffraction (XRD) pattern from phosphorus-doped ZnO NW arrays. Reproduced with permission [26] opyright 2009, American Chemical Society.

length. The width-to-thickness ratio normally falls between 5 and 10, with the width varying from 50 to 300 nm.

The ZnO nanowires, which were doped with phosphorus, were cultivated on a silicon substrate with a (001) orientation. These nanowires had a length of 1 μm and a diameter of roughly 50 nm, as depicted in Figure 14.2 (b) ((i)-(v)) [26]. The creation and administration of ZnO-based NGs require the specified area electron diffraction pattern and high-resolution transmission electron microscopy (HRTEM) photographs to confirm that these ZnO nanostructures are single-crystalline. ZnO NW arrays were first used to make a piezoelectric nanogenerator (NG) that turns mechanical energy into electricity. This is possible because the piezoelectric and semiconducting properties work together to make the NG more powerful [1].

Following the initial discovery [27], people have observed the development of both direct current and alternating current ZnO NGs. Graphene, silicon, polyester, indium tin oxide (ITO), and AlN/sapphire substrates are some substrates that can be used to incorporate ZnO-based NGs. Kim et al. [28] employed a low-temperature solution approach to cultivate one-dimensional ZnO nanorods on two-dimensional graphene, with the aim of producing a heterogeneous NG. The ZnO nanorods shown in Figure 14.3 (a) are single-crystalline and have a diameter smaller than 100 nm and a length of about 2 mm. Their density is around 20 mm2, and they exhibit a preference for developing in the [1] direction on the 2D graphene. This cutting-edge NG has the ability to operate consistently, even under continuous bending and rolling conditions. A team led by Zhu et al. [29] recently used photolithography to add a growth-limited NG made of piezoelectric zinc oxide (ZnO) nanowire (NW) arrays to a photonic crystal. Figure 14.3 (b) shows the results of the position-controlled preparation, which showed that ZnO nanowires only formed in the spots marked by photolithography. The maximum power density achieved was 0.78 W cm^3, while the open-circuit voltage and peak short-circuit current reached unprecedented levels of 58 V and 134 μA, respectively. This structure lets each unit work on its own, which improves the NG's ability to handle flaws and the ZnO NWs' ability to collect energy efficiently. NGs based on ZnO nanostructures have four distinct characteristics that differentiate them from their competitors: i) Excessive elastic deformation does not pose a challenge for 1D nanomaterials. ii) The narrow dimensions of 1D NWs make them theoretically free from dislocations, indicating that their resistance to fatigue is likely to prolong the lifespan of the device. iii) Because they have a low elastic modulus and can deform when mechanical loads are applied to them from the outside, 1D nanomaterials can efficiently store energy even when there are only small mechanical disturbances. ZnO nanoparticles in a one-dimensional array can be easily synthesized on various substrates. Substrates can exhibit the morphology of nanowires (NWs), nanorods, nanoribbons, or nanobelts. Possible substrates with different levels of crystallinity and morphologies include silicon, graphene, polymers, and other nanomaterials. In order for these materials to be utilized in technology, it is crucial that the synthetic approach be highly user friendly.

14.4.2 TMSs-Based NGs

Transition metal sulfides have proven to be beneficial in nanogenerators due to their unique properties and characteristics. These materials have been shown to exhibit high electrocatalytic activity for energy storage and energy generation applications. For example, transition metal-doped MnS nanoparticles have demonstrated satisfactory specific capacitances and superior performance for the water-splitting process. Metal-gallium arsenide tribovoltaic nanogenerators (MG-TVNGs) based on

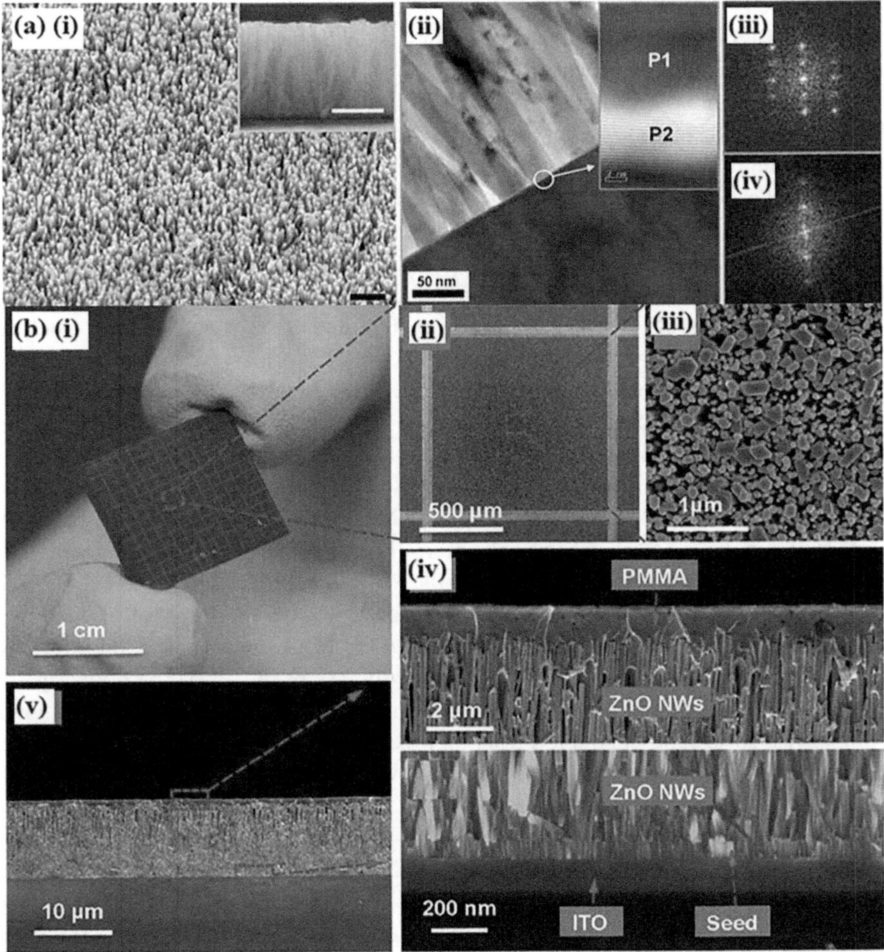

FIGURE 14.3 Panel (a):1D ZnO nanorods on 2D graphene sheets. (i) SEM image of the ZnO nanorods; the inset is a cross-section image of the sample. The entire scale bar is 1 µm; (ii) TEM and HRTEM (inset) image, the area is ZnO (P1) and graphene (P2) layers, respectively; (iii)-(iv) Fast Fourier transformation patterns of ZnO and graphene, respectively. Reproduced with permission. [28] Copyright 2010, Wiley. Panel (b): (i) Optical photograph of the ZnO NWs grown on graphene; and SEM images of (ii) one segment designated by photolithography; (iii) the prepared ZnO NWs from a top view; (iv) ZnO NWs spin-coated in a PMMA layer; (v) enlarged image of (iv) showing the interface between ZnO NWs and PMMA layer, and (vi) enlarged image of (iv) showing the interface between ZnO seeds and ITO layer. Reproduced with permission. [29] Copyright 2012, American Chemical Society.

transition metal sulfides have also shown significantly higher DC output compared to silicon-based devices. Additionally, compositing transition metal sulfides with highly conductive supports, such as reduced graphene oxide (rGO), has been found to enhance their oxygen evolution reaction (OER) catalytic activity. Furthermore, colloidal metal sulfide nanoparticles have been utilized in nanogenerators to overcome challenges related to parasitic reactions and improve cyclability, making them promising candidates for energy storage systems.

A transition metal dichalcogenide (TMD) is a type of two-dimensional semiconductor with the formula MX_2, where M stands for the transition metal's atomic number and X stands for the number of chalcogen atoms (such as S, Se, Te, etc.). The band gap of TMDs is modulated by their thickness, akin to the utilization of graphene as a semiconductor without a band gap. Nevertheless, the 2D materials possess the same advantageous characteristics as graphene, including transparency, stability, and flexibility. The thickness [30, 31] influences a change in the direct/indirect transition. The characteristics of TMDs make them well-suited for a range of uses, such as energy harvesting, integrated circuits, and flexible electronics. Seol et al. conducted the initial thorough examination of the triboelectric charging characteristics of various innovative triboelectric materials, such as WS_2, $MoSe_2$, WSe_2, and MoS_2, utilizing the concept of a triboelectric nanogenerator [32]. The polarity of the voltage was assessed by employing friction layers composed of nylon and MoS_2. According to their findings, the contact between nylon and MoS_2 resulted in the electrification of MoS_2 with a negative charge. Similarly, many TENG shapes were created by mixing MoS_2 with six various triboelectric materials: PC, mica, PET, PTFE, PDMS, PC, and nylon (+). While all the other pairs seemed identical to nylon, the MoS_2-PTFE pair exhibited a polarity that was opposite to that of the MoS_2-nylon pair. Consequently, MoS_2 is situated in the triboelectric series with PTFE and PDMS. Following the same method, the authors found the exact locations of 2D materials within the known triboelectric series and conducted tests on other substances (Figure 14.4(a)). All the 2D materials used in the study are located towards the lower end of the triboelectric series. It is important to mention that the 2D materials used in the study were obtained by chemically separating them from larger pieces in a liquid phase. These materials had consistent thickness and surface roughness to ensure that external factors did not affect the results. The authors used first-principles simulations and Kelvin probe force microscopy (KPFM) to figure out the effective work function, which adds to the reliability of the results. The determination of the triboelectric polarity of MoS_2 was found to be unaffected by factors such as thickness, manufacturing process, and polytype. Chemical doping enables precise control of its position within the friction series.

TMDs materials, such as MoS_2, possess the ability to accumulate charge and prevent the recombination of electrons and positive charges through drift and diffusion processes [33, 34]. Wu et al. did a study to see what would happen if MoS_2 was added to the friction layer as the triboelectric electron-acceptor layer [35]. The goal was to improve the output performance of TENGs. In order to produce a monolayer of MoS_2, the researchers utilized liquid-phase exfoliation while subjecting the material to ultrasonic vibrations. This procedure entailed extracting the majority of MoS_2 from an organic solvent. The experimental setup involved the employment of an aluminum electrode and a layered structure consisting of a polyimide (Kapton) layer, a molybdenum disulfide (MoS_2) layer embedded in polyimide (PI), and another polyimide (PI) layer as the negative friction layer (Figure 14.4(b)). The TENG, utilizing monolayer MoS_2 as the electron-acceptor layer, attained a power density of up to

25.7 W m^2, which is 120 times greater than the device without monolayer MoS_2. To showcase the capacity to trap charge, the authors additionally fabricated a metal-insulator semiconductor device (MIS) utilizing MoS_2 with a floating gate. Monolayer MoS_2 possesses a band gap of up to 1.8 eV, whereas rGO does not. Additionally, the trap states located at the interface are evenly distributed within this gap. In Figure 14.4(b), you can see that the C-V curves change as the voltage is increased, showing that the number of electrons trapped in the MoS_2 interface increases. The enhancement of TENGs' friction layer performance is significantly facilitated by the surface microstructure and the suitable level of friction. Conventional preparation processes like mechanical stripping and chemical vapor deposition (CVD) have limitations when it comes to regulating the surface form [36, 37]. Park et al. [38] utilized pulsed laser-directed thermolysis to create a surface-crumpled MoS_2, as depicted in Figure 14.4(c)), as a means of resolving this problem. This technique involves modifying the morphological configuration of a MoS_2 crystal to induce internal stress within the crystal. The smoothness of the MoS_2 surface (F-MoS_2) decreases and becomes more wrinkled (LC-MoS_2) with increasing laser irradiation fluence. In addition, the little creases merge together, thereby sealing the most wrinkled structure (MC-MoS_2). The power output of the flat F-MoS_2 device is approximately 40% lower than that of the MC-MoS_2 TENG device. The maximum open-circuit voltage recorded was 25 V, whereas the highest short-circuit current measured was 1.2 μA. Additionally, they predicted that the tensile strain on MoS_2 would result in an elevation in atomic distance and lattice expansion. Consequently, the valence bandwidth and Fermi level of MoS_2 shift towards a lower energy range, resulting in an increase in its work function. The combination of these factors enhances the effectiveness of surface-crumpled MoS_2 TENGs. The process of TMD preparation is susceptible to surface imperfections. WS_2 has sulfur vacancies on its basal planes or edges, which negatively impact carrier mobility and charge density [39]. This has a substantial impact on the efficiency of TENGs that depend on TMDs. Figure 14.5d illustrates the enhanced output performance of TENG devices recently developed by Kim et al. [40]. The surfaces of these devices were made with TMDs and thiolated ligands (LC-WS_2) that had different lengths of alkane chains attached to them. To alter the surface of WS_2 nanosheets with defects, thiol-containing ligands such as mercaptopropionic acid (MPA), mercaptohexanoic acid (MHA), mercapto-octanoic acid (MOA), and mercaptoundecanoic acid (MUA) were employed for functionalization. Using ATR-IR and XPS, the study looked at how the surface of WS_2 connected to different thiol-containing ligands. The LC-WS_2 TENGs exhibited an output voltage of 12.2 V and a power density of 138 mW m^{-2}, respectively. It was found that the LC-WS_2 had a better work function because of two things: a higher surface charge density and the p-type doping effect that comes from thiolate ligand conjugation. The scientists concluded that this better performance was primarily due to the latter. Furthermore, WS_2's stability in the presence of air was enhanced. TMDs possess adjustable band-gap properties, allowing them to be utilized for various detector preparations, including light, humidity, and biological signal detection. Additionally, they can be effectively combined with TENGs to expand their range of applications. Furthermore, graphene and its derivatives exhibit electron capture and surface modification capabilities. Both reduced graphene oxide (rGO) and MoS_2, which is a type of transition metal dichalcogenide (TMD), possess the capacity to store charges and inhibit the recombination of electrons and positive charges resulting from drift and diffusion [33, 34]. Wu et al. looked into what would happen if MOS_2 was used in the friction layer as the triboelectric electron-acceptor layer to make TENGs work better [35]. The researchers employed ultrasonic treatment in combination with liquid-phase peeling conduction to generate a single layer of MoS_2. During this process, an organic solvent was employed to isolate the molecular composition of MoS_2. Figure 14.4(b) illustrates the composition of the electrodes, which comprised an aluminum electrode and a layered negative friction layer consisting of PI (Kapton)/MoS_2:PI/PI. The TENG, employing monolayer MoS_2 as the electron-acceptor layer, achieved a power density of 25.7 W m^{-2}, surpassing the device without monolayer MoS_2 by a factor of 120. In addition, the authors employed MoS_2 to fabricate a floating-gate metal-insulator semiconductor device (MIS) capable of confining charges. The trap states within the band gap of monolayer MoS_2 are uniformly distributed, spanning a range of up to 1.8 eV. This is in contrast to rGO. The C-V curves in Figure 14.4(b) move as the number of voltage sweeps increases. This is because electrons are building up in the MoS_2 interface trap states. The surface microstructure and optimal friction coefficient play a crucial role in enhancing the performance of the friction layer in TENGs. Nevertheless, conventional techniques like mechanical stripping and CVD have inherent limitations in terms of their ability to precisely control the form of the surface [36, 37]. To solve this problem, Park et al. [38] used

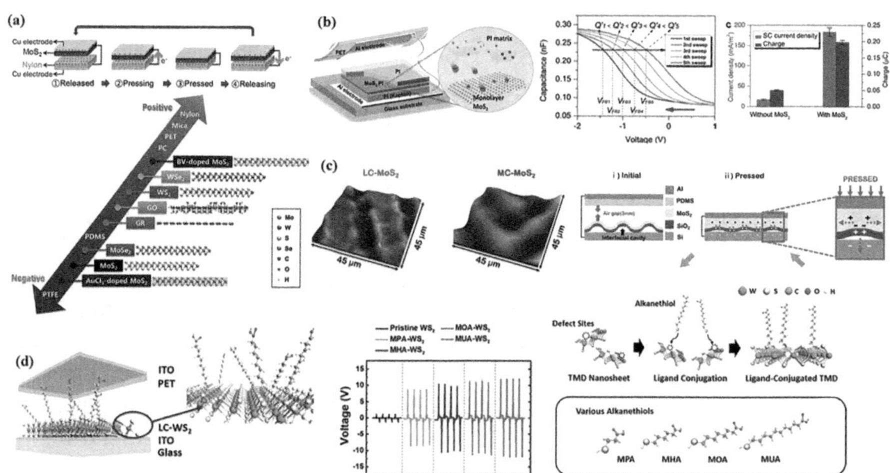

FIGURE 14.4 (a) Triboelectric series including 2D materials. Reproduced with permission from ref. [32]. Copyright © 2018 Wiley-VCH. (b) MoS2 acts as a charge-trapping layer to enhance the output performance of TENGs. Reproduced with permission from ref. [35]. Copyright © 2017 American Chemical Society. (c) The use of MC-MoS_2 as the friction layer increases the contact area. Reproduced with permission from ref. [38]. Copyright © 2020 Elsevier. (d) The surface defects of friction layer WS_2 were modified by thiol groups to improve the output of LC-WS2 TENGs. Reproduced with permission from ref. [40]. Copyright © 2021 American Chemical Society.

pulsed-laser-directed thermolysis to make surface-crumpled MoS2 (Figure 14.4(c)). In order to induce internal stress in a MoS_2 crystal, this technique entails altering the crystal's morphological configuration. Recently, Kim et al. designed the surface of TMDs utilizing thiolated ligands (LC-WS_2) of varying alkane chain lengths and used them in conjugation in order to create TMD based TENG devices with improved output performance (Figure 14.4(d)) [40].

14.4.3 TMPs-Based NGs

Transition-metal phosphide electrocatalyst is a very large and complex family with a variety of properties. With different transition metals, there are Ni_2P, $Ni_{12}P_5$, Ni_5P_4, Cu_3P, Fe_2P, FeP, Co_2P, CoP, InP, MoP, WP, W_2P, etc. with excellent HER catalytic activities. Metal phosphides in various forms can be synthesized using a variety of methods, including the metal-organic precursor decomposition method [41], solvothermal [42], solid phase reaction [43], electrodeposition [44], and several other synthesis methods. Overall, transition metal phosphides hold great promise for the development of efficient and cost-effective nanogenerators for future energy needs.

TMP materials have become promising candidates as highly efficient electrocatalysts owing to their better electrical conductivity, redox activity, and capacity than transition metal oxides/LDHs. [45] The electronegativity of phosphorus is lower than that of oxygen, which facilitates electron transport and redox reaction kinetics. Furthermore, TMP materials are considered to be stable electrodes due to their favorable resistance to the surrounding environment and good thermal stability.[46] Bimetallic phosphides, like NiCoP, [45, 47] Co_xNi_1–xP, [46] NiCuP, [48] and FeNiP, [49] have been prepared as flexible battery-type electrodes due to their large charge-storage ability. Intriguingly, Wang's group [45] prepared 3D cactus-like NiCoP/NiCo–OH on CC as a flexible SC electrode, which integrated both transition metal phosphide and hydroxide materials. The NiCoP/NiCo–OH showed a structure that combined 2D nanoflakes with 1D nanospines, and displayed polycrystalline features, which offered effective pathways for charge transfer, guaranteed structural stability, and provided substantial active sites. Because of the hierarchical nanostructure and low electronegativity of P, the NiCoP/NiCo–OH electrode delivered a high capacitance of 1100 F g^{-1} and a long cycle life (90% capacitance retention over 1,000 cycles).

Transition metal phosphides have proven to be beneficial in nanogenerators due to their unique properties and characteristics. These nanomaterials exhibit increased structural and thermal stability, good hardness, and electrical conductivity, making them suitable for energy generation systems. Additionally, transition metal phosphides have available surfaces with enriched active sites and strong metal-phosphorous bonds, which contribute to their excellent performance in energy conversion and storage systems, including hydrogen and oxygen evolution reactions, water splitting, metal-ion batteries, and supercapacitors. The short pathways of diffused ions, increased specific capacitance, and lowered charge-discharge voltage of these nanomaterials further enhance their effectiveness in nanogenerators. Furthermore, the use of transition metal phosphides in nanogenerators offers advantages such as high catalytic activity, good stability, simple preparation methods, and repeatability, making them suitable for industrial-scale production.

14.4.4 MXenes-Based NGs

The fundamental constituents of MXenes are nitrides and carbides derived from transition metals. The formula $M_{n+1}AX_n$ is a commonly used representation, where n can be 1, 2, or 3. In this formula, M represents an abundant metal element, A represents an element from group II or IV, and X can be either carbon or nitrogen. The M-A bond is replaced to make MXene, which is made up of $M_{n+1}X_nT_x$, where T_x is the surface termination. This is done by selectively etching away layer A. MXene, a unique two-dimensional layered material, has notable physical and chemical properties, including a robust negative zeta potential, abundant surface functional groups, excellent electrical conductivity (more than 20,000 S cm^{-1}), and high mechanical strength [50, 51]. It has the potential to be highly advantageous for energy devices [52]. Dong initially investigated the feasibility of employing MXene materials as the negative triboelectrification layer and electrode for a TENG. The $Ti_3C_2T_x$ MXene surface exhibits high electronegativity because of the presence of halogen groups (-F) and oxygen-containing terminal functional groups (Figure 14.5(a)). When a TENG works in a single-electrode mode with PTFE as the friction layer, the output voltage is compared. It is shown that $Ti_3C_2T_x$ MXene is the same in the triboelectric series as PTFE. A substitute has been found for non-conductive electronegative polymers (PTFE, PDMS, and FEP) that are limited to operating in single-electrode mode. A MXene-PET TENG with dimensions of 2.5 × 5 cm2 was capable of generating potential differences of roughly 650 V and achieving a peak power output of around 0.65 mW. Figure 14.5(a) shows that when the MXene-PET TENG was fixed at a 30° angle, it produced an output above 40 V when a force greater than 1 N was applied. Curved fingers enable this TENG to accumulate energy. Cao et al. employed identical ingredients to fabricate a creased MXene film through the process of pouring $Ti_3C_2T_x$ ink onto an elongated latex balloon. Once the ink had dried, the balloon underwent a steady reduction in size. Figure 14.6b served as the basis for the creation of a stretchable TENG that operates in a single-electrode mode [53]. The crumpled structure of the MXene TENG led to a maximum area strain of 2,150% and a linear tensile ratio of 400%. Although the scientists observed that both the fold structure and the enhanced MXene content in the ink had a good impact on production, the fold structure had a far greater influence. A very sensitive pressure sensor was developed with the stretchable TENG technology.

Figure 14.5(c) depicts the initial non-contact mode (CNM) TENG developed by Salauddin et al. [55]. They created this TENG by using lasers to carbonize MXene composites, resulting in the formation of an LC-MXene/ZiF-67 nanocomposite layer. This layer was subsequently employed as an intermediary layer to preserve electrons. The LC-MXene/ZiF-67 nanocomposite layer has a porous structure and a high charge density. This increases the surface potential and makes it easier for surface charges to stay in place for a long time. It creates a special porous structure framework when ZiF-67 is added to the nanocomposite. This framework has consistent porosity and high chemical stability. Carbonization significantly increases the electron-capturing properties of the nanocomposite, thereby facilitating the sustained maintenance of surface charge and potential over time. Scientists made tiny structures on the MXene/silicone nanocomposite's surface to increase the contact area. This surface was used as the charge-generating layer (CGL). By leveraging the surface textures of fabrics, they managed to enhance the power density of the output. The surface of the MXene/silicon nanocomposite in a CNM-TENG had a negative charge before a person touched it. The movement of the hand-generated electric impulses due to the separation between the CGL-containing intermediate layer and the user's hand. The relative distance between the two sites had an impact on the electric signals produced as a result of the hand movement. The researchers also investigated the impact of several parameters, such as distance, laser power, laser speed, and LC-MXene/ZiF-67 ratios, on the performance of the TENG. Luo et al. improved the properties of a hydrogel by adding $Ti_3C_2T_x$ MXene nanosheets. This led to the creation of a TENG that is very flexible, performs well, and heals itself very well (Figure 14.5(d)) [56]. The conductivity of hydrogels can

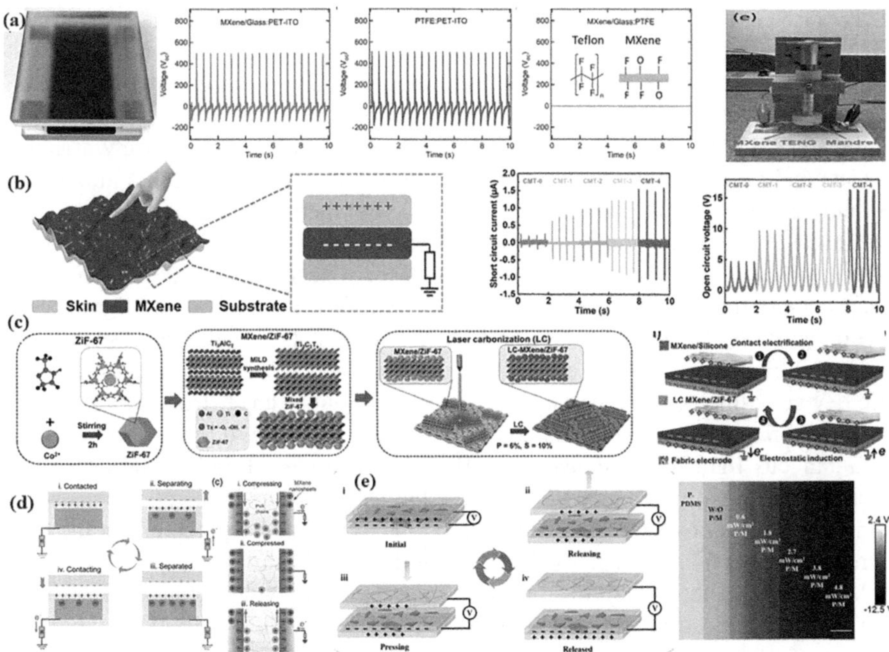

FIGURE 14.5 (a) $Ti_3C_2T_x$ (MXene) was first used as a friction layer in a TENG. Reproduced with permission from ref. [54]. Copyright © 2018 Elsevier. (b) Wrinkled MXene was used to increase the contact area of the friction layer. Reproduced with permission from ref. [53]. Copyright © 2022 Elsevier. (c) A non-contact TENG with LC-MXene/ZiF-67 as the electron capture layer. (1–4) represent the electricity generation of TENG. Reproduced with permission from ref. [55]. Copyright © 2022 Elsevier. (d) The SVP model (i–iii on right of picture) generated between MXene and hydrogel enhances the TENG output. (i–iv) on the left of picture represent the electricity generation of TENG. Reproduced with permission from ref. [56]. Copyright © 2021 Wiley-VCH. (e) Light enhanced the triboelectric properties of MXene. i–iv represent the electricity generation of TENG. Reproduced with permission from ref. [57] Copyright © 2021 Elsevier.

be increased by improving ion transport and enhancing the mechanical properties of composites with the addition of nanosheets. These nanosheets also create microchannels on the surface that come into contact with the hydrogels. Furthermore, the triboelectric effect can be intensified by utilizing a system that generates a vibrating potential through.

The researchers put the MXene/PVA hydrogel inside eco-flex silicone rubber to make a single-motor mode TENG out of Kapton. Subsequently, they used the hydrogel as an electrode. Figure 14.5(d) demonstrates that the MH-TENG exhibited various potential changes when exposed to pressure. The contact potential between the eco-flex and Kapton friction layers, the streaming vibration potential model made by the interface between MXene and the hydrogel, and the extra potential difference caused by compressing the material were all things that were looked at. The level of compression also augmented this possible discrepancy. The use of MXene nanosheets resulted in enhanced output performance, mechanical stability, and stretchability of the TENG. The TENG's capacity to generate electrical signals via compression and contact renders it a highly attractive contender for implementation in handwriting recognition systems. MXene has both photosensitivity and exceptional electronegative conductivity [58]. MXenes exhibit a crystal structure composed of several nanometer-thick layers of disordered material. By virtue of their densely arranged borders and minuscule gaps, they have the ability to mitigate plasmonic momentum limitations and facilitate the production of highly energetic electrons [59, 60]. Liu et al. created a light-enhanced stretchable TENG based on this property, as shown in Figure 14.5(e). The TENGs, composed of PDMS/MXenes with a solitary structural layer, possess the capability to simultaneously harvest mechanical and light energy. The integration of Ag nanowires and MXene nanosheets into the PDMS elastomer resulted in the formation of a conductive electrode and a friction layer, respectively. The device exhibited a significant rise in open-circuit voltage (from 145 to 453 V) and short-circuit current (from 27 to 131 μA) upon exposure to light, resulting in an instantaneous light power conversion efficiency of 19.6%. Heating and energizing electrons can greatly enhance the surface charge density and dielectric constant of composite films consisting of PDMS and MXene. The surface potentials of the films increased proportionally with the intensity of the light beams.

MXenes have shown significant progress in nanogenerators, offering advantages such as electronegativity, metallic conductivity, mechanical flexibility, and customizable surface chemistry. They have been explored for various applications, including energy harvesting, sensing, and electronic devices. MXenes can be used as active materials in nanogenerators, either alone or in combination with other material; by incorporating MXenes into nanogenerators, researchers have achieved enhanced output performance, such as increased surface potential and boosted electrical polarization. MXenes have also been used in the fabrication of all-MXene flexible and integrated thermoelectric nanogenerators, demonstrating promising output voltage and power. The use of MXenes in nanogenerators has gained significant attention in recent years, with researchers focusing on design strategies and internal improvement mechanisms to optimize their performance. Overall, MXenes-based materials show great potential for improving the output performance of nanogenerators and advancing their practical applications.

14.4.5 Metal-Organic Coordination Polymers (MOCPs)-Based NGs

Metal-organic coordination polymers, also known as MOCPs, are a type of crystalline material that combines inorganic metal nodes with organic linkers. They are among the countless potential substances that are appropriate for a wide variety of applications. [1–6] Among the application domains, there has been a significant amount of interest in the evolution of fuel cells, supercapacitors, thermo-electrics, and other applications of MOCPs in the field of electricity. As of late, porous coordination polymers, which are also known as metal-organic frameworks (MOFs), have been used as TENG materials in self-powered systems and sensor applications. Both the development of TENG materials and the breadth of applications for MOCPs have benefited from this development, which has been beneficial. In spite of this, there is still a significant amount of work to be done before MOCP-based TENGs that excel in output performance can be produced for a wide range of applications.

14.4.6 Metal-Organic Framework (MOF)-Based NGs

The coordination of inorganic metal ions (node components) and organic ligands (structural linkers) forms intricate MOFs. Importantly, MOFs possess the capability to form multi-layer structures in addition to three-dimensional ones. Layered MOFs possess an extraordinarily elevated surface area as a result of their broad nanoscale porosity and vast inner surfaces. Hence, the utilization of stratified MOFs in triboelectric materials enables the capture of electrons. The multi-layer MOF-based TENGs have the ability to sense and harvest energy in high-humidity settings due to their high dielectric constant in humid environments and remarkable sensitivity to pollutants.

MOFs have been shown to be beneficial in nanogenerators, specifically in TENGs. MOFs, such as UiO-66–4F and Cd-MOF, have been designed and synthesized to improve the performance of TENGs. The introduction of strong electron-withdrawing groups in MOFs, such as tetrafluoroterephthalic acid (TFA), has led to significant improvements in the power density and voltage output of MOF-based TENGs. Additionally, functionalized filler MOFs with large electron-withdrawing functional groups, such as UiO-66-NO_2, have been found to enhance the output performance of TENGs. Furthermore, the combination of NH_2-MOFs with cellulose nanofibers has been shown to improve the stability, positive triboelectric properties, and charge-trapping performance of TENGs. These findings demonstrate the potential of MOFs in enhancing the electrical properties and long-term operational durability of nanogenerators, particularly TENGs.

14.5 CHALLENGES IN THE DEVELOPMENT OF NGs

The discoveries presented above have sparked the interest of scientists worldwide due to the potential of TENGs as a highly efficient energy conversion mechanism. Compared to other renewable energy sources such as solar power, wind power, or nuclear power, TENGs possess a unique advantage in their ability to convert mechanical friction into usable energy with exceptional efficiency, regardless of location or time. In addition, it should be noted that TENGs do not require any additional structural components in comparison to PENGs. Several TENGs constructed using 2D materials have been created and introduced within the past ten years. Additional two-dimensional materials, such as MOFs and COFs, have been incorporated into TENGs following the initial use of high-conductivity MXene by graphene [61–63]. 2D materials are frequently employed to modify doping in order to enhance the output performance of TENGs. Additionally, they function as electrodes and friction layers. Researchers are currently prioritizing the exploration of practical applications for TENGs, such as energy harvesting, autonomous sensors, and HMI, instead of solely focusing on their performance analysis. Extensive research efforts have significantly improved the output performance of TENGs and expanded their range of applications. However, TENGs have their own set of anticipated outcomes and challenges, particularly when viewed in relation to the rapid expansion of the Internet of Things and electronic information technology.

It has been twenty years since 2D materials were successfully separated; nevertheless, the remarkable thermal, triboelectric, piezoelectric, and magnetic properties of materials that are just one atomic layer thick are still not fully comprehended [64, 65]. This is the primary obstacle that hinders the widespread utilization of these materials. The second concern pertains to the limited availability of techniques for preparing 2D materials, which poses a significant obstacle to the cost-effective, extensive, high-grade, and manageable production of crystal-phase 2D materials. 2D materials must adhere to specific electronegativity and conductivity criteria since they serve as electrodes and friction layers in TENGs. Hence, it is imperative to optimize the efficiency of TENGs through the development and utilization of innovative materials that exhibit similar properties to MXene, such as $Ti_3C_2T_x$. In order to fully achieve their potential, further investigation into the production and characteristics of two-dimensional materials is still required. According to the information presented in the previous summary, it is possible to use 2D materials as altered constituents in composites composed of primary materials.

For instance, they have the ability to manipulate the work function, enhance the dielectric properties of materials, and facilitate charge capture and interface lubrication. The aforementioned benefits have had a significant impact on TENGs. Many testing tools at the macroscopic level have confirmed these effects. However, we now lack a comprehensive understanding of the microscopic process responsible for these effects. To make TENG work better, we need a solid theoretical framework for studying the properties of 2D materials, how they interact with functionalization, and the base materials that they are built on. 2D materials are very suitable for constructing flexible TENGs due to their inherent pliability, transparency, and extremely thin nature. However, with the ongoing progress of the IoT and the increasing popularity of wearable gadgets, there is a growing demand for devices that are highly flexible, compact, self-sustaining, and capable of wireless communication. To meet specific needs like being stable in harsh environments and being safe for living things, TENGs need to be built with smart stacking structures and extra signal processing and transmission units added. To promote the use of TENGs utilising 2D materials and facilitate the detection of many signal modes, it is crucial to develop hybrid energy harvesters that harness both triboelectric effects and other phenomena [66].

Regarding TENGs, there is currently no universally accepted method for quantifying their efficacy in generating outcomes. The results of general tests, such as open-circuit voltage, short-circuit current, power density, and energy conversion efficiency, can differ based on many test parameters, including the test environment, applied loads, contact areas, and structural elements of the device [67, 68]. Due to the lack of comparative optimization of TENG performance, it is insufficient to construct a single evaluation criterion solely based on the quantity of LED lights and the capacity of a capacitor. The high impedance of TENGs, in addition to causing heat damage, also restricts their possible usage.

14.6 CONCLUSION AND PROSPECTS

This chapter provides a concise overview of the latest research findings on electrodes or NG materials. Upon examining the material aspect, we discovered that carbon nitrides such as MXene, TMOs, TMSs, and graphene possess the ability to function as electrodes, friction layers, or doping modifiers. NGs have multiple applications in various sectors, like energy harvesting, autonomous sensing, and HCI. The creation of NGs was ultimately defined by the identification of challenges and possibilities. The utilization of carbon nanotubes in NGs has created new opportunities for enhancing their efficiency and broadening their real-world uses, resulting in progress in self-powered sensing and wearable technology. To enhance the performance of TENGs and PENGs, it is imperative to broaden our existing theoretical knowledge, develop novel materials, and devise innovative structural configurations.

REFERENCES

[1] T. Park, N. Kim, D. Kim, S.-W. Kim, Y. Oh, J.-K. Yoo, J. You, M.-K. Um, An organic/inorganic nanocomposite of cellulose nanofibers and ZnO nanorods for highly sensitive, reliable, wireless, and wearable multifunctional sensor applications, *ACS Applied Materials & Interfaces*, 11 (2019) 48239–48248.

[2] Q. Shi, T. He, C. Lee, More than energy harvesting–Combining triboelectric nanogenerator and flexible electronics technology for enabling novel micro-/nano-systems, *Nano Energy*, 57 (2019) 851–871.

[3] S. Sripadmanabhan Indira, C. Aravind Vaithilingam, K.S.P. Oruganti, F. Mohd, S. Rahman, Nanogenerators as a sustainable power source: State of art, applications, and challenges, *Nanomaterials*, 9 (2019) 773.

[4] F. Ali, W. Raza, X. Li, H. Gul, K.-H. Kim, Piezoelectric energy harvesters for biomedical applications, *Nano Energy*, 57 (2019) 879–902.

[5] Z.L. Wang, Self-powered nanotech, *Scientific American*, 298 (2008) 82–87.

[6] Y. Hu, Y. Zhang, C. Xu, L. Lin, R.L. Snyder, Z.L. Wang, Self-powered system with wireless data transmission, *Nano Letters*, 11 (2011) 2572–2577.

[7] Y. Yang, K.C. Pradel, Q. Jing, J.M. Wu, F. Zhang, Y. Zhou, Y. Zhang, Z.L. Wang, Thermoelectric nanogenerators based on single Sb-doped ZnO micro/nanobelts, *ACS Nano*, 6 (2012) 6984–6989.

[8] F. Yi, Z. Zhang, Z. Kang, Q. Liao, Y. Zhang, Recent advances in triboelectric nanogenerator-based health monitoring, *Advanced Functional Materials*, 29 (2019) 1808849.

[9] T.P. Huynh, H. Haick, Autonomous flexible sensors for health monitoring, *Advanced Materials*, 30 (2018) 1802337.

[10] W. Ding, A.C. Wang, C. Wu, H. Guo, Z.L. Wang, Human–machine interfacing enabled by triboelectric nanogenerators and tribotronics, *Advanced Materials Technologies*, 4 (2019) 1800487.

[11] H. Askari, A. Khajepour, M.B. Khamesee, Z. Saadatnia, Z.L. Wang, Piezoelectric and triboelectric nanogenerators: Trends and impacts, *Nano Today*, 22 (2018) 10–13.

[12] Y. Yang, W. Guo, K.C. Pradel, G. Zhu, Y. Zhou, Y. Zhang, Y. Hu, L. Lin, Z.L. Wang, Pyroelectric nanogenerators for harvesting thermoelectric energy, *Nano Letters*, 12 (2012) 2833–2838.

[13] Z.L. Wang, J. Song, Piezoelectric nanogenerators based on zinc oxide nanowire arrays, *Science*, 312 (2006) 242–246.

[14] C. Zhou, G. Zeng, D. Huang, Y. Luo, M. Cheng, Y. Liu, W. Xiong, Y. Yang, B. Song, W. Wang, Distorted polymeric carbon nitride via carriers transfer bridges with superior photocatalytic activity for organic pollutants oxidation and hydrogen production under visible light, *Journal of Hazardous Materials*, 386 (2020) 121947.

[15] W. Wang, Q. Niu, G. Zeng, C. Zhang, D. Huang, B. Shao, C. Zhou, Y. Yang, Y. Liu, H. Guo, 1D porous tubular g-C3N4 capture black phosphorus quantum dots as 1D/0D metal-free photocatalysts for oxytetracycline hydrochloride degradation and hexavalent chromium reduction, *Applied Catalysis B: Environmental*, 273 (2020) 119051.

[16] W. Wang, M. Chen, D. Huang, G. Zeng, C. Zhang, C. Lai, C. Zhou, Y. Yang, M. Cheng, L. Hu, An overview on nitride and nitrogen-doped photocatalysts for energy and environmental applications, *Composites Part B: Engineering*, 172 (2019) 704–723.

[17] K.-A.N. Duerloo, M.T. Ong, E.J. Reed, Intrinsic piezoelectricity in two-dimensional materials, *The Journal of Physical Chemistry Letters*, 3 (2012) 2871–2876.

[18] W. Wu, L. Wang, Y. Li, F. Zhang, L. Lin, S. Niu, D. Chenet, X. Zhang, Y. Hao, T.F. Heinz, Piezoelectricity of single-atomic-layer MoS2 for energy conversion and piezotronics, *Nature*, 514 (2014) 470–474.

[19] C. Yan, W. Deng, L. Jin, T. Yang, Z. Wang, X. Chu, H. Su, J. Chen, W. Yang, Epidermis-inspired ultrathin 3D cellular sensor array for self-powered biomedical monitoring, *ACS Applied Materials & Interfaces*, 10 (2018) 41070–41075.

[20] L. Jin, B. Zhang, L. Zhang, W. Yang, Nanogenerator as new energy technology for self-powered intelligent transportation system, *Nano Energy*, 66 (2019) 104086.

[21] Z.L. Wang, Triboelectric nanogenerators as new energy technology for self-powered systems and as active mechanical and chemical sensors, *ACS Nano*, 7 (2013) 9533–9557.

[22] Y.S. Zhou, K. Wang, W. Han, S.C. Rai, Y. Zhang, Y. Ding, C. Pan, F. Zhang, W. Zhou, Z.L. Wang, Vertically aligned CdSe nanowire arrays for energy harvesting and piezotronic devices, *ACS Nano*, 6 (2012) 6478–6482.

[23] G.-C. Yi, C. Wang, W.I. Park, ZnO nanorods: Synthesis, characterization and applications, *Semiconductor Science and Technology*, 20 (2005) S22.

[24] Y. Heo, D. Norton, L. Tien, Y. Kwon, B. Kang, F. Ren, S. Pearton, J. LaRoche, ZnO nanowire growth and devices, *Materials Science and Engineering: R: Reports*, 47 (2004) 1–47.

[25] Y. Heo, V. Varadarajan, M. Kaufman, K. Kim, D. Norton, F. Ren, P. Fleming, Site-specific growth of ZnO nanorods using catalysis-driven molecular-beam epitaxy, *Applied Physics Letters*, 81 (2002) 3046–3048.

[26] M.-P. Lu, J. Song, M.-Y. Lu, M.-T. Chen, Y. Gao, L.-J. Chen, Z.L. Wang, Piezoelectric nanogenerator using p-type ZnO nanowire arrays, *Nano Letters,* 9 (2009) 1223–1227.

[27] M.Y. Choi, D. Choi, M.J. Jin, I. Kim, S.H. Kim, J.Y. Choi, S.Y. Lee, J.M. Kim, S.W. Kim, Mechanically powered transparent flexible charge-generating nanodevices with piezoelectric ZnO nanorods, *Advanced Materials*, 21 (2009) 2185–2189.

[28] D. Choi, M.Y. Choi, W.M. Choi, H.J. Shin, H.K. Park, J.S. Seo, J. Park, S.M. Yoon, S.J. Chae, Y.H. Lee, Fully rollable transparent nanogenerators based on graphene electrodes, *Advanced Materials*, 22 (2010) 2187–2192.

[29] G. Zhu, A.C. Wang, Y. Liu, Y. Zhou, Z.L. Wang, Functional electrical stimulation by nanogenerator with 58 V output voltage, *Nano Letters*, 12 (2012) 3086–3090.
[30] Q. Yun, L. Li, Z. Hu, Q. Lu, B. Chen, H. Zhang, Layered transition metal dichalcogenide-based nanomaterials for electrochemical energy storage, *Advanced Materials*, 32 (2020) 1903826.
[31] W. Choi, N. Choudhary, G.H. Han, J. Park, D. Akinwande, Y.H. Lee, Recent development of two-dimensional transition metal dichalcogenides and their applications, *Materials Today*, 20 (2017) 116–130.
[32] M. Seol, S. Kim, Y. Cho, K.E. Byun, H. Kim, J. Kim, S.K. Kim, S.W. Kim, H.J. Shin, S. Park, Triboelectric series of 2D layered materials, *Advanced Materials*, 30 (2018) 1801210.
[33] J. Liu, Z. Zeng, X. Cao, G. Lu, L.H. Wang, Q.L. Fan, W. Huang, H. Zhang, Preparation of MoS2-polyvinylpyrrolidone nanocomposites for flexible nonvolatile rewritable memory devices with reduced graphene oxide electrodes, *Small*, 8 (2012) 3517–3522.
[34] G.H. Shin, C.-K. Kim, G.S. Bang, J.Y. Kim, B.C. Jang, B.J. Koo, M.H. Woo, Y.-K. Choi, S.-Y. Choi, Multilevel resistive switching nonvolatile memory based on MoS2 nanosheet-embedded graphene oxide, *2D Materials*, 3 (2016) 034002.
[35] C. Wu, T.W. Kim, J.H. Park, H. An, J. Shao, X. Chen, Z.L. Wang, Enhanced triboelectric nanogenerators based on MoS2 monolayer nanocomposites acting as electron-acceptor layers, *ACS Nano*, 11 (2017) 8356–8363.
[36] R. Kumar, S. Sahoo, E. Joanni, R.K. Singh, R.M. Yadav, R.K. Verma, D.P. Singh, W.K. Tan, A. Perez del Pino, S.A. Moshkalev, A review on synthesis of graphene, h-BN and MoS 2 for energy storage applications: Recent progress and perspectives, *Nano Research*, 12 (2019) 2655–2694.
[37] Z. Ji, L. Zhang, G. Xie, W. Xu, D. Guo, J. Luo, B. Prakash, Mechanical and tribological properties of nanocomposites incorporated with two-dimensional materials, *Friction*, 8 (2020) 813–846.
[38] S. Park, J. Park, Y.-g. Kim, S. Bae, T.-W. Kim, K.-I. Park, B.H. Hong, C.K. Jeong, S.-K. Lee, Laser-directed synthesis of strain-induced crumpled MoS2 structure for enhanced triboelectrification toward haptic sensors, *Nano Energy*, 78 (2020) 105266.
[39] S. Wang, A. Robertson, J.H. Warner, Atomic structure of defects and dopants in 2D layered transition metal dichalcogenides, *Chemical Society Reviews*, 47 (2018) 6764–6794.
[40] T.I. Kim, I.-J. Park, S. Kang, T.-S. Kim, S.-Y. Choi, Enhanced triboelectric nanogenerator based on tungsten disulfide via thiolated ligand conjugation, *ACS Applied Materials & Interfaces*, 13 (2021) 21299–21309.
[41] C.G. Read, J.F. Callejas, C.F. Holder, R.E. Schaak, General strategy for the synthesis of transition metal phosphide films for electrocatalytic hydrogen and oxygen evolution, *ACS Applied Materials & Interfaces,* 8 (2016) 12798–12803.
[42] E.J. Popczun, J.R. McKone, C.G. Read, A.J. Biacchi, A.M. Wiltrout, N.S. Lewis, R.E. Schaak, Nanostructured nickel phosphide as an electrocatalyst for the hydrogen evolution reaction, *Journal of the American Chemical Society*, 135 (2013) 9267–9270.
[43] J. Li, J. Li, X. Zhou, Z. Xia, W. Gao, Y. Ma, Y. Qu, Highly efficient and robust nickel phosphides as bifunctional electrocatalysts for overall water-splitting, *ACS Applied Materials & Interfaces*, 8 (2016) 10826–10834.
[44] N. Jiang, B. You, M. Sheng, Y. Sun, Electrodeposited cobalt-phosphorous-derived films as competent bifunctional catalysts for overall water splitting, *Angewandte Chemie*, 127 (2015) 6349–6352.
[45] X. Li, H. Wu, A.M. Elshahawy, L. Wang, S.J. Pennycook, C. Guan, J. Wang, Cactus-like NiCoP/NiCo-OH 3D architecture with tunable composition for high-performance electrochemical capacitors, *Advanced Functional Materials*, 28 (2018) 1800036.
[46] N. Zhang, Y. Li, J. Xu, J. Li, B. Wei, Y. Ding, I. Amorim, R. Thomas, S.M. Thalluri, Y. Liu, High-performance flexible solid-state asymmetric supercapacitors based on bimetallic transition metal phosphide nanocrystals, *ACS Nano,* 13 (2019) 10612–10621.

[47] P. Sun, M. Qiu, J. Huang, J. Zhao, L. Chen, Y. Fu, G. Cui, Y. Tong, Scalable three-dimensional Ni3P-based composite networks for flexible asymmertric supercapacitors, *Chemical Engineering Journal*, 380 (2020) 122621.
[48] L. Naderi, S. Shahrokhian, Nickel vanadium sulfide grown on nickel copper phosphide Dendrites/Cu fibers for fabrication of all-solid-state wire-type micro-supercapacitors, *Chemical Engineering Journal*, 392 (2020) 124880.
[49] L. Wan, D. Chen, J. Liu, Y. Zhang, J. Chen, M. Xie, C. Du, Construction of FeNiP@CoNi-layered double hydroxide hybrid nanosheets on carbon cloth for high energy asymmetric supercapacitors, *Journal of Power Sources*, 465 (2020) 228293.
[50] Z. Ling, C.E. Ren, M.-Q. Zhao, J. Yang, J.M. Giammarco, J. Qiu, M.W. Barsoum, Y. Gogotsi, Flexible and conductive MXene films and nanocomposites with high capacitance, *Proceedings of the National Academy of Sciences*, 111 (2014) 16676–16681.
[51] M. Naguib, O. Mashtalir, J. Carle, V. Presser, J. Lu, L. Hultman, Y. Gogotsi, M.W. Barsoum, Two-dimensional transition metal carbides, *ACS Nano*, 6 (2012) 1322–1331.
[52] B. Anasori, M.R. Lukatskaya, Y. Gogotsi, 2D metal carbides and nitrides (MXenes) for energy storage, *Nature Reviews Materials*, 2 (2017) 1–17.
[53] Y. Cao, Y. Guo, Z. Chen, W. Yang, K. Li, X. He, J. Li, Highly sensitive self-powered pressure and strain sensor based on crumpled MXene film for wireless human motion detection, *Nano Energy*, 92 (2022) 106689.
[54] M. Li, Y. Jie, L.-H. Shao, Y. Guo, X. Cao, N. Wang, Z.L. Wang, All-in-one cellulose based hybrid tribo/piezoelectric nanogenerator, *Nano Research*, 12 (2019) 1831–1835.
[55] M. Salauddin, S.S. Rana, M. Sharifuzzaman, S.H. Lee, M.A. Zahed, Y. Do Shin, S. Seonu, H.S. Song, T. Bhatta, J.Y. Park, Laser-carbonized MXene/ZiF-67 nanocomposite as an intermediate layer for boosting the output performance of fabric-based triboelectric nanogenerator, *Nano Energy*, 100 (2022) 107462.
[56] X. Luo, L. Zhu, Y.C. Wang, J. Li, J. Nie, Z.L. Wang, A flexible multifunctional triboelectric nanogenerator based on MXene/PVA hydrogel, *Advanced Functional Materials,* 31 (2021) 2104928.
[57] A.R. Chowdhury, A.M. Abdullah, I. Hussain, J. Lopez, D. Cantu, S.K. Gupta, Y. Mao, S. Danti, M.J. Uddin, Lithium doped zinc oxide based flexible piezoelectric-triboelectric hybrid nanogenerator, *Nano Energy*, 61 (2019) 327–336.
[58] D.B. Velusamy, J.K. El-Demellawi, A.M. El-Zohry, A. Giugni, S. Lopatin, M.N. Hedhili, A.E. Mansour, E.D. Fabrizio, O.F. Mohammed, H.N. Alshareef, MXenes for plasmonic photodetection, *Advanced Materials,* 31 (2019) 1807658.
[59] H. Xu, A. Ren, J. Wu, Z. Wang, Recent advances in 2D MXenes for photodetection, *Advanced Functional Materials*, 30 (2020) 2000907.
[60] Z. Liu, H.N. Alshareef, MXenes for optoelectronic devices, *Advanced Electronic Materials*, 7 (2021) 2100295.
[61] G. Khandelwal, A. Chandrasekhar, N.P. Maria Joseph Raj, S.J. Kim, Metal–organic framework: A novel material for triboelectric nanogenerator–based self-powered sensors and systems, *Advanced Energy Materials*, 9 (2019) 1803581.
[62] S. Hajra, M. Sahu, A.M. Padhan, I.S. Lee, D.K. Yi, P. Alagarsamy, S.S. Nanda, H.J. Kim, A green metal–organic framework-cyclodextrin MOF: A novel multifunctional material based triboelectric nanogenerator for highly efficient mechanical energy harvesting, *Advanced Functional Materials*, 31 (2021) 2101829.
[63] L. Zhai, W. Wei, B. Ma, W. Ye, J. Wang, W. Chen, X. Yang, S. Cui, Z. Wu, C. Soutis, Cationic covalent organic frameworks for fabricating an efficient triboelectric nanogenerator, *ACS Materials Letters*, 2 (2020) 1691–1697.
[64] M. Gibertini, M. Koperski, A.F. Morpurgo, K.S. Novoselov, Magnetic 2D materials and heterostructures, *Nature Nanotechnology*, 14 (2019) 408–419.
[65] M. Long, P. Wang, H. Fang, W. Hu, Progress, challenges, and opportunities for 2D material based photodetectors, *Advanced Functional Materials*, 29 (2019) 1803807.

[66] C. Wu, A.C. Wang, W. Ding, H. Guo, Z.L. Wang, Triboelectric nanogenerator: A foundation of the energy for the new era, *Advanced Energy Materials,* 9 (2019) 1802906.
[67] J.-H. Zhang, Y. Li, X. Hao, A high-performance triboelectric nanogenerator with improved output stability by construction of biomimetic superhydrophobic nanoporous fibers, *Nanotechnology,* 31 (2020) 215401.
[68] J.-H. Zhang, Y. Li, J. Du, X. Hao, H. Huang, A high-power wearable triboelectric nanogenerator prepared from self-assembled electrospun poly (vinylidene fluoride) fibers with a heart-like structure, *Journal of Materials Chemistry A*, 7 (2019) 11724–11733.

15 Recycling of Coordination Nanomaterials-Based Energy Devices

Harimohan Sharma, Monika Singh, Jasvinder Kaur, Anuj Kumar and Ghulam Yasin

15.1 INTRODUCTION

Over the past few years, the entire world has been confronted with imminent threats of energy crisis and environmental degradation. During the UN climate change conference in Glasgow in 2021, a proposal was made to minimize the utilization of fossil fuels and provide reimbursement for more investments in clean energy. This initiative has undeniably expedited the advancement of clean and sustainable energy sources. Nevertheless, the majority of contemporary renewable energy sources lack a stable and sustained output, hence raising significant concerns regarding the development of energy storage and conversion devices [1]. Within this framework, the potential of coordination nanomaterials (CNs)-based energy gadgets, including fuel cells, electrolyzers, batteries, and supercapacitors, is highly intriguing as their promising candidature to meet the global energy demand. Consequently, the concerned scientific community has been continually developing different categories of CNs, including transition metal oxides (TMOs), transition metal nitrides (TMNs), transition metal carbides (TMCs), transition metal sulfides (TMSs), and others [2]. However, the fundamental issue lies in the fact that the wide range of applications and excellent characteristics of these materials are intricately connected to their special structures [3–5]. Due to their small size, high specific surface area, quantum effect, and unique electrical structure, nanomaterials based on CNs have applications in energy devices. Moreover, the CNs-based materials possess exceptional catalytic and electrical characteristics compared to other comparable compound nanomaterials, because of their unique crystal structure. Moreover, CNs have demonstrated significant commercial potential for catalytic applications, particularly in noble metal catalysts, due to their abundant availability and inexpensive raw material costs. However, their facile preparation, tuning, recycling, and further utilization of energy devices are still under debate for their improvement, benefiting the modern generation.

 DOI: 10.1201/9781003345886-17

15.1.1 Coordination Nanomaterials (CNs)

The CNs are a fascinating class of materials that have been extensively studied in recent years due to their unique structural and functional properties. These materials are characterized by the presence of coordination bonds between metal ions or clusters and organic ligands, which give rise to a diverse range of structures and properties that are highly tunable for various applications [6]. One of the most promising applications of CNs is in energy storage systems, where their unique electronic, optical, and catalytic properties can be leveraged to improve energy conversion and storage efficiency. For example, metal-organic frameworks (MOFs) are highly porous CNs composed of metal nodes connected by organic ligands, which form a crystalline structure with a high surface area, making them promising candidates for energy storage applications. Similarly, coordination polymers are extended structures formed by the coordination of metal ions with organic ligands, which can create a variety of frameworks with unique properties for energy storage. Metal nitride, metal phosphide, and metal sulfide nanomaterials are other examples of CNs that have been explored for their high electrical conductivity and catalytic activity in energy storage systems. Overall, the diverse range of structures and properties exhibited by CNs makes them highly versatile for various energy storage applications, and ongoing research in this field is expected to lead to further advancements in energy conversion and storage technologies.

15.1.2 Importance of Recycling of CNs

Recycling is converting waste material into a new material showing physicochemical properties suitable for a particular application. Billion tons of waste, such as glass, biomass, electronics, metals, paper, plastic, textiles, batteries, electronics, and many more, are generated every day due to the population's growing needs and consumption. Therefore, recycling these spent materials into the manufacture of new products would significantly impact the environment and the economy. The recycling of CNs is an essential practice that aligns with the principles of the circular economy (Figure 15.1 (a)) and promotes the efficient use of resources, environmental stewardship, sustainable development, and mitigating potential risks associated with the disposal of nanomaterials. Recycling CNs is significant due to several key aspects. First, the synthesis of CNs typically involves the use of rare and valuable elements, and recycling facilitates the recovery of these materials, contributing to the conservation of scarce resources. Second, CNs, with their intricate structures, may pose challenges in waste disposal, and recycling minimizes the volume of waste generated, reducing the environmental burden and addressing concerns related to the accumulation of nanomaterials in landfills [7]. Third, recycling CNs typically requires less energy than the production of new materials, and it contributes to reducing energy consumption associated with the extraction and synthesis processes. Fourth, recycling ensures that these materials are handled responsibly, minimizing the potential ecological and human health risks associated with their release into the environment. The improper disposal of nanomaterials may lead to environmental contamination. Fifth, the recycling of CNs aligns with the principles of the circular economy, wherein materials

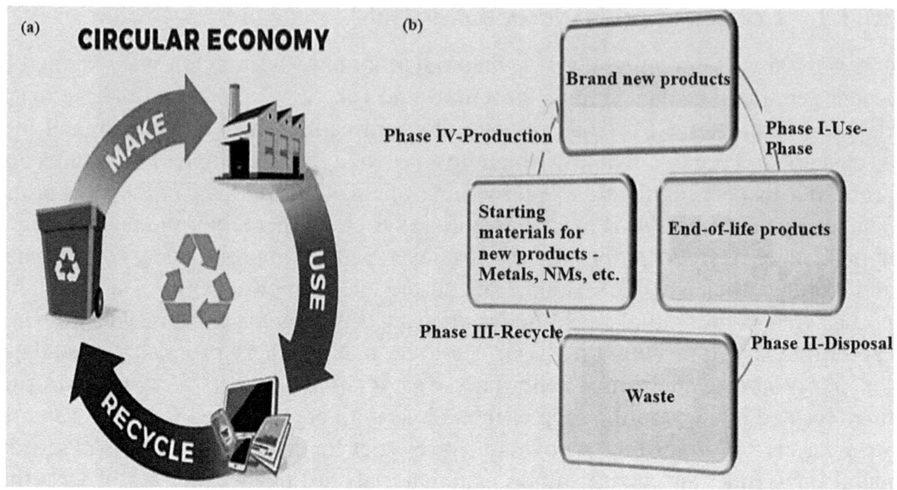

FIGURE 15.1 (a) A closed-loop of circular economy and (b) closed-loop materials cycle.

are reused, remanufactured, and recycled to create a closed-loop system (Figure 15.1 (b)) fostering sustainable resource management and reducing dependence on primary raw materials. [8] Sixth, responsible recycling practices for CNs enhance public perception and corporate responsibility, demonstrating a commitment to ethical and sustainable practices. [9]

The recycling of CNs is in line with existing and evolving regulations governing the disposal of nanomaterials. Compliance with these regulations ensures responsible and ethical practices in the management of nanomaterial waste [10]. Green chemistry principles emphasize the design and use of materials with minimal environmental impact. Recycling CNs supports the tenets of green chemistry by reducing the need for new material synthesis and minimizing waste generation. Recycling CNs contributes to economic sustainability by reducing the costs associated with raw material extraction and processing. It promotes the development of cost-effective and resource-efficient technologies [11]. Overall, the recycling of CNs is crucial for ensuring the long-term sustainability of nanomaterial applications and intersects environmental, economic, and societal considerations. Recycling of CNs-based energy devices is a highly required process due to the high initial cost of nanomaterials and the environmental problem of waste accumulation. Various techniques have been investigated for the efficient separation and recovery of nanomaterials from waste sources. These techniques include magnetic separation, solvents, aqueous dispersion, colloidal solvents, centrifugation/solvent evaporation, and other methods. The recycling of nanomaterials from energy devices has been a topic of interest in recent research.

15.1.3 Role of CNs in Energy Devices

CNs are a diverse group of materials with potential applications in various fields including the energy sector due to their unique characteristics like tunable structures, porosity, and conductivity. Metal-organic frameworks (MOFs) with high porosity and tunable structures have garnered significant attention for their potential application in fabrication of energy devices [12]. MOFs have been investigated as electrode materials in fuel cells and water electrolyzers or hosts for active materials in lithium-ion batteries, leading to improved their performance [13]. Redox-active metal centers in coordination polymers have been explored for supercapacitors, offering high capacitance and energy storage capabilities [14]. In addition, coordination polymers have been found to serve as catalysts for oxygen reduction reactions in metal-air batteries, contributing to enhanced performance [15]. Transition metal nitride nanomaterials have also shown promise for supercapacitor applications, owing to their high conductivity and electrochemical stability. Furthermore, metal nitrides, such as titanium nitride, have been studied as anodes for lithium-ion batteries, exhibiting high capacity and cycling stability. Metal phosphide nanomaterials, such as nickel phosphide, have been researched for their suitability as supercapacitor electrodes, providing high specific capacitance. Similarly, some metal phosphides have been found to serve as anode materials in sodium-ion batteries, contributing to improved electrochemical performance [16]. Finally, metal sulfide nanomaterials, such as iron sulfide, have been investigated for their potential application in lithium-sulfur batteries, owing to their high capacity and improved cycling stability. Additionally, some metal sulfides have been identified as promising electrode materials for supercapacitors, providing a high specific capacitance.

15.2 RECENT ADVANCEMENTS IN CNS-BASED ENERGY DEVICES

Recent advancements in CNs for energy devices have been the focus of research. This section is dedicated to the latest developments in CNs-based energy devices like fuel cells, water electrolyzers, batteries, and supercapacitors, correlating their performance with the structural and electronic features.

15.2.1 Fuel Cells

Fuel cells utilize oxygen as an oxidizing agent to transform the chemical energy contained in biomass fuel into electrical energy. Fuel cells undergo electrochemical reactions involving the oxidation of fuel at the anode and the reduction of oxygen at the cathode, known as the oxygen reduction reaction (ORR; Figure 15.2 (a)) The sluggish reaction kinetics of the cathode reaction is a major factor that restricts the extensive commercial utilization of fuel cells. Platinum and its alloys have been demonstrated to be the most effective electrocatalysts for ORR [17, 18]. However, platinum (Pt) is not only expensive, but it also has limited availability and a lack of durability. Furthermore, Pt, while catalyzing the ORR process, is also susceptible to the drawback of being easily poisoned. Consequently, extensive research efforts have been focused on investigating novel catalysts that provide both economic value and exceptional performance. Transition metal-based materials, such as TMOs, TMNs, TMCs, and TMSs, show potential as electrocatalysts that could replace noble metal Pt-based materials [19–21] For instance, Varga et al. [22] produced N-doped graphene sheets that were altered with cobalt nitride (Co_4N) nanoparticles (NPs).

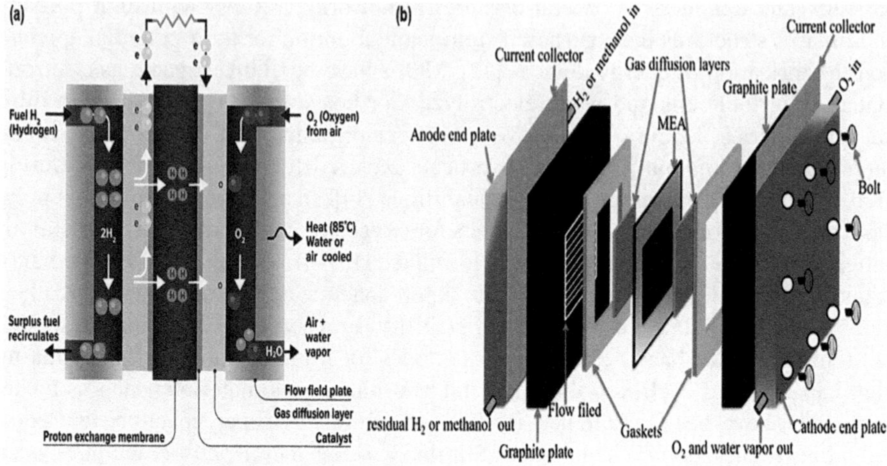

FIGURE 15.2 (a) Schematic illustration of fuel cell and (b) schematic diagram of operational parts of a single PEMFC. Reprinted with permission from ref. [40]. Copyright 2021 Royal Society of Chemistry.

They achieved this by using freeze-dried GO nanosheets and cobalt acetate (II) in an ammonia environment at a temperature of 600°C. The average particle size of Co_4N grows from 14 to 201 nm with the higher cobalt content. The novel catalyst exhibits oxygen reduction activity that is equivalent to that of the widely used Pt/C catalyst. The reduction current density reaches its maximum value of 4.1 mA cm^{-2} under alkaline circumstances. In another work, Rui et al. [23] synthesized a range of metal nitrides supported on carbon (M_xN/C, M=Ti, V, Cr, Mn, Fe, Co, Ni, $x = 1$ or 3) using the process of nitriding using ammonia. The use of carbon carriers enhances the dispersal of transition metal nitrides, hence optimizing the utilization efficiency of nitrogen molecules. Upon exposure to air, the material undergoes oxidation, resulting in the formation of a thin layer of oxygen-based oxides. This layer catalyzes facilitating chemical processes on the material's surface. Electrochemical investigations conducted in alkaline electrolytes indicate that Co_3N/C, MnN/C, and Fe_3N/C display favorable ORR capabilities. Among them, Co_3N/C demonstrates the highest ORR performance, which is comparable with commercially available Pt/C. When Co_3N/C was employed as the cathode catalyst in anion exchange membrane fuel cells (AEMFCs), it achieved a peak power density (PPD) of 700 mW cm^{-2}. This PPD is the highest reported performance among nitride cathode catalysts for membrane electrode assemblies (MEAs).

TMOs provide features of cost-effectiveness, high reactivity, and eco-friendliness, making them reliable cathode catalyst materials for fuel cells. Among all the options, Mn- and Co-based oxides exhibit the highest catalytic activity for ORR. Manganese oxides have garnered significant interest due to their notable advantages, including their affordability, eco-friendliness, ability to exist in multivalent states, and plentiful crystal structure [24]. The results indicate that the ORR catalytic activity of Mn oxide catalysts is positively correlated with the Mn valence state. Furthermore, Mn oxide catalysts synthesized at high potential exhibit significantly higher ORR catalytic activity compared to those synthesized at low potential. For instance, Zhang et al. [25] synthesized manganese-deficient Mn_3O_4 by subjecting manganese glycerate to calcination. The Mn defect in Mn_3O_4 alters the electronic structure, enhances electrical conductivity and electron delocalization, and facilitates the exposure of more surface Mn^{3+} as the main active site. Consequently, this promotes the activation of O_2 and the desorption of OH^*, leading to a significant reduction in the limiting rate of Gibb's free energy and the theoretical overpotential change of the material. The Mn_3O_4 with Mn defects exhibits superior onset potential, half-wave potential, and limiting current density compared to ordinary Mn_3O_4. Specifically, the onset potential is 0.87 V, the half-wave potential is 0.65 V, and the limiting current density is 5.0 $mA{\cdot}cm^{-2}$, whereas ordinary Mn_3O_4 has an onset potential of 0.77 V, a half-wave potential of 0.62 V, and a limiting current density of 2.6 $mA{\cdot}cm^{-2}$. The properties of cobalt oxide, including its excellent performance, affordability, and remarkable stability, have generated significant interest. Ma et al. [26] employed Ar plasma etching technology to fabricate cobalt oxide with a high concentration of oxygen vacancies (referred to as Co_3O_{4-x}), resulting in the creation of a novel zinc-cobalt oxide and zinc-air hybrid battery. Co_3O_{4-x} exhibits proficiency in several electrochemical reactions, such as ORR/OER electrocatalytic reaction and Faraday Co –O↔Co –O–OH redox reaction. The novel zinc-based power cell exhibits a remarkable power density of 3,200 $W{\cdot}kg^{-1}$, along with a high energy density of 1,060 $Wh{\cdot}kg^{-1}$. Furthermore, it demonstrates excellent electrochemical stability even after undergoing 1,500 cycles lasting 440 hours each. Furthermore, the solid-state hybrid battery exhibited remarkable safety, exceptional water resistance, and the capability to be washed after being immersed in water for 20 hours or washed for 1 hour, while retaining an electrochemical performance of approximately 90% or higher. Furthermore, the novel hybrid zinc-based power battery possesses the capability to autonomously regenerate power output upon exposure to air. Transition metal carbides are cheap and readily accessible materials that possess excellent electrical conductivity and unique electronic structure. They have the potential to serve as electrocatalysts, potentially replacing precious metals.

Further, Xiao et al. [27] reported to have developed a N-doped carbon material by polymerizing 1,8-diaminonaphthalene with ferric chloride, followed by pyrolysis and etching. The resulting material contained Fe_3C nanoparticles enclosed within a graphite layer and exhibited a porous structure. Out of all the samples, the Fe_3C/NG-800 sample heated to 800°C has the highest specific surface area. Materials with a high specific surface area have more active sites available. Additionally, having an adequate pore structure and distribution may accelerate up the rate of mass transfer and enhance catalytic efficiency. The Fe_3C/NG-800 exhibited an onset potential of 0.92 V and a half-wave potential of 0.77 V in a 0.1 mol/L $HClO_4$ solution. Additionally, the Fe_3C/NG-800 demonstrated significantly greater resistance to methanol and carbon monoxide compared to commercial platinum carbon. The Fe_3C/NG-800 exhibits an onset potential of 1.03 V and a half-wave potential of 0.86 V in a 0.1 mol/L KOH solution. These values are even higher than those of commercial platinum carbon. Additionally, the stability of Fe_3C/NG-800 is similarly superior to that of commercial platinum carbon. Furthermore, WC exhibits exceptional catalytic efficacy as a result of its characteristics as a covalent molecule, an ionic crystal, and a transition metal material. WC has excellent resistance to acid and possesses high catalytic activity for the electro-oxidation of methanol. Nevertheless, WC produced through high-temperature carbonization is prone to becoming coated with elemental carbon on the catalytic active site's surface, resulting in a decrease in its catalytic activity. Prior to utilization in WC catalysis, it is necessary for the agent to eliminate the carbon that covers the surface, hence exposing the active sites of the catalyst and enabling the manifestation of its utmost catalytic activity. In this context, Weigert et al. [28] synthesized WC and Pt-modified WC catalysts for use as anodes in DMFCs and assessed their catalytic activity and long-term reliability. WC has higher reactivity in the dissociation of methanol and water compared to Pt and Ru, as evidenced by the findings of temperature step-up desorption and high-resolution electron energy loss spectroscopy. The desorption temperature of WC for CO is at a minimum of 100 K lower compared to Pt and Ru. WC exhibits greater resistance to CO poisoning during the electro-oxidation of catalytic methanol. Further, Sheng et al. [29] synthesized a range of Pt-modified WC catalysts. The DFT calculation findings indicated that the double-layer Pt-modified WC and Pt (111) catalyst exhibited comparable starting potential for methanol electro-oxidation. Furthermore, the reaction activity of the double-layer Pt-modified WC catalyst was 2.4 times greater than that of pure Pt. The combined impact of Pt and WC also offers a novel medium for the anodic catalyst.

15.2.2 Water Electrolyzers

Electrolysis shows great potential as a method for producing hydrogen without carbon emissions, utilizing electric current to decompose water into its constituent elements H_2 and O_2. Typically, water electrolyzation involves hydrogen evolution reaction (HER) on the cathode and oxygen evolution reaction (OER) on the anode. The corresponding water-splitting reactions (i.e., alkaline solution) are as follows [30].

Various types of electrolyzers, such as polymer electrolyte membrane (PEM) electrolyzers (Figure. 15.2(b)), alkaline electrolyzers, and solid oxide electrolyzers, operate differently primarily because of the variations in the electrolyte material and the ionic species they facilitate. To separate water into H_2 and O_2, a standard thermodynamic potential of 1.23 V (vs. reversible hydrogen electrode (RHE)) is necessary, regardless of the electrolyte media. Nevertheless, it is important to note that both the cathode and anode reactions consist of multiple steps of electron transfer. Furthermore, additional energy is required to overcome obstacles and facilitate electron transfer, leading to the phenomenon known as overpotential (η).

Therefore, to achieve cost-effective water-splitting, it is necessary to use electrocatalysts that are both extremely active and robust, since they can greatly enhance the kinetics of the two essential reactions that occur in each half-cell. Essentially, a superior catalyst must fulfil two fundamental requirements: (i) the catalyst for each half-reaction must possess excellent efficiency, enabling it to create a high current density with minimal overpotential. (ii) additionally, the catalyst must exhibit remarkable durability. The development of highly efficient and durable electrocatalysts for

HER/OER is crucial for reducing costs in commercial applications. The manufacturing of an efficient and durable electrocatalyst requires the following important characteristics [31–39]: (i) large number of surface-active sites, (ii) excellent electrical conductivity, (iii) high intrinsic activity, and (iv) suitable specific adsorption/desorption energy of the intermediate species.

In this context, the transition metal-hydroxide phosphates (TMHP) being a CNs-based material were found to be more durable and efficient electrocatalysts in water electrolyzers. For instance, Mani et al. reported to have prepared an orthorhombic iron hydroxy phosphate ($Fe_5(PO_4)_4(OH)_3{\cdot}2H_2O$ (FeHP)) using a hydrothermal method. They used metal chloride precursors and disodium hydrogen phosphate as a source of phosphate ligand precursor (Figure 15.3 (a)) [41]. To improve the behavior of OER, researchers added Sn into FeHP to create Sn-FeHP, which weakens the M^{3+}-OH^- bonding. Upon the incorporation of Sn, the nanostructure of FeHP transformed, resulting in the formation of a star-shaped nanostructure (Figure 15.3 (b–g)).

Significantly, Sn–FeHP has demonstrated catalytic efficiency with a small onset overpotential of 0.35 V at a current density of 10 mA cm^{-2}, exhibiting a Tafel slope of 81 mV dec^{-1} in a 1 M KOH solution. In addition, the SN–FeHP exhibited exceptional stability in alkaline conditions for 800 minutes at a potential of 1.65 volts vs

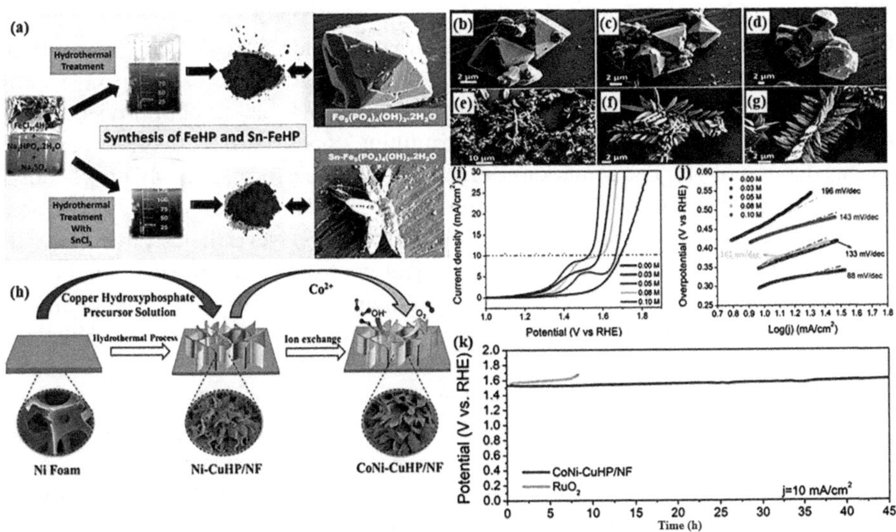

FIGURE 15.3 (a) The synthesis of FeHP and Sn-FeHP compounds by combining iron (III) chloride, tin (II) chloride, and disodium hydrogen phosphate. (b–d) FE-SEM images of octahedral crystals composed of FeHP. (e-–g) self-assembled needle-like microcrystals composed of Sn-FeHP. Taken with permission from ref. [41]. Copyright 2017, Royal Society of Chemistry. (h) Depicts an illustration of the synthesis process for CoNi–CuHP/NF, (i) displays the LSV curves, (j) shows the related Tafel curves for CoNi–CuHP/NF and different catalysts, and **(K)** displays the results of long-term cycling tests conducted on the CoNi–CuHP/NF catalyst and a commercial RuO_2 catalyst. Taken with permission from ref. [42] copyright 2017, Royal Society of Chemistry.

the RHE. Furthermore, the electrocatalytic activity of the FeHP electrode can be enhanced through a synergistic effect of including Ni or Co. Fe (Ni/Co) hydroxyphosphate/NF was reported as a bifunctional electrocatalyst [43]. In summary, the overall water-splitting process using Fe (Ni/Co)HP/NF as both the anode and cathode in an alkaline solution required a cell voltage of just 1.65/1.67 V at a current density of 10 mA cm^{-2}. Furthermore, the construction of trimetallic (Co/Ni/Cu) HP NSs on Ni foam has also been published alongside the quaternary compound. In another work, Zhang et al. investigated the fabrication of CuHP-loaded CoNi hybrid nanostructures using an ion exchange method. The remarkable OER catalytic activity, characterized by a low overpotential of 370 mV at 50 mA cm^{-2}, can be primarily due to the formation of high-valent cobalt species during the electrooxidation process. Simultaneously, the incorporation of cobalt and nickel augmented the electrochemically active surface area and furnished a greater number of efficient active sites for the OER. Although these types of materials showed great potential towards electrolysis, their catalytic activity is still limited.

To address this issue, several efforts have been made to develop oxysulfides (OS) to attain exceptional performance in electrocatalysis [44, 45]. Oxygen anions are typically (hard) non-polarized. On the other hand, sulfur anions are considered soft (polarized in sulfides). The O and S, which belong to the same VIA group of the periodic table, can be readily combined to alter the properties of compounds. Various TM compounds containing oxysulfides, such as $(CoNi)O_xS_y$, Si-MW_s@MoO_xS_y, $CoO_{0.87}S_{0.13}$/GN, $Ni_yCo_{1-y}O_xS_z$, and Mn(Zn, Ge) oxysulfide, [46, 47] have been investigated as catalysts for electrochemical applications including supercapacitors [48], PEC [46], zinc-air batteries [49], PEMWE [50], and ORR [51]. Later on, Suntivich's group examined the introduction of a partial substitution of S ions in CoO NPs to enhance the production of H_2. The remarkable electrocatalytic performance of CoO_xS_y can be related to the substitution of anion S^{2-} and O^{2-} in CoO NPs, resulting in a decrease in the energy required for the intermediate step in the HER. Nevertheless, increasing the exchange of sulfur atoms would elevate the energy of the H^* intermediate and reduce the kinetics of the HER, ultimately leading to a drop in performance. Cai et al. used an anion exchange strategy to create amorphous $CoO_{0.6}S_{4.6}$ porous nanocubes PNCs from a precursor called CoFe Prussian blue analog (PBA) in another example [48NC]. The precursor was based on TM-oxysulfide. The introduction of O atoms surrounding the Co center resulted in the formation of a local disordered structure, demonstrated by a notable reduction in intensity beyond the peaks of the first shell. In addition, the Co atom is encompassed by 4.6 S atoms and 0.6 O atoms, which possess a significant number of Co-S dangling bonds. The presence of Co–S dangling bonds and the inclusion of oxygen into the CoS_x host can dramatically augment the adsorption of O^* molecules, thus leading to a substantial enhancement in activity at a single site. The $CoO_{0.6}S_{4.6}$ PNCs that were produced had remarkable performance in water oxidation, with a low onset potential of 1.52 V (1 M KOH) and 1.50 V (0.1 M PBS), enabling a current density (j) of 10 mA cm^{-2}. Self-supporting structures possess the benefits of a high current density and a simple preparation process. Li et al. fabricated 3D NiCo sulfoxide ($NiCoS_xO_y$) supported on Ni foam by electrodeposition using Ni, Co, and S precursors (Figure 15.4a) [52]. Consequently, a cell voltage of 1.64 V was required to achieve a current density of 20 mA cm^{-2} (Figure 15.4b).

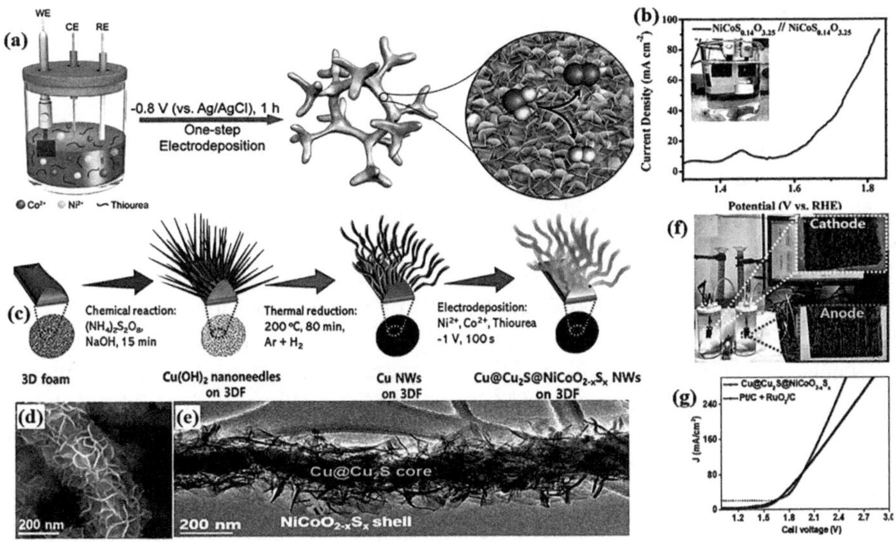

FIGURE 15.4 (a) The diagram depicts the synthesis of 3D arrays of NiCoSxOy nanosheets on Ni foam, and their application as catalysts for both OER and HER in the process of water splitting. (b) The graph shows the OER polarization curves of $NiCoS_{0.14}O_{3.25}$ NSs/NF, without iR compensation, at a scan rate of 1 mV s^{-1}. A Photograph of the electrolyzer composed of arrays of Ni-Co nanosheets. Reproduced with permission from [52] copyright 2018. Elsevier. (c) The fabrication process of $Cu@Cu_2S@NiCoO_{2-x}S_x$ NWs is schematically illustrated. (d) FE-SEM, and (e) TEM images show the $Cu@Cu_2S@NiCoO_{2-x}S_x$ NWs on 3DF. (f) An electrolyzer is fabricated using cathodic and anodic $Cu@Cu_2S@NiCoO_{2-x}$Sx NW electrodes. (g) The overall water-splitting performance of the electrolyzer devices based on $Cu@Cu_2S@NiCoO_{2-x}S_x$ NW electrodes and (Pt/C + RuO_2/C) electrodes is measured using LSV. Taken with permission from ref. [53] copyright 2020. Royal Society of Chemistry.

The electronegativity of the S ligand was comparable to OH species in alkaline electrolyte, which made the OH species attract electrons intensively, inhibiting the transformation electron from the M–S to M–O bond, which resulted in a shorter and more stable M-S bond in $NiCoS_{0.14}$ $O_{3.25}$ than the other TM sulfides. On the other hand, the nanosheet arrays constructed from O^{2-} and large S^{2-} anions with Ni and Co cations at different valence states could help diverse active sites achieve lower activation potential. NiFe oxysulfide is also very active as a bifunctional catalyst. For example, Liu et al. reported the in-situ grown fullerene-like nickel oxysulfide hollow nanospheres on 3D Ni foam by a solvothermal method [53]. The as-prepared catalyst showed a low overpotential of 0.29 V and 0.14 V at 10 mA cm^{-2} for OER and HER, respectively. Similarly, Li et al. fabricated NiFe oxysulfide ($NiFeS^{-2}$) by the vulcanization treatment of NiFe layered double hydroxide (LDH) precursor using thioacetamide [59NC]. The catalyst exhibited superior OER performance (η_{10} = 0.286 V) because the polarized S and non-polarized O anions equally adjusted the electronic configuration of the active sites in the catalyst. For overall water-splitting, a j of 10 mA cm^{-2} was achieved at 1.64 V in N_2 saturated 1 M KOH solution with

$NiFeS^{-2}$/NF electrode serving as anode and cathode. Furthermore, Tran et al. designed a novel catalyst derived from a porous interconnected network of nickel cobalt oxysulfide interfacial assembled Cu@Cu_2S nanowires (NWs) via a three-step process (Figure 15. 4 (c–e)) [54]. In 1.0 M KOH medium, the catalyst only required an overpotential of 203 mV to achieve a current response of 295 mV to reach 50 mA cm^{-2} for the OER and 20 mA cm^{-2} for the HER. Besides, a developed electrolyzer enabled a small cell voltage of 1.61 V at 20 mA cm^{-2} without performance decay upon long-term operation (Figure. 15.4 (f–g)).

Currently, TM phosphosulfides (TMPS) have remarkable catalytic efficiency. The electrocatalytic activity of S entirely depends on the electron-donation of chalcogen ligands (X_2^{2-}). The presence of P atoms in X_2^{2-} increases its electron-donating capacity, hence promoting the redox reactions of metal atoms. Recently, various reports are available in the literature for TM-based phosphosulfides such as MoPS, PdPS, CoPS, $FePS_3$, $NiPS_3$, and so on [55–63]. Kibsgaard et al. devised a wet chemical method to produce MoPS using sulfidation treatment. The MoPS catalyst, with a mixture of anions, exhibited a significantly lower overpotential of 0.064 V at a current density of 10 mA cm^{-2} in acidic conditions for the HER. This performance is superior to that of the MoP/Ti-based catalyst. Pd combined with phosphosulfide also exhibits exceptional electrochemical characteristics. The report by Sampath's group indicates that the PdPS/rGO composites had outstanding HER activity, with a minimal overpotential of around 0.09 V compared to the RHE and a narrow Tafel slope of 46 mV dec^{-1} [64]. The overall structure was determined to be a "layered type" consisting of pentagons joined together to form wrinkled two-dimensional sheets. It was observed that four layers are required to create the one-unit cell. The P and S atoms are predicted to combine and create polyanions with a structure of $[S–P–P–S]^{4-}$. The aforementioned group utilized few-layered 2D $FePS_3$ nanosheets as the composite material in the fabrication of rGo-$FePS_3$ composite electrocatalysts [65]. It was proposed that a direct bond between P and S in the $FePS_3$ could boost its HER activity. The augmented HER activity was additionally substantiated by DFT calculations, which verified that the inclusion of the $[P_2S_6]^{4-}$ (phosphorus and sulfur sites) unit may engage in the adsorption and desorption of the hydrogen atom. Flexible carbon fiber (CF) conductive substrates have been employed in several studies to facilitate two-dimensional (2D) development. The effectively synthesized $WP_{2x}S_{2(1-x)}$ nanoribbons (NRs) on a CF substrate by subjecting WO_3 NWs to phosphatization and sulfidation reactions [66]. Remarkably, the introduction of P atoms into $WP_{2x}S_{2(1-x)}$ resulted in electronic perturbation in WS_2, which significantly enhanced the activity, stability, and durability of the material for HER. Pyrite-type transition metal dichalcogenides (TMDs) that incorporate non-metal anions are widely recognized for their superior catalytic activity compared to other transition metal compounds. According to Jin's group, electrodes made of ternary pyrite-type cobalt phospho-sulfide (CoPS; nanowires, nanoparticles) exhibit better performance in the hydrogen evolution reaction (HER) compared to CoS_2 NWs and cobalt selenide nanoparticles (CoSe NPs). The CoPS electrode demonstrated a reduced overpotential of 48 mV at a current density of 10 mA cm^{-2}, a Tafel slope of 56 mV dec^{-1}, and maintained operational stability for 36 hours in an acidic solution. The researcher successfully employed a partial sulfurization/phosphorization method to create yolk-shell spheres of pyrrhotite-type cobalt mono-phosphosulfate ($Co_{0.9}P_{0.42}S_{0.58}$) with a hexagonal close-packed phase. (Figure 15.5 (a–f)). It is important to note that the combination of non-stoichiometric properties and the ability to vary the P/S ratio leads to the strengthening of Co^{3+}/Co^{2+} pairs and the ability to modify the electronic structure. Co^{3+} ions effectively enhance the OER activity due to their higher valence 3d electron orbits and more electron-accepting features than Co^{2+} ions. For HER, it showed a low operating overpotential of around 140 mV with small Tafel slopes around 70 mV dec^{-1} in both alkaline and acidic media (Figure 15.5 (g–j)). Coupled with the high HER activity of $Co_{0.9}S_{0.58}P_{0.42}$, the overall water-splitting was demonstrated with a low η_{10} at 1.59.

Moreover, layered nickel phosphosulfide represents an additional type of novel electrocatalyst. As an illustration, Ren's group synthesized highly pure layered $NiPS_3$ and $NiPS_3$ NSs/Graphene (G) composites for use as electrocatalysts in the OER [67]. It was believed that non-metallic components such as P and S typically do not directly enhance the performance of OER. Nevertheless, their significant electronegativity can influence the nearby electronic structure of Ni, thus facilitating the oxidation of Ni^{2+} to $Ni^{3+.}$ In addition, the $NiP_{0.62}S_{0.38}$/NiOOH heterostructure exhibited enhanced performance in the OER. The $NiP_{0.62}S_{0.38}$ nanosheets exhibited a low overpotential of 52 mV at a current density of 10 mA cm^{-2} for HER, while an overpotential of 240 mV was detected at the same current density for the OER. It is important to mention

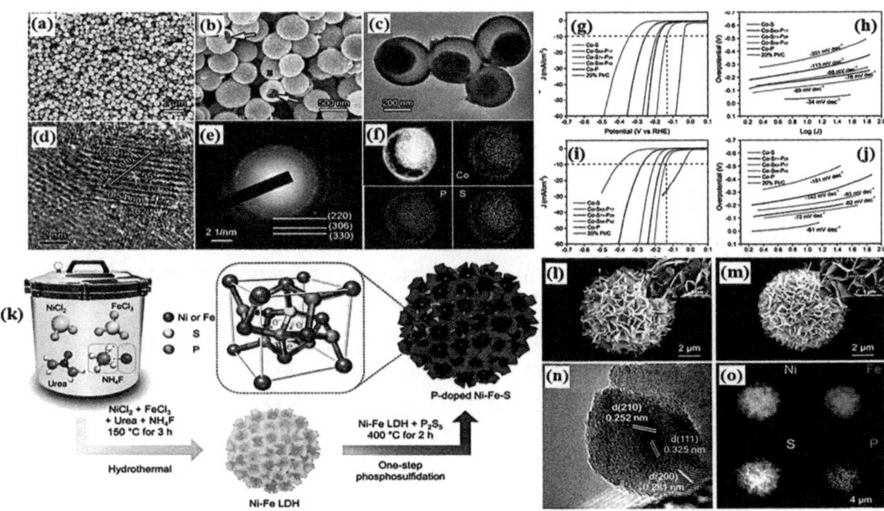

FIGURE 15.5 (a) SEM, (b) expanded SEM view. (c) TEM image. (d) High-resolution TEM image. (e) SAED pattern. (f) STEM image and elemental mapping of the $Co_{0.9}S_{0.58}P_{0.42}$ yolk-shell spheres. (g–j) Electrocatalytic performance of different catalysts for HER in alkaline and acidic electrolytes. (g) Polarization curves and (h) Tafel plots in 0.5 M H_2SO4. (i) Polarization curves and (j) Tafel plots in 1 M KOH. Reprinted with permission from ref. [52] copyright 2017 American Chemical Society. (k) Schematic illustration of the preparation of the P-doped Ni–Fe–S microspheres. High-magnification SEM images of (l) Ni–Fe–S and (m) P-doped Ni–Fe–S, (n) HRTEM image of the P-doped Ni–Fe–S, (o) EDS elemental mapping of the P-doped Ni–Fe–S. Reprinted with permission from ref. [69] copyright 2021 Elsevier.

that the phosphosulfide species with high electron affinity shifted ΔG_{H*} to the thermally neutral position and achieved a balanced hydrogen adsorption/desorption at the nickel site. In addition to ternary PS compounds, researchers have also studied the HER and OER activities of multicomponent mixed PS-based materials [68]. For instance, by utilizing P_2S_5 as a starting material, researchers created 3D hierarchical flower-like microspheres of P-doped Ni-Fe disulfide through a combination of hydrothermal and phosphosulfidation processes (Figure 15.5 (k-–o)) [69]. The microspheres exhibit exceptional performance in terms of OER, with an overpotential of 264 mV, enabling a *j* of 10 mA cm^{-2}. Furthermore, they demonstrate great long-term durability, maintaining their performance for over 50 hours without any significant degradation in alkaline conditions.

15.2.3 Batteries

The battery concept relies on the electrochemical storage of energy in electrochemical cells that are interconnected either in series or in parallel [70]. An electrochemical cell consists of a cathode and an anode, which are positive and negative electrodes, respectively. These electrodes are separated by a separator, which prevents short circuits and allows for the movement of ions. Electrochemical energy storage is the conversion of electrical energy into stored chemical energy and vice versa, through a redox reaction occurring between the positive and negative electrodes. Batteries can be categorized into two distinct types: primary batteries – which cannot be recharged – and secondary batteries, which can be recharged. In addition, secondary batteries can be categorized into i) alkali metal ion secondary batteries, which involve alkali metals such as Li^+, Na^+, K^+, Zn^{2+}, and Mg^{2+} [71]; ii) conventional secondary batteries like lead–acid batteries and nickel-electrode batteries [72, 73]; iii) molten salt batteries such as sodium–nickel chloride batteries and sodium–sulfur batteries; and iv) flow batteries including polysulfide–bromide batteries, vanadium redox batteries, and zinc-bromine batteries.

The global demand for lithium-ion batteries is experiencing tremendous growth due to the increasing popularity of electric vehicles (EVs) in recent years. Currently, the primary anode material used in commercial applications is graphite. However, graphite is constrained by its low theoretical capacity and poor rate of lithium-ion transfer. These limitations significantly hinder the storage performance of lithium-ion batteries [74–76]. Hence, the investigation and implementation of novel anode materials hold significant importance. CNs possess a significant theoretical capacity and demonstrate exceptional cycle performance and Coulomb efficiency when prepared with unique morphologies or composites. As a result, they are considered strong contenders for next-generation anode materials. Transition metal oxides utilized as anode active materials exhibit three distinct lithium-ion storage mechanisms: insertion, alloying, and transformation. Luo et al. [77] described a technique for effectively attaching a continuous mesoporous nanostructure of Fe_3O_4 onto 3D graphene foams using atomic layer deposition. The graphene's loose and porous nature mitigates the issue of shedding or pulverization of active components resulting from volume expansion. The composite material is utilized directly as the negative electrode in lithium-ion batteries, exhibiting a high reversible capacity and rapid discharge capability. The maximum capacity is 785 milliampere-hours per gram at a discharge rate of 1C, and it maintains this capacity consistently across 500 charge and discharge cycles. Larson et al. [78] effectively created nanoporous molybdenum oxide asymmetric membranes using spontaneous phase separation caused by a non-solvent. The presence of a three-dimensional nanoporous structure can mitigate the significant alteration in volume that occurs during the process of charging and discharging, hence enhancing the long-term durability of lithium-ion battery anodes. The experimental findings further demonstrate that the lithium-ion battery, which consists of the MoO_2 planar asymmetric film anode material, can retain 97% of its original capacity after 165 cycles at a current density of 120 mAh g^{-1}. Currently, there is ongoing research on mixed transition metal oxides. When compared to typical metal oxides, the two metal elements have distinct expansion coefficients, which leads to a synergistic action that helps to mitigate volume expansion. Furthermore, both components are metals that exhibit electrochemical activity, enabling them to create alloys with a greater number of lithium ions and exhibit enhanced electrochemical capabilities.

Xiao et al. [79] described a method for producing hollow structures of carbon-free CoS_x using a straightforward solvothermal reaction followed by annealing. The nanostructure in question possesses a distinctive composition, consisting of a hollow core and a porous shell. This composition is similar to a solid structure that has demonstrated enhancements in both rate capability and cycling stability. As an illustration, the initial discharge capacity is remarkably high at 1,059.7 mAh g^{-1}. Additionally, the Coulombic efficiency is 77.2%, and after 100 cycles at a current of 500 mA g^{-1}, the reversible capacity reaches 1,012.1 mAh g^{-1}. In a single-step process, Fayed et al. [80] synthesized nanosheets of molybdenum disulfide that were co-doped with nitrogen and carbon. The MoS_2 samples facilitate the movement of ions by means of disorganized nanosheet-like clusters, hence improving the electrochemical efficiency. The material used in lithium-ion batteries exhibits an initial discharge capacity of 1,280 mAh g^{-1} at a current density of 100 mA g^{-1}, and its rate performance is satisfactory. Transition metal nitrides exhibit superior performance as lithium-ion anode materials for other transition metal nanomaterials, primarily due to their higher conversion reaction sites compared to transition metal nitrides. Lai et al. [81] described a universal technique for producing transition metal nitride nanoparticles by attaching them to conductive graphene. These nanoparticles serve as anode materials in Li-ion batteries. This technique is appropriate for immobilizing FeN, FeCoN, CoN, NiN, and other similar compounds onto conductive graphene that has been treated with nitrogen. This hybrid material exhibits a high specific capacity and demonstrates excellent cycle performance. Despite a decrease in loading, the utilization of graphene can enhance the storage capacity of lithium-ion batteries. For instance, when the nanoparticles consist of FeN, the hybrid material exhibits an initial reversible capacity of around 665 mAh g^{-1} at a current density of 50 mA $g^{-1.}$ Furthermore, the capacity progressively increases to 698 mAh g^{-1} during further cycling, indicating excellent cycle performance. Wang et al. [82] employed a two-dimensional titanium carbide (MXene) metal conductive film as the current collector layer for the electrode material. They also utilized multi-layer $Ti_3C_2T_x$ (where T_x represents the surface functional group) as the negative electrode material. The investigation revealed that the implementation of the $Ti_3C_2T_x$ MXene current collector effectively decreases the weight and thickness of the device while maintaining its performance. This study expands the utilization of transition metal carbides in Li-ion batteries as current collectors, eliminating the need for extra surface treatment. It has the potential to be a promising option for the advancement of flexible and lightweight energy storage systems.

Crystallized Co_3O_4 with different nanostructures such as nanocube (NC), nanotruncated octahedron (NTO), and nanopolyhedron (NP) have been prepared via a facile and template-free hydrothermal strategy (Figure. 15.6a). [83] It is noteworthy that these three cobalt oxides expose [12], [12] +{111}, and {112} planes, respectively. Generally, different Miller-index planes contain different atomic arrangements and electronic structures. For {001} and {111} planes, Co ions only occupy the tetrahedral coordination (Co^{2+}Td). The {112} planes can provide not only Co ions in tetrahedral coordination but also in octahedral coordination (Co^{3+}Oh). According to the ligand field theory, Co^{3+}Oh has a strong electron-donating ability, which is beneficial to the displacement of OH^- and O_2^{2-} in the OER or ORR process. Moreover, as shown in Figure. 15.6 (b–d), DFT calculations reveal that {112} surface can elongate O=O bonds to a more favorable length (1.380 Å) compared with {001} (1.349 Å) or {111} (1.513 Å), which results in a moderate oxygen binding ability. Consequently, {112} faceted Co_3O_4 NP displays excellent bifunctional electroactivity for OER and ORR during the charging-discharging tests (1.18 and 1.95 V at 10 mA cm ^{2}).

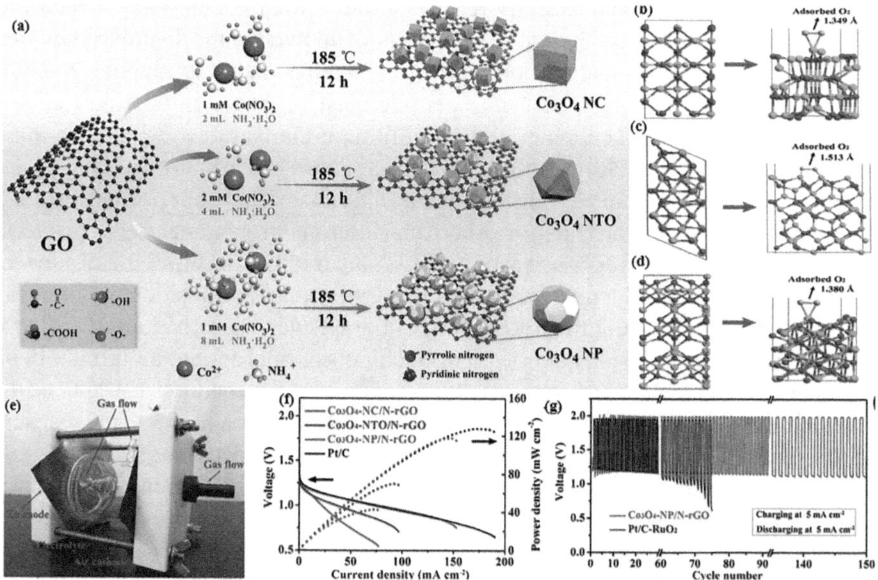

FIGURE 15.6 (a) Schematic illustration of Co_3O_4 with different nanostructure and (b–d) calculated O_2-adsorption on (001), (111), and (112) facets in Co_3O_4. (e) Photograph. (f) Discharging polarization curves and corresponding power densities and (g) cycling performance of Zn-air battery with Co_3O_4 air-cathode. Taken from [85] copyright Wiley, 2018.

Applying {112} faceted Co_3O_4 NP in Zn-air battery configuration as shown in Figure 15.6e, this battery can obtain a high-power density peak and favorable long-term stability as shown in Figure. 15.6 (f–g). In order to improve the exposure of highly active Co^{3+}Oh on the Co_3O_4 surface, ultrathin Co_3O_4 nanofilm was prepared by a surfactant- and template-free hydrothermal strategy. [84] Both XPS and electron energy loss spectroscopy (EELS) confirm that the Co_3O_4 nanofilm contains a higher ratio of Co^{3+}/Co^{2+} compared with that of Co_3O_4 nanoparticles. As a result, Co_3O_4 nanofilm requires only 461 mV to reach 40 mA cm^2 for OER and shows an outstanding ORR activity via the $4e^-$ pathway.

15.2.4 Supercapacitors

In recent decades, supercapacitors have been widely studied because of their fast charging and discharging speed, long service life, and high power density [86, 87]. According to the different charge storage mechanisms, supercapacitors are divided into two categories, electric double-layer capacitors (EDLCs) and pseudocapacitors (PCs). [88, 89] The field of energy development and advancement is undergoing intensive study focused on the fabrication of electrode materials. In recent times, there has been considerable interest in the use of CNs for supercapacitor applications due to their exceptional conductivity and remarkable physicochemical features, which are a result of their inherent structures and unique chemical bonding. By synthesizing transition metal materials with diverse nanostructures, one can harness the cost-effectiveness and stability inherent in transition metals, while also leveraging the substantial surface area provided by the nanostructure to buffer the volume fluctuations of active materials. By combining with other elements, the electrochemical energy storage performance can be further enhanced. Hence, it is imperative to methodically conduct the investigation and state-of-the-art modification of materials based on TMNs. Feng et al. [90] described a technique called polymer intercalation for producing three-dimensional composites of MoS_2 with nitrogen-doped carbon. Figure 15.7 demonstrates that the MoS_2/nitrogen-doped carbon material exhibited distinct interlayer spacing in comparison to the pure MoS_2 nanosheets following a sequence of processing. The interlayer spacing of pure MoS_2 nanosheets after calcination was 0.62 nm, while that of MoS_2/nitrogen-doped carbon composite was 0.98 nm. The observed value aligns with the interlayer spacing of graphene inside the MoS_2 structure, suggesting the successful embedding of graphene in the MoS_2 layer. This embedding is a result of the in-situ carbonization of PEI within the MoS_2 layer. The electrode material possessed the following advantages: i) the three-dimensional material structure facilitated the fast transport of electrons and ions, ii) the intercalation of a graphene-like monolayer carbon between the layers of MoS_2 increased the availability of active sites for redox reactions and opened up new opportunities for ion/electron transport, resulting in each monolayer of MoS_2 becoming electrochemically active, iii) the intrinsic electrical conductivity of MoS_2 is low. However, in MoS_2/nitrogen-doped carbon composites, electrons are transferred through metallic bond-like interfacial Mo-N bonds, which are more efficient than covalent bond-like Mo-O bonds in the MoS_2/C hetero-aerogel. Hence, the MoS_2/nitrogen-doped carbon material had exceptional electrochemical performance, with a mass-specific capacitance

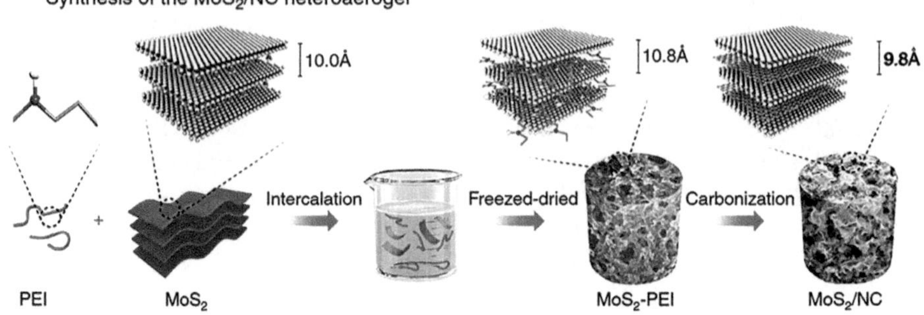

FIGURE 15.7 Synthesis of the MoS_2/NC hetero-aerogel [90].

of 4,144 F g^{-1} at a current density of 1 A g^{-1}. Even at a higher current density of 10 A g^{-1}, the mass-specific capacitance remained high at 2483 F g^{-1}.Supercapacitors exhibit superior power density, cycle efficiency, and energy density in comparison to regular batteries and are capable of enduring millions of charge-discharge cycles. The features of stretchable supercapacitors have significant promise for use in wearable and implantable energy systems. McPherson et al. [91] created a nano-porous electrode made of copper and copper oxide by dissolving Al in $Cu_{17.5}Al_{82.5}$ with NaOH and formed a supercapacitor with an aqueous electrolyte. This device is both secure and low-cost, exhibiting minimal ohmic losses and the potential for enhanced electrode surface area through geometric optimization. Zhou et al. investigated the use of two-dimensional (2D) titanium carbide ($Ti_3C_2T_x$) Mxene for creating flexible and printable energy storage devices. They achieved this by developing a strong and stretchable supercapacitor with great performance. The supercapacitor was made by combining reduced graphene oxide (RGO) with $Ti_3C_2T_x$ to generate a composite electrode. The composite electrode, consisting of $Ti_3C_2T_x$ and RGO, effectively integrates the excellent electrochemical and mechanical characteristics of $Ti_3C_2T_x$ with the robustness of RGO, which is achieved by strong interactions between nanosheets, increased nanoflake size, and mechanical flexibility. The study revealed that the incorporation of 50 wt% RGO in the $Ti_3C_2T_x$/RGO composite electrodes effectively reduced the formation of cracks caused by extreme strains. The composite electrodes demonstrated a significant capacitance of 49 mF cm^{-2} (~ 490 F cm^{-3} and ~ 140F g^{-1}) and exhibited greater electrochemical and mechanical stability when exposed to cyclic uniaxial strains of 300% or biaxial strains of 200%. The symmetric supercapacitor, when assembled, exhibited a specific capacitance of 18.6 mF cm^{-2} (~ 90 F cm^{-3} and ~ 29F g^{-1}) and stretchability up to 300%. The implemented technique offered an alternative method to construct stretchable energy storage devices using MXene-based materials and has the potential to be applied to other members of the extensive MXene family.

The synthesis of TMNs, which have been thoroughly investigated for their potential use in supercapacitors, is a highly challenging and time-consuming procedure. TMOs and TMCs exhibit great potential as electrode materials for advanced ultracapacitors owing to their ability to exist in many oxidation states. Hence, it poses a significant challenge to enhancing the electrochemical performance of pseudocapacitive materials by controlling their structure, morphology, and surface area. Hussain et al. [92] produced a zinc cobalt sulfide (ZCS) electrode without using a binder, using a facile single-step hydrothermal technique. The $Zn_{0.76}Co_{0.24}S$ produced can be utilized as a binder-free electrode for supercapacitor applications. Furthermore, an investigation was conducted to analyze the influence of morphology on the electrochemical performance. Out of the four electrodes, the 12-h ZCS that was synthesized exhibited the highest specific capacitance of 2,417 F g^{-1} (967 C g^{-1}) at 1 A g^{-1}. Additionally, it demonstrated 83% strong cycling stability after undergoing 10,000 charge/discharge cycles. In addition, a type of supercapacitor called an asymmetric supercapacitor (ASC) was used in this study. The positive electrode of the ASC was made of $Zn_{0.76}Co_{0.24}S$, while the negative electrode was made of activated carbon (AC). The ASC had a high energy density of 51 W h kg^{-1} when operated at a power density of 400 W kg^{-1}. Furthermore, it exhibited good cycling stability, retaining 82% of its initial capacity after 5,000 cycles. Manganese sulfide (MnS) is a promising metal sulfide contender due to its advantageous characteristics, including affordability, exceptional electrochemical performance, eco-friendliness, and abundant availability. This electrode material exhibits great potential for next-generation supercapacitors because of its theoretical SC, which can reach up to 1,370 F g^{-1} in aqueous electrolyte. In comparison to ternary transition metal oxides, transition metal sulfides typically demonstrate enhanced conductivity and ionic diffusivity as a result of the substitution of oxygen atoms with sulfur atoms. They also possess a narrower band gap, more anionic polarizability, and a bigger size of the S^{2-} ion. Rao et al. [93] described the facile synthesis of a novel Cu-MnS structure using PVP. This structure combines the advantageous properties of MnS, such as high theoretical capacitance, with the low cost and high electrical conductivity of Cu. Additionally, the structure has a significant surface area and benefits from the high thermal and mechanical conductivity of PVP. These features make it suitable for creating high-performance electrodes for supercapacitors on a single entity on nickel foam. The dynamic states of Cu and PVP in the presence of MnS facilitate the movement of ions at the interfaces, while the void spaces between particles are capable of enduring alterations in volume during extended periods of the cycle. The electrochemical measurements demonstrated that the Cu-MnS material, synthesized with the 2PVP electrode, displayed an exceptional SC of 833.58 F g^{-1} at a current density of 1 A g^{-1}. Additionally, it exhibited a remarkably high cycling stability of 96.95% over 1,000 cycles in a 2 M KOH solution. Hence, it has the potential to offer an alternative pathway for the commercialization of supercapacitors made from transition metal-based nanomaterials. Molybdenum disulfide (MoS_2), a type of TMDC, is extensively researched due to its notable benefits. These include exceptional electrocatalytic activity, a 2D layered crystal structure that enables ion intercalation, a large surface area, and superior electronic conductivity compared to transition metal oxides (TMOs). Researchers began considering the use of TMDC for supercapacitors only after the capacitive behavior of MoS_2 nanowall films was first described in 2007. Balasingam et al. [94] utilized the facile hydrothermal technique to deposit an amorphous MoSx thin film onto the carbon fiber paper. The electrode material, produced without the use of a binder, showed exceptional capacity for storing electrochemical energy. Additionally, it formed a symmetric device with a specific capacitance value of 41.96 mF cm^{-2}, indicating a high level of performance. The presence of a sulfuric acid medium led to a significant increase in the capacitance retention of amorphous MoS_x, reaching up to 600%, as a result of the process known as "electro-activation." Li et al. [95] fabricated a sophisticated supercapacitor electrode by electrochemically depositing MoS_2 onto TiN nanotube arrays (NTAs) on a Ti mesh substrate. The sample containing 900 pulses exhibited the largest specific capacitance, measuring 353.2 F/g at an applied current density of 0.6 A/g. Additionally, it displayed the lowest internal resistance of 2.195 Ω and interface resistance of 0.005 Ω. The preservation of specific capacitance was 63% after 700 cycles of galvanostatic charge and discharge, with a current density of 0.8 A/g. The hybrid electrode was anticipated to deliver exceptional electrochemical performance for supercapacitors due to the combination of TiN nanotubes' excellent conductivity and MoS_2 nanoparticles' high theoretical specific capacitance. The predominant phase of MoS_2 is 2H owing to its stability; yet, its semiconducting qualities render it unsuitable for energy storage applications. The electrical conductivity of metallic phase 1 T-MoS_2 is 107 times greater than that of 2H-MoS_2, rendering it highly desirable for energy storage applications. Consequently, it has gained recent use in batteries and supercapacitors.

Zhuo et al. [96] synthesized composite electrodes by combining aryl diazonium functionalized graphene with chemically exfoliated 1 T- MoS_2. The resulting electrodes are depicted in Figure 15.8. The inherently unstable 1 T- MoS_2 was rendered stable through its interaction with fct-graphene, hence inhibiting its transition to the 2H phase. The graphene-MoS_2 structure, when used as an alternate layer, exhibited a significant specific capacitance of 290 F cm^{-3} at an applied current density of 0.5 A g^{-1}. Furthermore, it demonstrated an impressive capacitance retention of 90% even after undergoing 10,000 cycles. The advantageous electrochemical characteristics ascribed to the hydrophilic nature, high electronic conductivity, expanded interlayer space, and reduced ion diffusion length demonstrated the potential for the advancement of energy storage with high-performance. MnO_2 is a preferred choice for supercapacitors among the

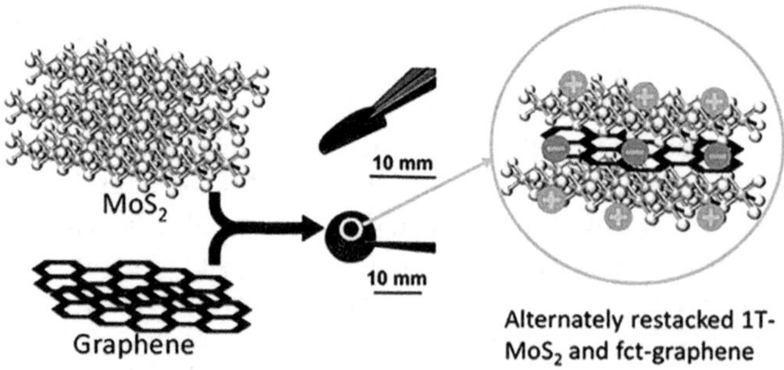

FIGURE 15.8 Schematic showing the 1 T-MoS_2/functionalized-graphene composite production procedure [96].

different transition metal oxides explored. This is because MnO_2 is low-cost, non-toxic, readily available in nature, and has a high theoretical pseudo capacitance of approximately 1370 F g^{-1}. Additionally, MnO_2 exhibits favorable interfacial properties when combined with carbon materials. The ε-phase of MnO_2 is recognized for its electrochemical reactivity. Rakhi et al. [97] constructed symmetric supercapacitors utilizing composites of ε-MnO_2/Ti_2CT_x and ε-MnO_2/$Ti_3C_2T_x$. The ε-MnO_2/MXene supercapacitor, along with its improved pseudo capacitance, demonstrated remarkable cycling stability, maintaining around 88% of its initial specific capacitance even after undergoing 10,000 cycles. The proposed electrode structure capitalized on the high specific capacitance of MnO_2 and the ability of MXenes to enhance conductivity and cycling durability.

15.3 CONCLUSIONS AND FUTURE PROSPECTS

This chapter primarily focuses on the recent progress in recycled carbon nanomaterials- (CNs)-based energy systems, such as fuel cells, water electrolyzers, batteries, and supercapacitors. Various materials have been extensively examined and highlighted for their performance in these energy devices. In addition, some research studies and the conclusions of review articles have been thoroughly explored, emphasizing their novel concepts and potential for future development. Recycled CNs-based energy devices present a complex and evolving landscape of sustainable materials management. Understanding and addressing the challenges and prospects of this process are crucial for realizing the full potential of recycling in the context of energy device technologies. CNs often have complex structures and compositions that make the separation and recovery of individual components challenging during recycling processes. Thus, the development of efficient and economically viable technologies for the recovery of valuable materials from CNs-based energy devices is a significant and ongoing challenge. Recycling processes may introduce impurities or contaminants, which can affect the purity and performance of recovered materials, particularly in the case of CNs. In addition, the lack of standardized regulatory frameworks for the recycling of CNs may hinder the development and implementation of effective recycling strategies. Therefore, it is essential to carefully assess the economic feasibility of recycling CNs, considering factors such as collection, transportation, and processing costs. Research and development of advanced separation technologies can enhance the efficient recovery of CNs from energy devices, improving recycling yields. Moreover, implementing Design for Recycling principles during the manufacturing of CNs-based devices can enhance recyclability by considering materials' ease of separation and recovery. Conducting life cycle assessments can help evaluate the environmental impacts of CNs-based energy devices and guide decisions to optimize recycling processes. Finally, the continued development and adoption of circular economy policies and initiatives can provide a supportive framework for the recycling of CNs in energy devices. Indeed, understanding and addressing the challenges and prospects of recycling CNs-based energy devices requires a multidisciplinary approach that encompasses materials science, engineering, economics, and policy-making.

REFERENCES

[1] A.G. Olabi, M. Mahmoud, B. Soudan, T. Wilberforce, M. Ramadan, Geothermal based hybrid energy systems, toward eco-friendly energy approaches, *Renewable Energy*, 147 (2020) 2003–2012.

[2] R. Qin, P. Wang, C. Lin, F. Cao, J. Zhang, L. Chen, S. Mu, Transition metal nitrides: Activity origin, synthesis and electrocatalytic applications, *Acta Physico-Chimica Sinica*, 37 (2021).

[3] Y. Luo, H. Jin, Y. Lu, Z. Zhu, S. Dai, L. Huang, X. Zhuang, K. Liu, L. Huang, Potential gradient-driven fast-switching electrochromic device, *ACS Energy Letters*, 7 (2022) 1880–1887.

[4] L. Jiao, J. Wu, H. Zhong, Y. Zhang, W. Xu, Y. Wu, Y. Chen, H. Yan, Q. Zhang, W. Gu, Densely isolated FeN4 sites for peroxidase mimicking, *ACS Catalysis*, 10 (2020) 6422–6429.

[5] K. Liu, H. Jin, L. Huang, Y. Luo, Z. Zhu, S. Dai, X. Zhuang, Z. Wang, L. Huang, J. Zhou, Puffing ultrathin oxides with nonlayered structures, *Science Advances*, 8 (2022) eabn2030.

[6] L. Jiao, J.Y.R. Seow, W.S. Skinner, Z.U. Wang, H.-L. Jiang, Metal–organic frameworks: Structures and functional applications, *Materials Today*, 27 (2019) 43–68.

[7] A.D. Maynard, R.J. Aitken, 'Safe handling of nanotechnology' ten years on, *Nature Nanotechnology*, 11 (2016) 998–1000.

[8] E. MacArthur, Towards the circular economy, *Journal of Industrial Ecology*, 2 (2013) 23–44.

[9] Y. Zheng, B. Nowack, Size-specific, dynamic, probabilistic material flow analysis of titanium dioxide releases into the environment, *Environmental Science & Technology*, 55 (2021) 2392–2402.

[10] J.C. Warner, Green chemistry: Theory and practice, *Papers of the American Chemical Society*, 244 (2012).

[11] D.J. Glass, Government regulation of bio-based fuels and chemicals, *Biobased Products and Industries*, Elsevier, 2020, pp. 79–123.

[12] H. Li, X. Zhang, Z. Zhao, Z. Hu, X. Liu, G. Yu, Flexible sodium-ion based energy storage devices: Recent progress and challenges, *Energy Storage Materials*, 26 (2020) 83–104.

[13] D. Sheberla, J.C. Bachman, J.S. Elias, C.-J. Sun, Y. Shao-Horn, M. Dincă, Conductive MOF electrodes for stable supercapacitors with high areal capacitance, *Nature Materials*, 16 (2017) 220–224.

[14] L. Jiao, G. Wan, R. Zhang, H. Zhou, S.H. Yu, H.L. Jiang, From metal–organic frameworks to single-atom Fe implanted N-doped porous carbons: Efficient oxygen reduction in both alkaline and acidic media, *Angewandte Chemie,* 130 (2018) 8661–8665.

[15] R. Nivetha, J. Jana, S. Ravichandran, H.N. Diem, T. Van Phuc, J.S. Chung, S.G. Kang, W.M. Choi, S.H. Hur, Two-dimensional bimetallic Fe/M- (Ni, Zn, Co and Cu) metal organic framework as efficient and stable electrodes for overall water splitting and supercapacitor applications, Journal of Energy Storage, 61 (2023) 106702.

[16] T.-F. Yi, J.-J. Pan, T.-T. Wei, Y. Li, G. Cao, NiCo2S4-based nanocomposites for energy storage in supercapacitors and batteries, *Nano Today,* 33 (2020) 100894.

[17] C. Chen, Y. Kang, Z. Huo, Z. Zhu, W. Huang, H.L. Xin, J.D. Snyder, D. Li, J.A. Herron, M. Mavrikakis, Highly crystalline multimetallic nanoframes with three-dimensional electrocatalytic surfaces, *Science*, 343 (2014) 1339–1343.

[18] D. Wang, H.L. Xin, R. Hovden, H. Wang, Y. Yu, D.A. Muller, F.J. DiSalvo, H.D. Abruña, Structurally ordered intermetallic platinum–cobalt core–shell nanoparticles with enhanced activity and stability as oxygen reduction electrocatalysts, *Nature Materials*, 12 (2013) 81–87.

[19] K.P. Singh, E.J. Bae, J.-S. Yu, Fe–P: A new class of electroactive catalyst for oxygen reduction reaction, *Journal of the American Chemical Society*, 137 (2015) 3165–3168.
[20] S. Dou, L. Tao, J. Huo, S. Wang, L. Dai, Etched and doped Co 9 S 8/graphene hybrid for oxygen electrocatalysis, *Energy & Environmental Science*, 9 (2016) 1320–1326.
[21] H.F. Wang, C. Tang, B. Wang, B.Q. Li, Q. Zhang, Bifunctional transition metal hydroxysulfides: Room-temperature sulfurization and their applications in Zn–air batteries, *Advanced Materials*, 29 (2017) 1702327.
[22] T. Varga, G. Ballai, L. Vásárhelyi, H. Haspel, Á. Kukovecz, Z. Kónya, Co4N/nitrogen-doped graphene: A non-noble metal oxygen reduction electrocatalyst for alkaline fuel cells, *Applied Catalysis B: Environmental,* 237 (2018) 826–834.
[23] R. Zeng, Y. Yang, X. Feng, H. Li, L.M. Gibbs, F.J. DiSalvo, H.D. Abruña, Nonprecious transition metal nitrides as efficient oxygen reduction electrocatalysts for alkaline fuel cells, *Science Advances*, 8 (2022) eabj1584.
[24] K.A. Stoerzinger, M. Risch, B. Han, Y. Shao-Horn, Recent insights into manganese oxides in catalyzing oxygen reduction kinetics, *ACS Catalysis*, 5 (2015) 6021–6031.
[25] Y.-C. Zhang, S. Ullah, R. Zhang, L. Pan, X. Zhang, J.-J. Zou, Manipulating electronic delocalization of Mn3O4 by manganese defects for oxygen reduction reaction, *Applied Catalysis B: Environmental*, 277 (2020) 119247.
[26] L. Ma, S. Chen, Z. Pei, H. Li, Z. Wang, Z. Liu, Z. Tang, J.A. Zapien, C. Zhi, Flexible waterproof rechargeable hybrid zinc batteries initiated by multifunctional oxygen vacancies-rich cobalt oxide, *ACS Nano*, 12 (2018) 8597–8605.
[27] M. Xiao, J. Zhu, L. Feng, C. Liu, W. Xing, Meso/macroporous nitrogen-doped carbon architectures with iron carbide encapsulated in graphitic layers as an efficient and robust catalyst for the oxygen reduction reaction in both acidic and alkaline solutions, *Advanced Materials*, 27 (2015) 2521–2527.
[28] E.C. Weigert, M.B. Zellner, A.L. Stottlemyer, J.G. Chen, A combined surface science and electrochemical study of tungsten carbides as anode electrocatalysts, *Topics in Catalysis*, 46 (2007) 349–357.
[29] T. Sheng, X. Lin, Z.-Y. Chen, P. Hu, S.-G. Sun, Y.-Q. Chu, C.-A. Ma, W.-F. Lin, Methanol electro-oxidation on platinum modified tungsten carbides in direct methanol fuel cells: A DFT study, *Physical Chemistry Chemical Physics*, 17 (2015) 25235–25243.
[30] R. Boppella, J. Tan, J. Yun, S.V. Manorama, J. Moon, Anion-mediated transition metal electrocatalysts for efficient water electrolysis: Recent advances and future perspectives, *Coordination Chemistry Reviews,* 427 (2021) 213552.
[31] Y. Yan, B.Y. Xia, B. Zhao, X. Wang, A review on noble-metal-free bifunctional heterogeneous catalysts for overall electrochemical water splitting, *Journal of Materials Chemistry A*, 4 (2016) 17587–17603.
[32] N.K. Chaudhari, H. Jin, B. Kim, K. Lee, Nanostructured materials on 3D nickel foam as electrocatalysts for water splitting, *Nanoscale*, 9 (2017) 12231–12247.
[33] N.-T. Suen, S.-F. Hung, Q. Quan, N. Zhang, Y.-J. Xu, H.M. Chen, Electrocatalysis for the oxygen evolution reaction: Recent development and future perspectives, *Chemical Society Reviews,* 46 (2017) 337–365.
[34] V. Vij, S. Sultan, A.M. Harzandi, A. Meena, J.N. Tiwari, W.-G. Lee, T. Yoon, K.S. Kim, Nickel-based electrocatalysts for energy-related applications: Oxygen reduction, oxygen evolution, and hydrogen evolution reactions, *Acs Catalysis*, 7 (2017) 7196–7225.
[35] R. Gusmão, Z. Sofer, M. Pumera, Metal phosphorous trichalcogenides (MPCh3): From synthesis to contemporary energy challenges, *Angewandte Chemie International Edition*, 58 (2019) 9326–9337.
[36] H.-F. Wang, C. Tang, B.-Q. Li, Q. Zhang, A review of anion-regulated multi-anion transition metal compounds for oxygen evolution electrocatalysis, *Inorganic Chemistry Frontiers,* 5 (2018) 521–534.

[37] Y. Yan, T. He, B. Zhao, K. Qi, H. Liu, B.Y. Xia, Metal/covalent–organic frameworks-based electrocatalysts for water splitting, *Journal of Materials Chemistry A*, 6 (2018) 15905–15926.

[38] G. Zhao, K. Rui, S.X. Dou, W. Sun, Heterostructures for electrochemical hydrogen evolution reaction: A review, *Advanced Functional Materials*, 28 (2018) 1803291.

[39] P. Li, W. Chen, Recent advances in one-dimensional nanostructures for energy electrocatalysis, *Chinese Journal of Catalysis*, 40 (2019) 4–22.

[40] M. Vinothkannan, A.R. Kim, D.J. Yoo, Potential carbon nanomaterials as additives for state-of-the-art Nafion electrolyte in proton-exchange membrane fuel cells: A concise review, *RSC Advances*, 11 (2021) 18351–18370.

[41] V. Mani, S. Anantharaj, S. Mishra, N. Kalaiselvi, S. Kundu, Iron hydroxyphosphate and Sn-incorporated iron hydroxyphosphate: Efficient and stable electrocatalysts for oxygen evolution reaction, *Catalysis Science & Technology*, 7 (2017) 5092–5104.

[42] Y. Zhang, T. Qu, F. Bi, P. Hao, M. Li, S. Chen, X. Guo, M. Xie, X. Guo, Trimetallic (Co/Ni/Cu) hydroxyphosphate nanosheet array as efficient and durable electrocatalyst for oxygen evolution reaction, *ACS Sustainable Chemistry & Engineering*, 6 (2018) 16859–16866.

[43] P. Babar, A. Lokhande, E. Jo, B. Pawar, M. Gang, S. Pawar, J. Kim, Facile electrosynthesis of Fe (Ni/Co) hydroxyphosphate as a bifunctional electrocatalyst for efficient water splitting, *Journal of Industrial and Engineering Chemistry*, 70 (2019) 116–123.

[44] X. Yu, Z.-Y. Yu, X.-L. Zhang, P. Li, B. Sun, X. Gao, K. Yan, H. Liu, Y. Duan, M.-R. Gao, Highly disordered cobalt oxide nanostructure induced by sulfur incorporation for efficient overall water splitting, *Nano Energy,* 71 (2020) 104652.

[45] C.-X. Zhao, J.-N. Liu, J. Wang, D. Ren, B.-Q. Li, Q. Zhang, Recent advances of noble-metal-free bifunctional oxygen reduction and evolution electrocatalysts, *Chemical Society Reviews,* 50 (2021) 7745–7778.

[46] X.-Q. Bao, D.Y. Petrovykh, P. Alpuim, D.G. Stroppa, N. Guldris, H. Fonseca, M. Costa, J. Gaspar, C. Jin, L. Liu, Amorphous oxygen-rich molybdenum oxysulfide decorated p-type silicon microwire arrays for efficient photoelectrochemical water reduction, *Nano Energy*, 16 (2015) 130–142.

[47] D.T. Tran, T. Kshetri, N.D. Chuong, J. Gautam, H. Van Hien, N.H. Kim, J.H. Lee, Emerging core-shell nanostructured catalysts of transition metal encapsulated by two-dimensional carbon materials for electrochemical applications, *Nano Today*, 22 (2018) 100–131.

[48] L. Liu, Nano-aggregates of cobalt nickel oxysulfide as a high-performance electrode material for supercapacitors, *Nanoscale*, 5 (2013) 11615–11619.

[49] J. Fu, F.M. Hassan, C. Zhong, J. Lu, H. Liu, A. Yu, Z. Chen, Defect engineering of chalcogen-tailored oxygen electrocatalysts for rechargeable quasi-solid-state zinc–air batteries, *Advanced Materials*, 29 (2017) 1702526.

[50] H. Kim, J. Kim, S.-K. Kim, S.H. Ahn, A transition metal oxysulfide cathode for the proton exchange membrane water electrolyzer, *Applied Catalysis B: Environmental,* 232 (2018) 93–100.

[51] Y. Zhang, X. Wang, D. Hu, C. Xue, W. Wang, H. Yang, D. Li, T. Wu, Monodisperse ultrasmall manganese-doped multimetallic oxysulfide nanoparticles as highly efficient oxygen reduction electrocatalyst, *ACS Applied Materials & Interfaces*, 10 (2018) 13413–13424.

[52] C. Li, X. Zhao, Y. Liu, W. Wei, Y. Lin, 3D Ni-Co sulfoxide nanosheet arrays electrodeposited on Ni foam: A bifunctional electrocatalyst towards efficient and stable water splitting, *Electrochimica Acta*, 292 (2018) 347–356.

[53] J. Liu, Y. Yang, B. Ni, H. Li, X. Wang, Fullerene-like nickel oxysulfide hollow nanospheres as bifunctional electrocatalysts for water splitting, *Small*, 13 (2017) 1602637.

[54] D.T. Tran, H.T. Le, N.H. Kim, J.H. Lee, Highly efficient overall water splitting over a porous interconnected network by nickel cobalt oxysulfide interfacial assembled Cu@ Cu 2 S nanowires, *Journal of Materials Chemistry A,* 8 (2020) 14746–14756.
[55] D. Goossens, D. James, J. Dong, R. Whitfield, L. Norén, R. Withers, Local order in layered NiPS3 and Ni0. 7Mg0. 3PS3, *Journal of Physics: Condensed Matter*, 23 (2011) 065401.
[56] N. Ismail, M. Madian, A. El-Meligi, Synthesis of NiPS3 and CoPS and its hydrogen storage capacity, *Journal of Alloys and Compounds*, 588 (2014) 573–577.
[57] J. Li, Z. Xia, X. Zhou, Y. Qin, Y. Ma, Y. Qu, Quaternary pyrite-structured nickel/cobalt phosphosulfide nanowires on carbon cloth as efficient and robust electrodes for water electrolysis, *Nano Research*, 10 (2017) 814–825.
[58] P. He, X.Y. Yu, X.W. Lou, Carbon-incorporated nickel–cobalt mixed metal phosphide nanoboxes with enhanced electrocatalytic activity for oxygen evolution, *Angewandte Chemie International Edition*, 56 (2017) 3897–3900.
[59] J. Chang, Y. Ouyang, J. Ge, J. Wang, C. Liu, W. Xing, Cobalt phosphosulfide in the tetragonal phase: A highly active and durable catalyst for the hydrogen evolution reaction, *Journal of Materials Chemistry A*, 6 (2018) 12353–12360.
[60] Y. Li, S. Niu, D. Rakov, Y. Wang, M. Cabán-Acevedo, S. Zheng, B. Song, P. Xu, Metal organic framework-derived CoPS/N-doped carbon for efficient electrocatalytic hydrogen evolution, *Nanoscale*, 10 (2018) 7291–7297.
[61] J. Li, C. Zhang, H. Ma, T. Wang, Z. Guo, Y. Yang, Y. Wang, H. Ma, Modulating interfacial charge distribution of single atoms confined in molybdenum phosphosulfide heterostructures for high efficiency hydrogen evolution, *Chemical Engineering Journal*, 414 (2021) 128834.
[62] K. Maiti, K. Kim, K.-J. Noh, J.W. Han, Synergistic coupling ensuing cobalt phosphosulfide encapsulated by heteroatom-doped two-dimensional graphene shell as an excellent catalyst for oxygen electroreduction, *Chemical Engineering Journal*, 423 (2021) 130233.
[63] Y. Tong, P. Chen, L. Chen, X. Cui, Dual vacancies confined in nickel phosphosulfide nanosheets enabling robust overall water splitting, *ChemSusChem,* 14 (2021) 2576–2584.
[64] S. Sarkar, S. Sampath, Equiatomic ternary chalcogenide: PdPS and its reduced graphene oxide composite for efficient electrocatalytic hydrogen evolution, *Chemical Communications*, 50 (2014) 7359–7362.
[65] D. Mukherjee, P.M. Austeria, S. Sampath, Two-dimensional, few-layer phosphochalcogenide, FePS3: A new catalyst for electrochemical hydrogen evolution over wide pH range, *ACS Energy Letters*, 1 (2016) 367–372.
[66] T.A. Shifa, F. Wang, K. Liu, Z. Cheng, K. Xu, Z. Wang, X. Zhan, C. Jiang, J. He, Efficient catalysis of hydrogen evolution reaction from WS2 (1− x) P2x nanoribbons, *Small*, 13 (2017) 1603706.
[67] S. Xue, L. Chen, Z. Liu, H.-M. Cheng, W. Ren, NiPS3 nanosheet–graphene composites as highly efficient electrocatalysts for oxygen evolution reaction, *ACS Nano,* 12 (2018) 5297–5305.
[68] L. Yin, X. Ding, W. Wei, Y. Wang, Z. Zhu, K. Xu, Z. Zhao, H. Zhao, T. Yu, T. Yang, Improving catalysis for electrochemical water splitting using a phosphosulphide surface, *Inorganic Chemistry Frontiers*, 7 (2020) 2388–2395.
[69] D. Lim, C. Lim, M. Hwang, M. Kim, S.E. Shim, S.-H. Baeck, Facile synthesis of flower-like P-doped nickel-iron disulfide microspheres as advanced electrocatalysts for the oxygen evolution reaction, *Journal of Power Sources*, 490 (2021) 229552.
[70] K.M. Winslow, S.J. Laux, T.G. Townsend, A review on the growing concern and potential management strategies of waste lithium-ion batteries, *Resources, Conservation and Recycling*, 129 (2018) 263–277.

[71] K.H. Prasad, S. Vinoth, A. Ratnakar, M. Venkateswarlu, N. Satyanarayana, Structural and electrical conductivity studies of spinel LiMn2O4 cathode films grown by RF sputtering, *Materials Today: Proceedings*, 3 (2016) 4046–4051.
[72] T. Sakai, I. Uehara, H. Ishikawa, R&D on metal hydride materials and Ni–MH batteries in Japan, *Journal of Alloys and Compounds*, 293 (1999) 762–769.
[73] A. Lewandowski, A. Świderska-Mocek, Ionic liquids as electrolytes for Li-ion batteries – An overview of electrochemical studies, *Journal of Power Sources*, 194 (2009) 601–609.
[74] K.T. Lee, Y.S. Jung, S.M. Oh, Synthesis of tin-encapsulated spherical hollow carbon for anode material in lithium secondary batteries, *Journal of the American Chemical Society*, 125 (2003) 5652–5653.
[75] Y.-P. Wu, E. Rahm, R. Holze, Carbon anode materials for lithium ion batteries, *Journal of Power Sources*, 114 (2003) 228–236.
[76] L. Ji, Z. Lin, M. Alcoutlabi, X. Zhang, Recent developments in nanostructured anode materials for rechargeable lithium-ion batteries, *Energy & Environmental Science*, 4 (2011) 2682–2699.
[77] J. Luo, J. Liu, Z. Zeng, C.F. Ng, L. Ma, H. Zhang, J. Lin, Z. Shen, H.J. Fan, Three-dimensional graphene foam supported Fe3O4 lithium battery anodes with long cycle life and high rate capability, *Nano Letters*, 13 (2013) 6136–6143.
[78] E. Larson, L. Williams, C. Jin, X. Chen, J. DiCesare, O. Sheppard, S. Xu, J. Wu, Molybdenum oxide nanoporous asymmetric membranes for high-capacity lithium ion battery anode, *Journal of Materials Research*, 37 (2022) 2204–2215.
[79] Y. Xiao, J.-Y. Hwang, I. Belharouak, Y.-K. Sun, Superior Li/Na-storage capability of a carbon-free hierarchical CoSx hollow nanostructure, *Nano Energy*, 32 (2017) 320–328.
[80] M.G. Fayed, S.Y. Attia, Y.F. Barakat, E. El-Shereafy, M. Rashad, S.G. Mohamed, Carbon and nitrogen co-doped MoS2 nanoflakes as an electrode material for lithium-ion batteries and supercapacitors, *Sustainable Materials and Technologies*, 29 (2021) e00306.
[81] L. Lai, J. Zhu, B. Li, Y. Zhen, Z. Shen, Q. Yan, J. Lin, One novel and universal method to prepare transition metal nitrides doped graphene anodes for Li-ion battery, *Electrochimica Acta*, 134 (2014) 28–34.
[82] C.-H. Wang, N. Kurra, M. Alhabeb, J.-K. Chang, H.N. Alshareef, Y. Gogotsi, Titanium carbide (MXene) as a current collector for lithium-ion batteries, *ACS Omega*, 3 (2018) 12489–12494.
[83] J. Yang, C. Hu, H. Wang, K. Yang, J.B. Liu, H. Yan, Review on the research of failure modes and mechanism for lead–acid batteries, *International Journal of Energy Research*, 41 (2017) 336–352.
[84] F. Feng, M. Geng, D. Northwood, Electrochemical behaviour of intermetallic-based metal hydrides used in Ni/metal hydride (MH) batteries: A review, *International Journal of Hydrogen Energy*, 26 (2001) 725–734.
[85] X. Han, G. He, Y. He, J. Zhang, X. Zheng, L. Li, C. Zhong, W. Hu, Y. Deng, T.Y. Ma, Engineering catalytic active sites on cobalt oxide surface for enhanced oxygen electrocatalysis, *Advanced Energy Materials*, 8 (2018) 1702222.
[86] C. Largeot, C. Portet, J. Chmiola, P.-L. Taberna, Y. Gogotsi, P. Simon, Relation between the ion size and pore size for an electric double-layer capacitor, *Journal of the American Chemical Society*, 130 (2008) 2730–2731.
[87] S. Kandalkar, D. Dhawale, C.-K. Kim, C. Lokhande, Chemical synthesis of cobalt oxide thin film electrode for supercapacitor application, *Synthetic Metals*, 160 (2010) 1299–1302.
[88] B. Babakhani, D.G. Ivey, Improved capacitive behavior of electrochemically synthesized Mn oxide/PEDOT electrodes utilized as electrochemical capacitors, *Electrochimica Acta*, 55 (2010) 4014–4024.

[89] S. Sarangapani, B. Tilak, C.P. Chen, Materials for electrochemical capacitors: Theoretical and experimental constraints, *Journal of the Electrochemical Society*, 143 (1996) 3791.
[90] N. Feng, R. Meng, L. Zu, Y. Feng, C. Peng, J. Huang, G. Liu, B. Chen, J. Yang, A polymer-direct-intercalation strategy for MoS2/carbon-derived heteroaerogels with ultrahigh pseudocapacitance, *Nature Communications*, 10 (2019) 1372.
[91] D.J. McPherson, A. Dowd, M.D. Arnold, A. Gentle, M.B. Cortie, Electrochemical energy storage on nanoporous copper sponge, *Journal of Materials Research*, 37 (2022) 2195–2203.
[92] I. Hussain, C. Lamiel, S.G. Mohamed, S. Vijayakumar, A. Ali, J.-J. Shim, Controlled synthesis and growth mechanism of zinc cobalt sulfide rods on Ni-foam for high-performance supercapacitors, *Journal of Industrial and Engineering Chemistry*, 71 (2019) 250–259.
[93] S.S. Rao, I. Kanaka Durga, B. Naresh, B. Jin-Soo, T. Krishna, C. In-Ho, J.-W. Ahn, H.-J. Kim, One-pot hydrothermal synthesis of novel Cu-MnS with PVP cabbage-like nanostructures for high-performance supercapacitors, *Energies*, 11 (2018) 1590.
[94] S.K. Balasingam, A. Thirumurugan, J.S. Lee, Y. Jun, Amorphous MoS x thin-film-coated carbon fiber paper as a 3D electrode for long cycle life symmetric supercapacitors, *Nanoscale*, 8 (2016) 11787–11791.
[95] J. Li, R. Wu, X. Yang, MoS2 modified TiN nanotube arrays for advanced supercapacitors electrode, *Physica E: Low-dimensional Systems and Nanostructures*, 118 (2020) 113951.
[96] Y. Zhuo, E. Prestat, I.A. Kinloch, M.A. Bissett, Self-assembled 1T-MoS2/functionalized graphene composite electrodes for supercapacitor devices, *ACS Applied Energy Materials*, 5 (2022) 61–70.
[97] R.B. Rakhi, B. Ahmed, D. Anjum, H.N. Alshareef, Direct chemical synthesis of MnO2 nanowhiskers on transition-metal carbide surfaces for supercapacitor applications, *ACS Applied Materials & Interfaces*, 8 (2016) 18806–18814.

Index

Note: Numbers in *italics* indicate a figure and numbers in **bold** indicate a table on the corresponding page.